Recent Results in Cancer Research 55

Fortschritte der Krebsforschung
Progrès dans les recherches sur le cancer

G. Mathé

Cancer Active Immunotherapy

Immunoprophylaxis and Immunorestoration

An Introduction

With 123 Figures

Springer-Verlag
Berlin · Heidelberg · New York 1976

Georges Mathé, MD, Professor of Oncology,
University of Paris Sud,
Institut de Cancérologie et d'Immunogénétique,
Unité Fred Siguier, Hopital Paul Brousse,
and Division of Hematology, Institut Gustave Roussy,
Villejuif, France

Sponsored by the Swiss League against Cancer

ISBN-13:978-3-642-81019-0 e-ISBN-13:978-3-642-81017-6
DOI: 10.1007/978-3-642-81017-6

Library of Congress Cataloging in Publication Data. Mathé, Georges, 1922— Cancer active immunotherapy, immunoprophylaxis, and immunorestoration.
(Recent results in cancer research; 55.) Bibliography: p.
1. Immunotherapy. 2. Cancer-Immunological aspects. I. Title. II. Series.
[DNLM: 1. Neoplasms-Prevention and control. 2. Neoplasms-Therapy. 3. Immunotherapy. 4. Biological products-Therapeutic use. 5. Adjuvants, Immunologic-Therapeutic use. Wl RE106P v. 55/QZ266 M426i.] RC261.R35 vol. 55
[RC271.I45] 616.9'94'008s. [616.9'94'079] 76-33.

Softcover reprint of the hardcover 1st edition 1976

Toute théorie n'est bonne
que si elle permet non le repos
mais un plus grand travail.

A. Gide

A theory is only sound
if it smooths the way,
not to tranquillity
but to greater effort.

Preface

I would like to thank all my co-workers who have collaborated with me, from 1963 until now, in biological and clinical research in the field of cancer active immunotherapy, of its immunoprevention and immunorestoration. They will often be quoted in this book.
I am particularly grateful to those who have helped me to write it by reviewing some chapters: D. Belpomme, J. F. Doré, Irene Florentin, A. Goutner, I. J. Hiu, R. Huchet and Marie-Christine Simmler.
I also thank Nicole Vriz, Marie-Claude Schneider, Fenella Riseley and M. Juvet for their willing and efficient co-operation in the preperation of the manuscript.
I am finally grateful to all authors of books or articles who authorized me to reproduce their figures or tables.

Paris, April 1976 G. Mathé

Contents

Chapter 1
Introduction and Definitions

For decades, oncologists have seen many cancers locally eradicated by surgery or radiotherapy, which relapse in the form of one or more metastases, in site(s) distant from the primary one, and possibly recurring a long time after primary treatment. The fact that these metastases were not detectable before and after treatment means that surgeons and radiotherapists often leave a few malignant cells in a patient who has been locally well treated. Systematic complementary chemotherapy in such cases has recently been claimed to be an excellent means of preventing relapses (E.O.R.T.C., Radiochemoth. Group, 1972; JAFFE *et al.*, 1974; BERLIN, 1974). However, it has, in the past, been generally disappointing (MATHÉ and KENIS, 1975).

The same reason may explain its failure to cure patients with perceptible disseminated neoplasias, with the exception of certain leukaemias, haematosarcomas and exceptional solid tumours such as placental choriocarcinoma (MATHÉ and KENIS, 1975). The contrast between the rare incidence of cure and the high frequency of complete regressions or remissions, which mean retrogression of the population of residual tumour cells to a number which is imperceptible, is explained by the fact that chemotherapy obeys first order kinetics (SKIPPER *et al.*, 1964, 1965, 1967).

If this therapy always leaves a few cells, and if it nevertheless cures some patients, it is reasonable to wonder if the last cells, in the case of a cure, are not killed by immune reactions. Three questions arise:

a) Are immune reactions able to kill tumour cells? Experimentally, it has been known since the beginning of this century (LOEB, 1901; JENSEN, 1903) that grafts of incompatible tumours are rejected, and it was when he was studying transplantation of neoplasias to hosts of different genetic constitution, that SNELL (1953) determined the immunogenetics laws valuable for all tissues and/or organs.

Clinically, it was impossible not to correlate the fact that choriocarcinoma was the first human cancer to be cured by (not even intensive) chemotherapy, and its semi-allogenic relationship with its carrier. In the serum of the (female) patients, antibodies were demonstrated against the tissues of their respective husbands (MATHÉ *et al.*, 1964).

b) Are tumours, immunogenetically identical to their hosts, recognizable by their immunological machinery and sensitive to it? In other words, do cancer cells carry special antigens?

The answer is yes. Since FOLEY (1953) gave the first demonstration that murine tumours may be rejected by isogeneic hosts, a thousand papers have been devoted to "tumour-associated antigens" of experimental and human neoplasias (KOLDOVSKI, 1969).

c) Hence, if tumour cells are recognized by the immune machinery and sensitive to it, the third question arises: can we manipulate the latter in order to make it kill more cancer cells than it does naturally? The answer is yes.

Since 1955, in animal and in man (THOMPSON and MATHÉ, 1972), the reactions against leukaemias of incompatible grafted (with bone marrow) or transfused lymphocytes have been used, and this form of treatment has been called *adoptive immunotherapy*.

In the author's laboratory, MOTTA (1971) treated experimental leukaemias with manipulated sera from animals immunized with cells of these tumours *(passive immunotherapy)*.

After 1963, we were tempted to increase the number of tumour cells that a subject can destroy with its own immune machinery, and this manipulation was called *active immunotherapy)* (MATHÉ, 1968; MATHÉ, 1971): for this purpose the author used the administration of (sterilized) tumour cells, isogeneic to the neoplasia to be treated, or extract(s) containing their tumour-associated antigen(s) (MARTYRE, 1976) *(specific immunotherapy)* and/or agents known to be able to potentiate, in some conditions, the immune responses. In fact, they may modulate them in different ways (MATHÉ, 1973), hence the traditional name of *immunity "adjuvants"* that we have kept for these agents which are legion. The name of *nonspecific immunotherapy* is retained for their exclusive use.

Active immunotherapy has also to be divided into two categories according to the presentation of the disease to be treated: when there is a nonaccessible tumour, specially in the case of imperceptible residual neoplasia, only the *systemic* type can be applied.

When, on the other hand, a tumour is accessible to the injection of adjuvant(s), such an injection is recommended by some and this form of immunotherapy is called *local* (MORTON *et al.*, 1970).

The results obtained with cancer immunotherapy are inducing research on *immunorestoration* of immunodepressed cancer patients (SIMMLER *et al.*, 1976) and on tumour *immunoprophylaxis*.

Though there is apparently much less work devoted to the latter question than to immunotherapy, it should be remembered that adjuvants called at that time "stimulants of the reticuloendothelial system" (OLD *et al.*, 1961) were applied for the prevention of experimental grafted tumours particularly by HALPERN *et al.*,(1959), after DUBOS and SCHAEDLER (1957) had shown that they could prevent experimental infections.

References

BERLIN, N.I. (Ed.): A report to the profession from Breast Cancer Task Force, 30th September 1974. Bethesda: Department of Health Education and Welfare U.S.A.

DUBOS, R.J., SCHAEDLER, R.W.: J. Exp. Med. 106, 703 (1957).

European Organization for Research on the Treatment of Cancer (EORTC). Radiotherapy-chemotherapy Cooperative Group. Europ. J. Cancer 8, 353 (1972).

FOLEY, E.J.: Cancer Res. 13, 835 (1953).

HALPERN, B., BIOZZI, G., STIFFEL, C., MOUTON, D.: C.R. Soc. Biol. Paris 153, 919 (1959).

JAFFE, N., FREI, E. III, TRAGGIS, D., BISHOP, Y.: New England J. Med. 291, 994 (1974).

JENSEN, C.O.: Zentralbl. Bakt. 1 Abt. 34, 28, 122 (1903).

KOLDOVSKI, P.: Tumor specific transplantation antigen. 1 vol. Berlin-Heidelberg-New York: Springer-Verlag 1969.

LOEB, L.: J. Med. Res. 6, 28 (1901).

MARTYRE, M.C.: Biomedicine in press (1976).

MATHÉ, G.: Rev. Fr. Et. Clin. Biol. 13, 881 (1968).

MATHÉ, G.: Adv. Cancer Res. 14, 1 (1971).

MATHÉ, G.: p. 305 in "Immunopotentiation" (Ciba Foundation Symposium 18). 1 vol., Amsterdam: Elsevier Excerpta Medica 1973.

MATHÉ, G., DAUSSET, J., HERVET, E., AMIEL, J.L., COLOMBANI, J., BRULE, G.: J. Nat. Cancer Inst. 33, 193 (1964).

MATHÉ, G., KENIS, Y.: La chimiothérapie des cancers (leucémies, hématosarcomes et tumeurs solides). 3rd ed., Paris: Expansion Scientifique 1975.

MORTON, D.L., EILBER, F.R., JOSEPH, W.L., WOOD, W.C., TRAHAN, E., KETCHAM, A.S.: Ann. Surg. 172, 740 (1970).

MOTTA, R.: Adv. Cancer Res. 14, 161 (1971).

OLD, L.J., BENACERRAF, B., CLARKE, D.A., CARSWELL, E.A., STOCKERT, E.: Cancer Res. 21, 1281 (1961).

SIMMLER, M.C., SCHWARZENBERG, L., MATHÉ, G.:Cancer Immunol.Immunoth. in press (1976).

SKIPPER, H.E., SCHABEL, F.M., WILCOX, W.S.: Cancer Chemoth. Rep. 35, 1 (1964).

SKIPPER, H.E., SCHABEL, F.M., WILCOX, W.S.: Cancer Chemoth. Rep. 45, 5 (1965).

SKIPPER, H.E., SCHABEL, F.M., WILCOX, W.S.: Cancer Chemoth. Rep. 51, 125 (1967).
SNELL, G.D.: p. 338, in "The Physiopathology of Cancer" (F. Homburger & W.H. Fistman, eds.). 1 vol. New York: Hoeber-Harper 1953.
THOMPSON, R.B., MATHÉ, G.: Transplant. Rev. 9, 54 (1972).

Chapter 2
Biological Basis: Tumour Associated Antigens, the Immune Machinery, and Its Behaviour Concerning Cancer Cells

Only a brief summary will be given of the present concepts concerning these subjects.

2.1. Tumour Associated Antigens

A monograph (KOLDOVSKI, 1969) in the series to which this book belongs, has been devoted to the problem of tumour-associated antigens (TAA).

FOLEY (1953) opened the field with an in vivo method: he induced a sarcoma in mice with methylcholanthrene (MCA), and showed that, after the tumour had been removed, sufficient immunity had developed to inhibit reimplantation.

This fundamental discovery was confirmed by several groups, mainly by PREHN and MAIN (1957) and by KLEIN *et al.* (1960). The workers immunized a group of mice with an isogeneic MCA induced tumour, and another group with normal tissue, from the donor of the neoplasia: animals preimmunized with the tumour were resistant to challenge with their neoplasia, while those preimmunized with normal tissue presented a similar growth of the tumour as in the controls. The second group of researchers showed that, by immunizing syngeneic mice with UV-treated tumour cells from a primary MCA-induced sarcoma, they could be immunized against the corresponding line of neoplasia, but not against any other of the 12 lines of MCA sarcomas obtained in the same strain of mice.

Much complementary work has been devoted to *chemically induced tumour*-associated antigens. Many more carcinogens have been shown to induce TAA: benzopyrene (OLD *et al.*, 1962), dimethylbenzanthracene (PREHN, 1960; KOLDOVSKI, 1961), azodyes etc. (KLEIN, 1966; LAW, 1970).

Soon after, *virus induced tumours* were shown to also contain TAA (SJÖGREN *et al.*, 1961). In adult mice, infection with polyoma virus produced resistance to transplant of a tumour induced by the same virus. Mice immunized with a small number of polyoma cells were also resistant to further graft of a great number of tumour cells. Neoplasias induced by *DNA viruses* (polyoma, SV-40, Shope papillomas etc.) or *RNA viruses* (RAUSCHER, MOLONEY etc.) have been the object of extensive studies which have all shown that they contain TAA (KLEIN, 1966; LAW, 1970).

However, while chemically induced tumour-associated antigens vary from one tumour to another, even when the tumours are induced by the same carcinogen in the same strain or in the same animal, TAA of a neoplasias induced by a virus are the same in all tumours induced in different animals of the same strain, or of different strains and even of different species (KLEIN, 1966; LAW, 1970).

Although the histocompatible graft reaction use of tumour graft in isogeneic animals is the best method to demonstrate the presence of antigens concerned in immunotherapy, because it reveals those antigens which are responsible for the possible rejection of an autologous tumour, other methods have been applied on experimental and on human neoplasias. These *methods* have shown that there is not one TAA, but several, which can be distinguished according to their situation (a) on the cell surface (those which are involved in cancer rejection, hence in immunotherapy), (b) in the cytoplasm, or (c) in the nucleus (Fig. 2.1).

These methods may be based on antigen-antibody reactions (the antibody being autologous and spontaneous or produced by allogeneic or heterospecific immunization and must, in the latter case, be purified by absorption with normal tissues of the same strain of animal as that of tumour cells). Other techniques such as immunodiffusion, cytotoxicity, complement fixation etc. can be used (LEVY *et al.*, 1972; WINN, 1972) (see appendix).

In addition to the serological demonstration of TAA, there are methods that depend upon cell-mediated immune responses. The techniques commonly used are in vitro techniques: the lymphocyte blastogenic response to TAA in either cellular or soluble form, the evaluation of the cytotoxic activity of lymphocytes towards tumour cells,

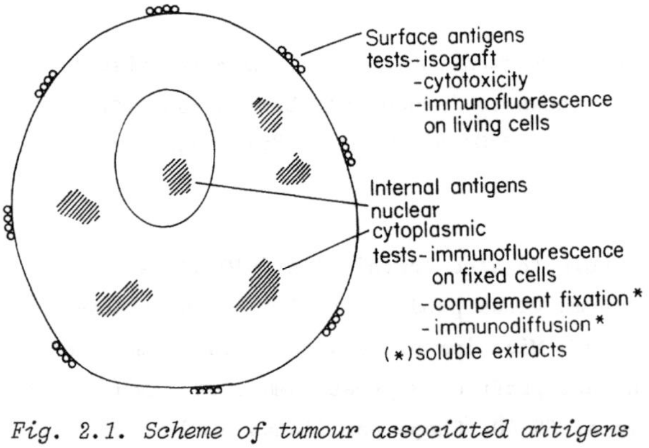

Fig. 2.1. Scheme of tumour associated antigens

the inhibition of leucocyte migration by tumoural extracts (in these two latter cases, lymphocytes should be sensitized against TAA) (BARTLETT, 1972; NORDIN, 1972) (see appendix).

In the light of present knowledge these tumour-associated antigens appear more complex than they seemed initially.

The following four examples which have been the object of intensive studies illustrate this idea: a) a murine neoplasia, the spontaneous leukaemia of the AkR strain (the inducing agent of which is the Gross RNA virus) and b) one human neoplasia, Burkitt's lymphosarcoma which is probably induced by a DNA virus, c) another neoplasia, the inducing agent (a RNA virus as found in animals) of which has not been demonstrated; d) bladder carcinomas.

Antigens Associated with Murine Spontaneous Leukaemia[a]. The antigenic system associated with the spontaneous leukaemia of AkR mice induced by the Gross virus (GROSS, 1970) was described as comprising two categories of antigens (Fig. 2.2): a) cell antigens induced by the virus, but not viral, and b) virion antigens, as indicated in Table 2.1 (BAUER, 1974). However, the term "virus-induced antigens" might turn out to be incorrect. It was coined at a time when it was not known whether these antigens were encoded in the viral or in the cell genomes. It seems, now, that the genetic information is brought in by the virus genome (LEVY, 1974).

More important than this classification based on semantic speculation is the distinction between the following types: a) a membrane located antigen called GCSA ("Gross cell surface antigen"), which may represent a transplantation antigen, and

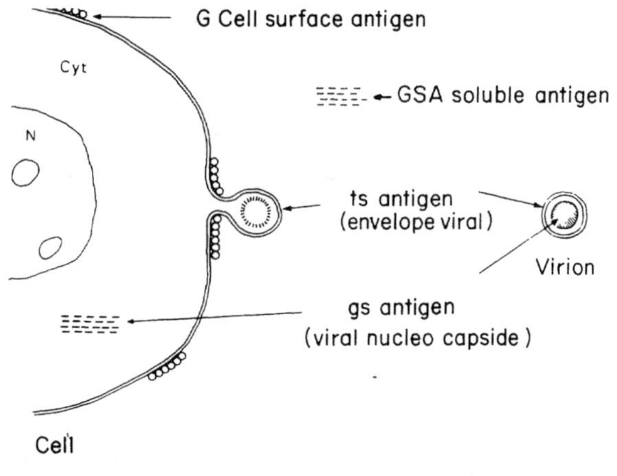

Fig. 2.2. Some antigens associated with murine spontaneous leukaemia

[a] Resistance to mice spontaneous leukaemia, which is known to be induced by Gross virus, is linked to a gene (called Rgv - 1) which behaves as a gene controlling immune resistance (Ir).

Table 2.I. Main antigens associated with murine spontaneous leukaemias (BAUER, 1974)

1. Cell surface "virus-induced antigens"[a]

Antigens	Immunological properties	Specificity for tumor cells
TSTA	Gross leukaemia specific	?
GCSAa (GSAa)	Gross leukaemia specific	-
GCSAb (GSAb)	Group specific (G and FMR leukaemias)	+
GCSAc (GSAc)	Gross leukaemia specific	?
VeA	ts and gs	-
EA	Specificity unknown	-
G_{IX}, G_L, G_T, TL	(Gross virus directed partly) act as allo-antigens in normal mice	- (TL 4 +)

2. Virion antigens

Antigens	Specificity	Localization[b] (virion proteins)
Envelope antigens (VeA)		
ts VeA	Type specific (Gross)	Virus surface projections (gp 45 and gp 70)
interspecific	gs	
haemagglutinin	ts	
H-2 ?		
Internal antigens		
MulV-gs 1	Group specific (G, FMR)	p 31
gs 3	Interspecific (= IGSA)	p 31
gs 2	Group specific	p 10

Table 2.I. (continued)

Reverse Transcriptase	Group specific	?

TSTA = Tumor specific transplantation antigen. Whether Gross virus directed TSTA is identical or not with the serologically defined GCSA is unknown.

GCSA = Gross cell surface antigen

VeA = Virus envelope antigens

EA = Embryonic antigens

G = Gross

FMR = Friend, Moloney, Rauscher

TL = Thymus-leukaemia

ts = type specific

gs = group specific

gp = glycoprotein

MulV = murine leukaemia virus

IGSA = interspecies group specific antigen

[a] See text for justification of this nomenclature

[b] Some of these antigens can be demonstrated in the cytoplasm of infected cells.

is different from viral antigens (this has been difficult to prove because oncorna viruses bud at the cell membrane). Three serologic specificities (a,b,c) of GCSA have been distinguished. This antigen is usually not specific for tumour cells (with the exception of GCSAb) and is found on all normal lymphoid cells naturally infected with the virus; b) a soluble antigen, localized in the plasma, called GSA (from "Gross soluble antigen"), serologically indentical to GCSA and which is probably the result of its exfoliation from the cell surface; c) a virus envelope antigen called ts (from "type specific"), carried by the surface of the virion. These last antigens are specific to each virus; hence a Gross virion is serologically different to a Friend's virion, another murine leukaemogenic virus, but which does not seem to play a role in spontaneous leukaemogenesis (FRIEND, 1957); d) antigens of the nucleocapside of the virus or gs (group specific) antigens: they represent internal components of the virus, and they are equally present in a soluble state in the cytoplasm of infected cells. They are called "group specific" because they are also found in all murine viruses which are leukaemogenic, spontaneously, or in the laboratory (GROSS, FRIEND, RAUSCHER, MOLONEY, GRAFFI etc.) (GROSS, 1970). Moreover, one of them, the antigen called Mulv-gs 3("murine leukaemia virus") is also found in viruses leukaemogenic for other mammal species, cats, guinea pigs etc. For this reason it has been given the name of "interspecific" antigen, or IGSA ("interspecies group specific antigen").

The available data on human tumours, though rare, are already very complex. They are summarized in Table 2.II. As indicated for a few tumours, one or several antigens have been detected or considered to have been detected.

Even if immunotherapists are probably not concerned with carcinoembryonic or foetal antigens (GOLD and FREEDMAN, 1965; Von KLEIST and BURTIN, 1966; URIEL *et al.*, 1967; TROUILLAS, 1971), and other internal (cytoplasmic or nuclear) antigens, the presence of which have been demonstrated serologically (using serums containing the specific antibodies), they already know of several human neoplasias in which cell surface antigens have been revealed, in particular Burkitt's African lymphosarcoma, leukaemias, other sarcomas, melanomas, some adenocarcinomas, neuroblastomas, and bladder carcinomas (Table 2.II).

Only three will be dealt with here: African Burkitt's lymphosarcoma and acute leukaemia, to show the complexity of the antigenic system of a tumour, the study of which is well advanced; and bladder carcinoma, to suggest that there may be some differences between human neoplasias and laboratory tumours.

Antigens Associated with Burkitt's (African) Lymphosarcoma. Though the role of Epstein-Barr-Virus (EBV) (EPSTEIN *et al.*, 1965) in the aetiology of the African Burkitt's lymphosarcoma (BURKITT and WRIGHT, 1970), is not definitely established,

Table 2.II. Antigens associated with some human tumours

Tumours	Antigens Denomination	Localisation	Cross reaction	Identification (methods, reaction techniques)	Authors
Burkitt's Lympho-sarcoma and carcinomas of the nasopharynx	MA Complex	Cell surface	Yes	Serum - autologous and/or allogeneic (IF)	KLEIN, 1966 PEARSON *et al.*, 1969
	-	Cell surface	Yes	Lymphocytes - autologous (SL)	STJERNSWARD, 1968
	VCA	Cytoplasm		Serum - autologous and/or allogeneic (from normal individuals or those with carcinoma of the upper respiratory digestive tract and infectious mononucleosis) (IF)	HENLE & HENLE 1966
	EA	Cytoplasm		Serum from patients (IF)	HENLE *et al.*, 1970
	EBNA	Nucleus		Serum from patients and normal individuals (ACIF)	REEDMAN and KLEIN 1973
Acute leukaemias	-	Cell surface	Yes	Serum - autologous (IA, IF)	DORÉ *et al.*, 1967 YOSHIDA and IMAI, 1970
	Embryonic	Cell surface	Yes	Hetero-immune serum (rabbit) (ID)	VIZA *et al.*, 1970
	-	Cell surface	Yes	Hetero-immune serum (rabbit) (TC)	MANN *et al.*, 1971
	-	Cell surface	Yes	Hetero-immune serum (monkey) (TC)	METZGAR *et al.*, 1972
	-	Cell surface	Yes	Serum from normal donors (TC)	BIAS et al., 1972
	Embryonic antigen(s) ?	Cell surface	Yes	Serum - multiparous (TC)	CULLEN and MASON, 1974
	-	Cell surface		Lymphocytes - autologous (SL) (LCL)	FRIDMAN and KOURILSKY, 1969; POWLES *et al.*,1971 LEVENTHAL *et al.*, 1971

Table 2.II. (continued)

Tumours	Antigens Denomination	Localisation	Cross reaction	Identification (methods, reaction techniques)	Authors
	–	Cell surface	YES	Cutaneous tests, soluble extract from membranes	OREN and HERBERMAN, 1971
Melanomas	–	Cell surface	No ? Yes ?	Serum - autologous (IF) Serum - autologous and/or allogeneic (IF)	LEWIS, 1967; LEWIS *et al.*, 1969 VRIES *et al.*, 1972 FOSSATI *et al.*, 1971
	–	Cell surface	Yes	Lymphocytes - autologous and allogeneic (CIT) (CMC)	HELLSTRÖM *et al.*, 1971a FOSSATI *et al.*, 1971 VRIES *et al.*, 1972
	–	Cytoplasm	Yes	Serum - autologous and/or allogeneic (IF)	LEWIS, 1967; LEWIS *et al.*, 1969 MORTON and MALMGREN, 1968 MUNA *et al.*, 1969
	–	Nucleus		Serum - autologous and/or allogeneic (IF)	LEWIS *et al.*, 1969
Sarcomas	–	Cell surface	Yes	Serum - autologous and/or allogeneic (IF)	MORTON and MALMGREN, 1968
	S	Cytoplasm		Serum - autologous and/or allogeneic (from normal individuals and parents of patients) (IF) idem (FC)	MORTON *et al.*, 1971 GIRALDO *et al.*, 1971 EILBER and MORTON, 1970a EILBER and MORTON, 1970b
Carcinoma of the colon	Carcino-embryonic antigen (CEA)	Cell surface and Plasma		Hetero-immune serum (ID) (IF)	GOLD and FREEDMAN, 1965 Von KLEIST and BURTIN, 1966

Table 2.II. (continued)

	NCA	Cell surface and plasma	with CEA Yes	Hetero-immune serum (IF)	Von KLEIST *et al.*, 1972
		Cell surface	(+ with colon and calf liver)	Lymphocytes - autologous and allogeneic (CIT)	HELLSTRÖM *et al.*, 1970
	Embryonic antigen (CEA)	Cell surface		Cutaneous tests, soluble extract from membranes	HOLLINSHEAD *et al.*, 1970
Carcinoma of the bladder	-	Cell surface	Yes	Autologous and/or allogeneic serum (GC)	BUBENIK *et al.*, 1970a
				Autologous and/or allogeneic lymphocytes (CMC)	BUBENIK *et al.*, 1970b
	TTSA	Cell surface	Yes	Allogeneic lymphocytes (CMC)	HELLSTROM and HELLSTROM, 1970a, 1970b

Abbreviations

IF : immunofluorescence
SL : stimulation in vitro of lymphocytes
IA : immunoadherence
ID : immunodiffusion
TC : test of cytotoxicity
MA : membrane antigen
EA: : early antigen
VCA : virus capside antigen
CEA : carcino-embryonic antigen

NCA : normal colon antigen
EBNA : EB nuclear antigen
ACIF : anticomplement immunofluorescence
LCL : Lymphocytolysis
FC : Fixation of complement
CIT : Colony inhibition test
CMC : Cell mediated cytotoxicity
TTSA : Tissue type specific antigen

its role is fairly certain (KLEIN, 1974), and experience of in vitro infection of human cells has demonstrated that antigens described in and on African Burkitt's lymphosarcoma cells are induced by EBV.

Four types of antigens and several subgroups have been individualized (Fig. 2.3). Three appear in EBV infected cells only during the lytic cycle which normally ends with virus production. They are the following: (a) "early antigens" (EA), demonstrated by immunofluorescence in the cytoplasm; the appearance of EA indicates the entry of the cell into the lytic cycle and is accompanied by a progressive inhibition of normal cell synthesis; (b) "membrane antigens" (MA) are also demonstrated by immunofluorescence on the membrane of the infected cells, and the immunoferritin method has shown that they are on the viral membrane; these MA appear early after infection of the cell; (c) "viral capside antigens" (VCA), also demonstrated by immunofluorescence; they only appear late, just before cell lysis and production of complete virions; (d) the nuclear antigen, which is complement fixing, and is also called "Epstein-Barr-Nuclear-Antigen" (EBNA); it is the most interesting as it is present in the nucleus of the cell after an abortive but transforming infection, while, in such a type of infection, the other antigens are absent. This EBNA is the most precious imprint of the presence of the viral genome in the cell.

Antigens Associated with Human Acute Leukaemias. In the case of acute lymphoid leukaemia (ALL), our present knowledge of tumour associated antigens is far less advanced (HARRIS, 1973; DORÉ and KOURILSKY, 1974). DORÉ *et al.*, (1967, 1970a) found, in the serum of ALL patients, antibodies reactive with autologous leukaemia cells. Similar findings were reported by MARUYAMA *et al.*, (1968) and by YOSHIDA and IMAI

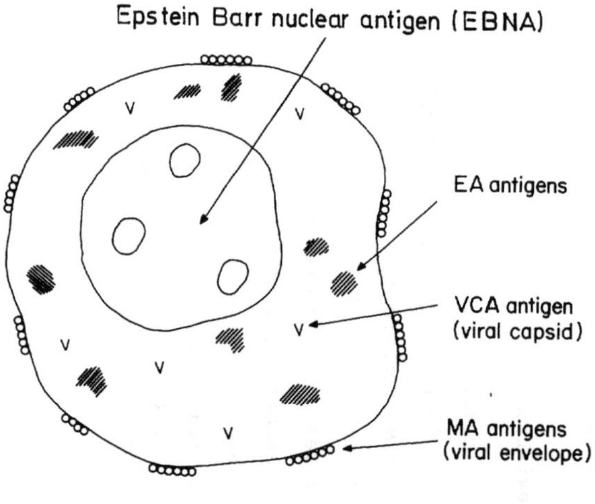

Fig. 2.3. Main antigens associated with Burkitt's lymphosarcoma

(1970). However, such antibodies are rather infrequent and very weak. Direct demonstration of IgG bound to the surface of leukaemic cells has also been reported (GUTTERMAN *et al.*,1973a; PEGRUM, 1973), but the specificity of these immunoglobulins is not yet clear. More recently, using permanent cell lines established from the peripheral blood of ALL or lymphosarcoma patients (BELPOMME *et al.*, 1969) and the corresponding autologous sera in an absorption typing assay, DORÉ and GUIBOUT (1974) identified a cell surface antigen associated with most of the ALL. Human leukaemia associated antigens have also been recognized using sera from normal individuals (BIAS *et al.*, 1972; GREENSPAN *et al.*, 1963), from immunized patients (MANN *et al.*, 1975), and from animals immunized with human leukaemia cells (METZGAR *et al.*, 1972; BAKER and TAUB, 1973) or subcellular fractions (HARRIS *et al.*, 1971; MANN *et al.*, 1971).

These serologic data are further supported by the demonstration that lymphocytes from ALL patients in remission may react against autologous blasts taken in a perceptible phase of the disease and stored in the frozen state (FRIDMAN and KOURILSKY, 1969; POWLES *et al.*, 1971; LEVENTHAL *et al.*, 1972), or against solubilized membrane components of ALL cells (GUTTERMAN *et al.*, 1972). It has also been claimed by several investigators that normal lymphocytes from identical twins of ALL patients, or from HL-A identical siblings, can react in vitro against ALL cells (BACH *et al.*, 1969; PEGRUM *et al.*, 1972; HAN and WANG, 1972; ROSENBERG *et al.*, 1972), although this is still controversial (SCHWEITZER *et al.*, 1973). Perhaps the most important work may be that of ANDERSON *et al.* (1974) concerning 38 adult acute leukaemia patients and 91 of their normal siblings who were studied for cell-mediated immune reaction to isogeneic leukaemic cells. HLA identical siblings (Joint Report Hist. Workshop, 1970, 1972) responded to leukaemic cells with a frequency of 84% in mixed leucocyte cultures (MLC), 42% in macrophage migration inhibition (MIF) test, and 73% in ^{51}Cr release cell-mediated immune cytotoxicity (CMIC) assay. For MLC, no evidence was found of stimulation due to antigens of the normal blasts present in normal bone marrow. MLC, MIF, and CMIC responses to autologous frozen-stored leukaemic cells were found when remission lymphocytes were tested. This data is very much in favour of the existence of leukaemia associated antigens and of the existence of cell mediated immune responses in many patients and siblings.

Antigens Associated with Chemically Induced Murine and Apparently Spontaneous Human Bladder Tumours. Human bladder cancer is specially interesting as far as comparative pathology is concerned. For example, a cross reaction has often been found in human neoplasias, while, in experimental neoplasias, it has only been described in virus-induced tumours. Do the cross reactions found in human cancers mean that most are virus-induced, although ecologists estimate that 80-90% are due to environmental factors (EPSTEIN, 1974)?

It may be that the scheme deduced from the first experimental data was too quickly accepted and too strongly imposed. HELLSTRÖM and HELLSTRÖM (1974a) showed that bladder carcinomas of rats or mice induced by 3-methylcholanthrene, which should not, theoretically, cross react, in fact do so. Finally, they showed that these bladder carcinomas and papillomas cross react. Thus the antigens found in malignant tumours may be found in benign ones.

These antigens are referred to as "tissue type specific antigens" (TTSA). They have also been detected in chemically-induced carcinomas of the colon (STEELE and SJÖGREN, 1974a) and suspected to be present in chemically-induced mammary carcimonas (ANKERST *et al.*, 1974).

A certain amount of cross reactivity has been detected between some tumour antigens and normal tissue antigens (MARTIN and MARTIN, 1970).

They are similar to organ antigens known in auto-immunization (see GLYNN and HOLBOROW, 1964), and they may not be very different from commonly found carcino-embryonic antigens (see above). Some have been shown to be expressed in embryos (STEELE and SJÖGREN, 1974b; MARTIN *et al.*, 1975).

In any case, new antigens are not peculiar to malignant tumours. They have been observed in preneoplastic tissues (WEISS, 1972; WEISS *et al.*, 1971).

However, it has been shown that pieces of skin into which Friend's virus (MATHÉ, 1967a) or dimethylbenzanthracene (MATHÉ, 1967b) were injected, were rejected when grafted into isogeneic hosts, while they were not transformed into malignant tissues. The author proposed, for these experiments, the concept by which antigens might be induced by the carcinogens, without or before malignant transformation, and this might explain the discovery of antigens in normal tissues, the persistence of antigens during remission, the presence of antibodies or lymphocyte reactivity against tumour cells in normal subjects, and also cross-reactivity.

All of these data allowed the author to use the term "tumour-associated antigens" (TAA) and to avoid the terms "tumour specific immunity" (BURNET, 1970), and "tumour specific transplantation antigens" (KOLDOVSKY, 1969). But they do not decrease the potential of immunotherapy. Even if some antigens in some tumours are due to contaminant viruses, and this is probably frequent, even in chemically induced tumours (BARBIERI *et al.*, 1971), they may still be good targets for immune reactions (BARBIERI *et al.*, 1971).

Hence, more important than specificity of the TAA are other characteristics.

The *structure* of (surface) tumour-associated transplantation antigens is the object of much study. There is a probable functional and/or anatomical relationship between them and some histocompatibility antigens (HA): the expression of the latter measured by a quantitative cytotoxicity test is modified in leukaemic cells compared to normal control cells, and varies also from one leukaemia to another of the same strain of mice as shown by MOTTA (1974) in the author's laboratory. Normal antigenicity can be decreased (HAYWOOD and McKHANN, 1970) or increased (PEACOCKE *et al.*, 1966); on the same cells, some specificities can be decreased, others increased (MOTTA, 1974).

TAA have been prepared in hydrosoluble extracts by the methods used for histocompatibility antigens (HA) in mice by DAVIES *et al.*,(1966) and, in the author's laboratory, by MARTYRÉ *et al.*,(1974). These authors showed that they are probably, like the HA, glycoproteins with a molecular weight of about 50.000 (Fig. 2.4). Some have already tried to extract TAA from human leukaemic cells and claim that they have been able to produce an antiserum specific for all leukaemic extracts which they have prepared (VIZA *et al.*, 1970). Further work of some of these authors showed

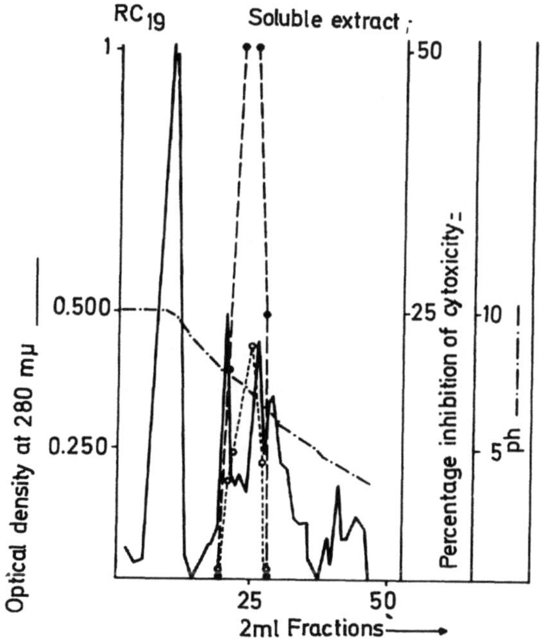

Fig. 2.4. Characterization and purification of Rauscher leukaemia-associated transplantation antigens: isoelectric focusing between pH3 and pH10 of the soluble fraction obtained by autodigestion from murine RC19 (Balb/c) tumour cell membrane. ——— absorbance at 280 mμ – – – – – pH •-------• inhibition of cytotoxicity of anti-FMR antiserum o--------o inhibition of cytotoxicity of anti-H-2^b antiserum (MARTYRE et al., 1974)

that this "anti-leukaemia" serum reacts with the serum of acute and chronic leukaemia patients, as well as that of subjects with other haematological disorders and against the sera of foetuses. Hence the antigen(s) detected may be embryonic ones.

TAA strength varies from tumour to tumour: for example the methylcholanthrene-induced sarcomas are more antigenic than these induced by dibenzanthracene (OLD *et al.*, 1962). Moloney virus murine leukaemia is more antigenic than the Gross virus leukaemia (KLEIN and KLEIN, 1964).

TAA are generally stable, but their expression is modified with the cell cycle. CIKES (1971) showed that Moloney virus-induced leukaemic cells are not sensitive to a cytotoxic serum during the S phase, while they are during the other phases. It seems that the antigenicity is maximally expressed during the G1 phase.

Another factor for variation of antigenicity of tumour cells is their modulation by specific antibodies (BOYSE *et al.*, 1968a).

Sialic acid (Fig. 2.5) residues have been considered as playing an important part in *antigenicity and/or immunogenicity* of tumour cells. They could be responsible for a steric hindrance to antigen recognition and for an increment of negative charge of the cell (SIMON-REUSS *et al.*, 1964), hence decreasing deformability (WEISS, 1965) and susceptibility to phagocytosis (WEISS *et al.*, 1966; LEE, 1968).

Relations at the cell surface between malignancy, antigenicity, and immunogenicity are suggested by the action of concanavalin. This lectin is known to react with polysaccharides and glycoproteins (GOLDSTEIN and IYER, 1966), and to agglutinate many types of tumour cells, but not normal cells unless they have been pre-exposed to trypsin (INBAR and SACHS, 1969). Furthermore, at high concentration, it is able to mask strong histocompatibility antigens on normal cells and to increase immunogenicity of tumour-associated antigens on neoplastic cells (SIMMONS and RIOS, 1974a,

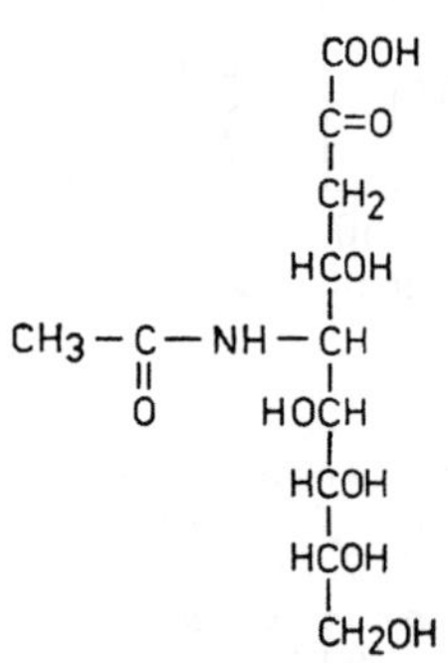

Fig. 2.5. Chemical formula of N-acetyl neuraminic (or sialic) acid

1974b); it shows that relations, which are not explained, do exist at the cell surface between malignancy, antigenicity, and immunogenicity.

LAPPE (1972) has shown that this escape to what he called "maladaptive host - immune responses" does exist also for preneoplastic tissues such as papillomas.

2.2. The Immune Machinery

Basic immunology has been the topic of several books (CAMPBELL *et al.*, 1970; BELLANTI, 1971; WEIR, 1973a, 1973b; ROITT, 1974; BIGLEY, 1975) during the last five years, and the following is only to give an idea of the present concept of the immune machinery, which has been elaborated, step by step, since the beginning of this century.

2.2.1. Humoral Mediated Immunity and Cell Mediated Immunity

The first important step was the distinction between (a) *humoral immunity* which was shown to be mediated by *antibodies* by Von Behring in 1890, identified first as serum globulins by TISELIUS and KABAT (1939) and later found to belong to different classes of *immunoglobulins* (Fig. 2.6) (EDELMAN and POULIK, 1961; PORTER 1959, 1962; Cold Spring Harbor Symposium Q.B., 1967), and (b) *cell-mediated immunity*, which is expressed in *delayed hypersensitivity* reactions.

Humoral immunity can give rise to *immediate hypersensitivity* reactions, which can be transferred "passively" to unsensitized individuals by means of the serum of sensitized donors. In contrast, delayed hypersensitivity, of which the typical example is the cutaneous reaction to intradermal injection of tuberculin in subjects sensitized to tubercle bacillus, can be transferred "adoptively" by injection of lymphoid cells from a sensitized donor, and not by the serum (LANDSTEINER and CHASE, 1942).

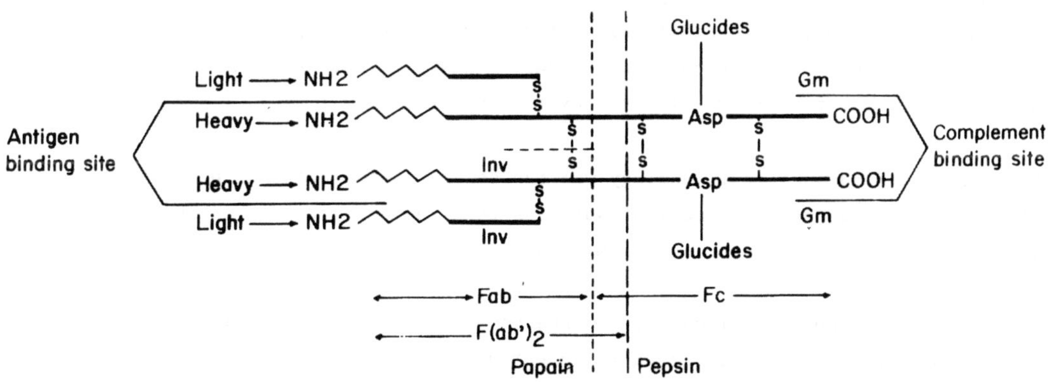

Fig. 2.6. Scheme of the present concept of the immunoglobulin molecule

The next major step came in 1963 with the identification of the *small lymphocyte* as the cell which initiates cell-mediated, as well as humoral immune reactions, and which carries *immunological memory* (GOWANS and McGREGOR, 1963; GOWANS *et al.*, 1963; GOWANS and McGREGOR, 1965; GOWANS and UHR, 1966). Moreover, recirculating properties of small lymphocytes were demonstrated (McGREGOR and GOWANS, 1963).

During the last decade, a revolution has taken place with the recognition that two classes of lymphocytes exist, one involved in cell-mediated immunity, the other in humoral immunity; one dependent of the presence of the *thymus*, and the other not.

This two-cell concept arises from experimental and clinical observations. In 1962, MILLER reported that neonatal thymectomy in mice causes a profound reduction in their subsequent immune development. Simultaneously, WARNER *et al.*,(1962) and later, COOPER *et al.*, (1966), showed that, while neonatal thymectomy in chickens as in mice, is followed by a profound defect in cell-mediated immunity, the removal of the *bursa of Fabricius* (also called "cloacal thymus") in chickens induced a state of humoral unresponsiveness.

Three congenital immunological diseases were described in human babies; a) the postbursectomy manifestations called *agammaglobulinaemia* (BRUTON, 1952); b) the postthymectomy immunological manifestations which are due to a congenital absence of thymus (Di GEORGE, 1968), and c) *Swiss-type agammaglobulinaemia* (GLANZMANN and RINIKER, 1950), combining the manifestations of the other two and supposedly due to a lack of (lymphocyte?) stem cells, curable by bone marrow graft (BUCKLEY, 1971).

All of these studies and observations demonstrate the existence of two types of lymphocytes (Fig. 2.7), according to the *primary lymphoid organ*, thymus or bursa of Fabricius, where they differentiate before migrating into the *secondary lymphoid organs* (spleen, lymph nodes, Peyer's patches) when they express their respective immunological functions. The thymus-derived lymphocytes, or *T-lymphocytes*, are responsible for cell-mediated immunity; the bursa-derived lymphocytes in avians, and the bursa-equivalent (bone marrow?) derived lymphocytes in mammals, or *B-lymphocytes*, are the precursors of antibody-forming cells.

The next major advance in immunology was the demonstration that the production of the antibodies against a variety of antigens requires co-operation between T- and B-lymphocytes (CLAMAN *et al.*, 1966; MILLER and MITCHELL, 1968).

A third type of mononuclear cells, which belongs to the *monocyte* series, participates in immunological reactions, either at an early stage in the cellular sequence resulting in antibody formation, and/or as nonspecific effector cells of cell-mediated immunity (Fig. 2.7).

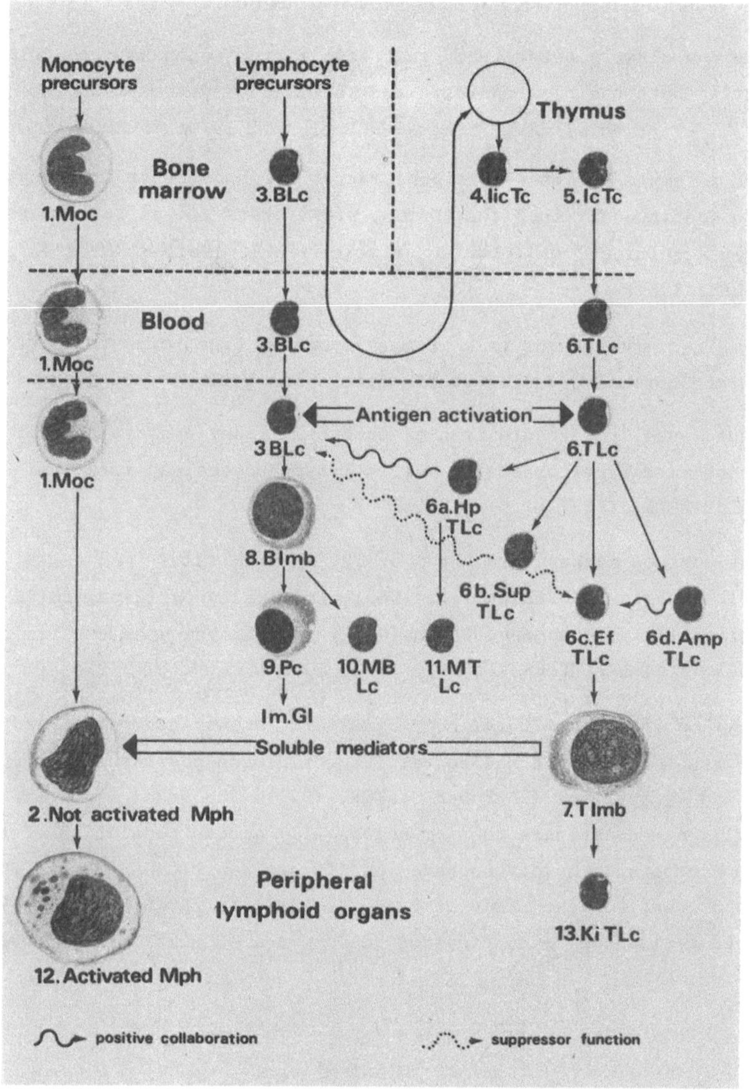

Fig. 2.7. 1. Moc = Monocyte, 2. Nonactivated MpH = Nonactivated macrophage, 3. BLc = B-lymphocyte, 4. Iic Tc = Immunoincompetent thymocyte, 5. Ic Tc = Immunocompetent thymocyte, 6. TLc = T-lymphocyte, 6a. Hp TLc = "Helper" T-lymphocyte, 6b. Sup. TLc = suppressor T-lymphocyte, 6c. Ef TLc = "Effector" T-lymphocyte, 6d. Amp TLc = "Amplifier" T-lymphocyte, 7. T Imb = T immunoblast, 8. B Imb = B immunoblast, 9. Pc = Plasma cell , 10. MBLc = Memory B-lymphocyte, 11. MTLc = Memory T-lymphocyte, 12. Activated Mph = Activated macrophagë, 13. Ki TLc = "Killer" T-lymphocyte.
Simplified scheme of physiological relationsship between different types of lymphocytes and between these and macrophages (kindly prepared by D. DANTCHEV)

2.2.2. T- and B-Lymphocytes and Monocyte Differentiation

It is now clearly established that bone marrow contains stem cells capable of repopulating both the hematopoietic and the immune systems (FORD *et al.*, 1956; TRENTIN *et al.*, 1967; WU *et al.*, 1968; NOWELL *et al.*, 1970; EDWARDS *et al.*, 1970).

There is ample evidence that bone marrow lymphocytes or lymphocyte precursors migrate continuously into the thymus, proliferate and differentiate there, concomitantly acquiring T-cell characteristics (POPP, 1961; FORD *et al.*, 1966; ORDER and WAKSMAN, 1969).

The role of the thymus as a primary lymphoid organ appears to be mediated through a thymic hormone produced by the epithelial cells of the gland (TRAININ, 1974).

In the mouse, the acquisition of surface alloantigens is part of the differentiation of bone marrow precursors to mature T-lymphocytes and marks this process (SCHLESINGER, 1972).

These surface markers are the θ or Thy antigen (REIF and ALLEN, 1964; RAFF, 1969), the TL antigens (BOYSE *et al.*, 1965), and the Ly antigens (BOYSE *et al.*, 1968b,1971) (Table 2.III). In humans, the majority of T-lymphocytes acquire receptors for sheep red blood cells (WYBRAN *et al.*, 1972; JONDAL *et al.*, 1972).

Ninety to 95% of the total *thymocytes* are immuno-incompetent and probably "immature". They are localized in the cortex (Fig. 2.8) and possess a high concentration of θ, TL, and Ly antigens (RAFF and CANTOR, 1971). The other 5-10% are thymocytes of the medulla and constitute the immunologically active population of the organ. These immunocompetent thymocytes have lost TL antigens (LECKBAND and BOYSE, 1971) and possess a lower concentration of θ and Ly antigens (KONDA *et al.*, 1973). They have acquired resistance to corticosteroid hormones (SCHLESINGER and GOLAKAI, 1967;

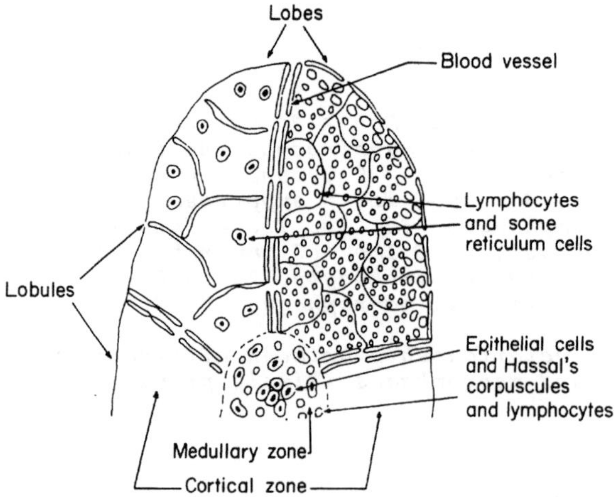

Fig. 2.8. Thymus structure

Table 2.III. Main immunological surface markers of lymphocytes and monocytes

a) Thymus-derived (T-) lymphocytes		
Mouse	θ antigen	expressed on mouse thymocytes and peripheral T-lymphocytes, a number of murine leukaemias and nervous tissue
	TL antigen	expressed on mouse thymocytes and on a number of mouse leukaemic cells (OLD *et al.*, 1968)
	Ly antigen	expressed on mouse thymocytes and peripheral T-lymphocytes
Human	Receptors for sheep erythrocytes present on thymocytes and peripheral T-lymphocytes	
b) Bone marrow-derived (B-) lymphocytes		
	Immunoglobulin receptors able to bind antigenic determinants	
	Receptors for the Fc region of IgG molecule able to bind antigen-antibody complexes	
	Receptors for the third component of complement (C'3) capable of binding antigen-antibody-complement complexes	
c) Monocytes		
	Receptors for the Fc fragment of IgG	
	Receptors for the third component of complement (C'3)	
d) Null cells		
	Receptors for the Fc portion of IgG molecule	
	?Receptors for the third component of the complement (C'3)	

BLOMGREN and ANDERSSON, 1969; COHEN *et al.*, 1970) and are capable of recirculation (LANCE and TAUB, 1969). There is strong evidence that these cells represent "mature" competent T-cells which are ready to migrate into the secondary lymphoid organs (WEISSMAN, 1967, 1973).

T-lymphocytes occupy predominantly paracortical regions of lymph nodes and periarteriolar zones of the splenic white pulp, which are consequently referred to as *thymus T-dependent areas* (Figs. 2.9 and 2.10). They are depleted after neonatal thymectomy and in Di George's disease (MILLER, 1962; Di GEORGE, 1968).

In secondary lymphoid organs, B-lymphocytes are found mostly in the cortical region including primary follicles and in the medullary cords of the lymph nodes, in the primary follicles and the red pulp of the spleen (PARROTT *et al.*, 1966; TURK and OORT, 1967). These *B-dependent areas* (Figs. 2.9 and 2.10) are depleted after bursectomy in avians and in children with agammaglobulinaemia (BRUTON, 1952). B-lymphocytes are distinguished from T-lymphocytes by the presence of surface receptors (Table 2.III) (FRÖLAND and NATVIG, 1973): a) detectable immunoglobulin-receptors, able to bind antigenic determinants specifically (RAFF, 1970; RABELLINO *et al.*, 1971; FRÖLAND and NATVIG, 1971); b) receptors for the Fc region of IgG molecule, able to bind antigen-antibody complexes (BASTEN *et al.*, 1972a, 1972b; FRÖLAND *et al.*, 1973); c) receptors for the third component of complement (C'3), capable of binding antigen-antibody-complement complexes (BIANCO *et al.*, 1970).

Little is known about the development of stem cells to immunocompetent B-lymphocytes. Bone marrow is the most likely site where this differentiation can take place in mammals (ABDOU and ABDOU, 1972; NOSSAL and PIKE, 1974). Recent studies have shown that bone marrow, in mice, contains B-cell precursors which are nondividing Ig-baering cells with already restricted specificity. They give rise to proliferating large transitional B-cells which differentiate into non-dividing mature B-lymphocytes (LAFLEUR *et al.*, 1972, 1973). Other authors have reported that a large fraction of bone marrow small lymphocytes bear neither T- nor B-lymphocyte surface markers. These cells arise from dividing large lymphoid cells, which appear similar to the transitional B-cells of LAFLEUR *et al.* (1972, 1973), and they rapidly acquire B-cell surface markers (RYSER and VASSALI, 1974; OSMOND and NOSSAL, 1974).

The third type of mononuclear cell which participates in immunological reactions belongs to the *monocyte series*. The monocyte which is supposed to derive from a

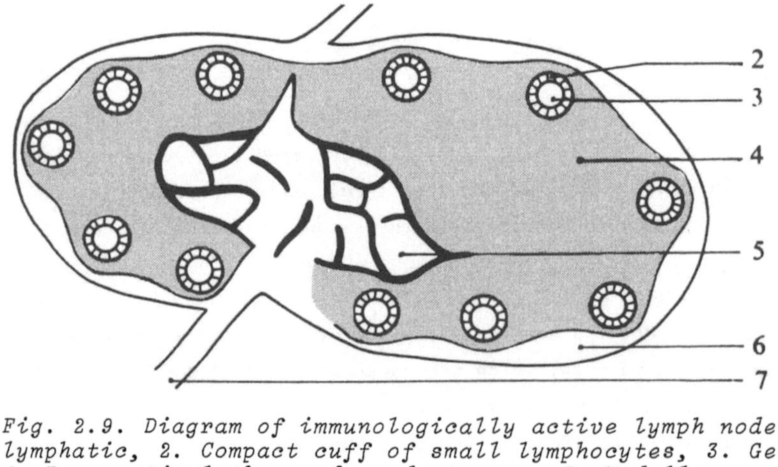

Fig. 2.9. Diagram of immunologically active lymph node. 1. Afferent lymphatic, 2. Compact cuff of small lymphocytes, 3. Germinal centre, 4. Paracortical thymus dependent area, 5. Medullary cords (plasma cells and macrophages), 6. Marginal sinus, 7. Efferent lymphatic (Taken from TURK, 1969)

Fig. 2.10. Diagram showing the thymus dependant area in the spleen (Taken from TURK, 1969)

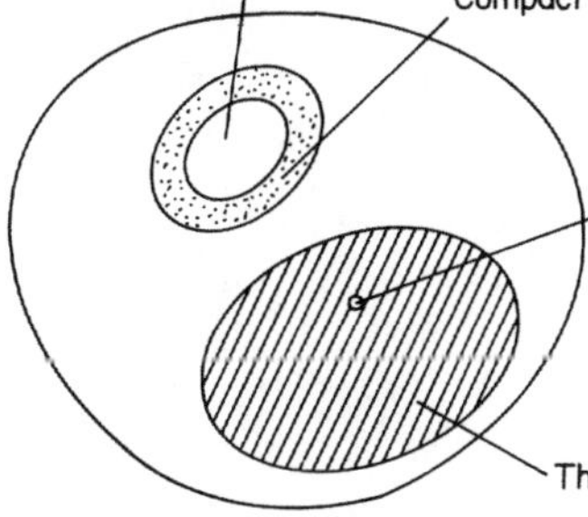

rapidly proliferating precursor cell in bone marrow, called "promonocytes", enters the blood, and migrates into connective tissue spaces where it differentiates into macrophage (VAN FURTH, 1970). The monocytes possess receptors for the third complement component (HUBER *et al.*, 1968; LAY and NUSSENZWEIG, 1968) (Table 2.III).

A fourth type of cell has recently been described in human peripheral blood. Unclassifiable as either T or B lymphocytes by the conventional surface markers (receptors for sheep erythrocytes, membrane-bound immunoglobulins), they are called *null cells*. They possess receptors for the Fc portion of IgG molecules and probably receptors for the third component of the complement. They have not yet been classified as belonging to B-lymphocyte or monocyte subpopulations (JOSEPH *et al.*, 1976; BELPOMME *et al.*, 1976).

T-lymphocytes, B-lymphocytes, and monocytes were said to be distinguishable morphologically by scanning electron microscopy (POLLIACK *et al.*, 1973). T-cells were described as smooth, B-cells as villous, and monocytes as ondulated, but, in fact, transitional aspects can be seen (Fig. 2.11), and some villous lymphocytes can make sheep red blood cell rosettes as do smooth cells (Fig. 2.12). According to DANTCHEV *et al.* (1976), thymocytes and T-lymphocytes are slightly ondulated, B-lymphocytes are covered by very small villosities (Fig. 2.11), and long villosities are functional aspects or artefacts of preparation. These are the different components of the immune machinery; a summary of their respective immunological functions follows.

2.2.3. T-Lymphocyte Functions

In response to antigenic stimuli leading to cell-mediated reactions, T-lymphocytes are transformed from small lymphocytes into large dividing cells (DAVIES *et al.*, 1966) with pyrininophylic cytoplasm and called *immunoblasts* by DAMESHEK (1963).

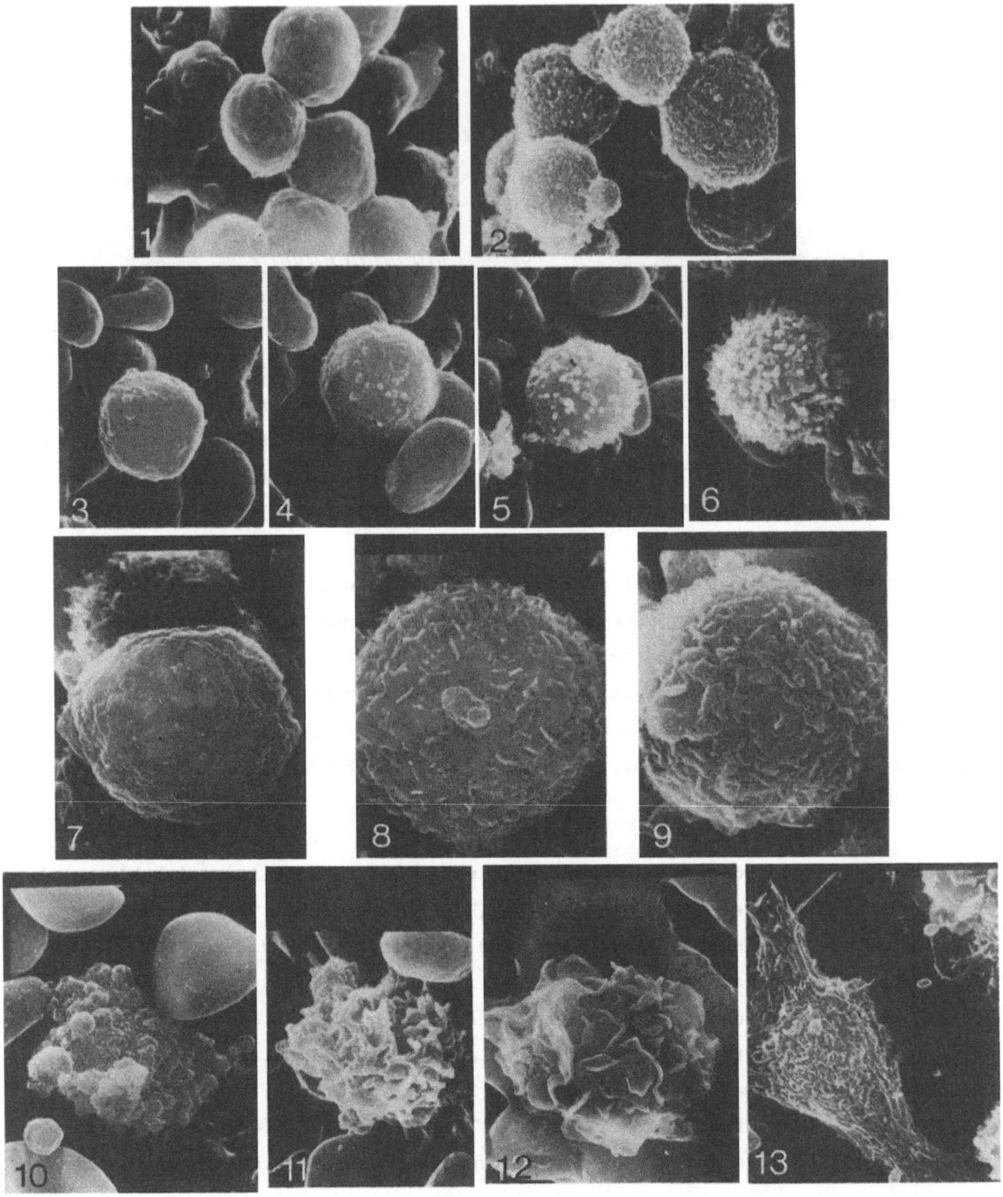

Fig. 2.11. Blood or lymphoid tissue mononuclear cells at scanning electron microscopy. 1. Thymus cell, 2. Bursa of Fabricius cell, 3. Smooth blood lymphocyte, 4. Poorly villous blood lymphocyte, 5. Averagely villous blood lymphocyte, 7. Thymus-dependent immunoblast, 8. and 9. Other types of immunoblasts, 10, Plasma cell, 11. Blood monocyte, 12. Tissue histiocyte, 13. Activated histiocyte (DANTCHEV, 1976)

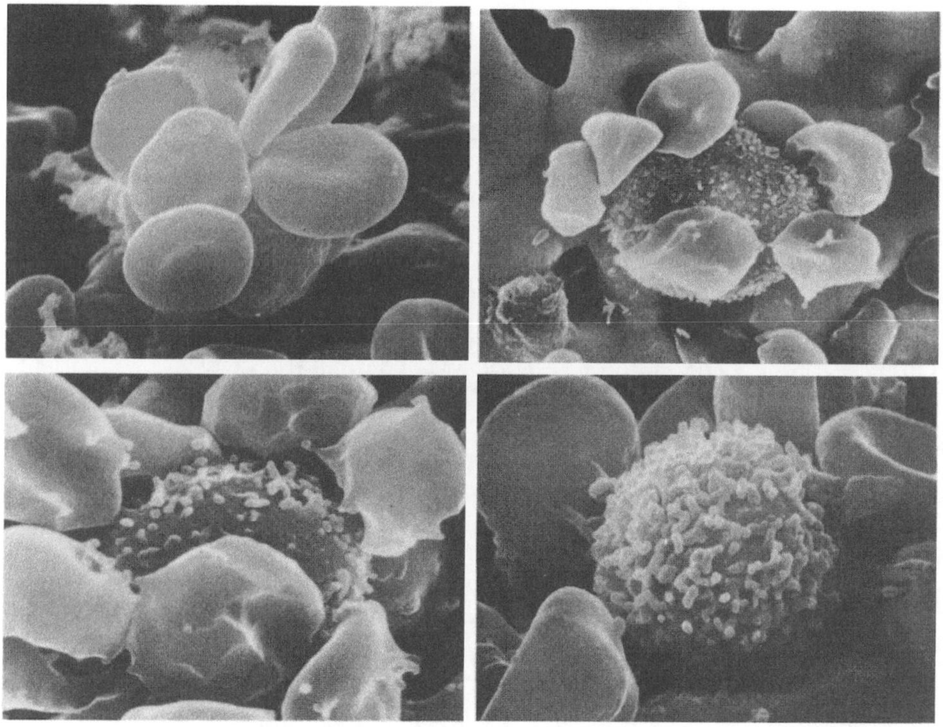

Fig. 2.12. Though smooth cells have been considered initially as T-lymphocytes and villous as B-lymphocytes, sheep red blood cell rosettes are made by smooth lymphocytes and by lymphocytes with long villosities (DANTCHEV, 1976)

These cells have been described in the case of the graft-versus-host reaction, and the absence of endoplasmic reticulum in their cytoplasm has been shown contrasting with the presence of numerous ribosomes and polyribosomes (BINET and MATHÉ, 1962) (Fig. 2.13).

In vitro, some lectins (phytohemagglutinin, concanavalin A) induce T-lymphocyte transformation into cells morphologically similar to immunoblasts (IMMAN and COOPER, 1963; CHALMERS *et al.*, 1966; JANOSSY and GREAVES, 1972).

T-lymphocytes from sensitized individuals, placed in vitro in the presence of the sensitizing antigen, are able to synthetize and excrete soluble factors, *lymphokines*, possessing a spectrum of activities, and which are considered as mediators of cell-mediated immunity (LAWRENCE and LANDY, 1969). Some of them act on macrophages (migration inhibiting factor: MIF; macrophage activating factor: MAF etc.); others on lymphocytes (transfer factor, mitogenic factor); others on neutrophils (chemotactic

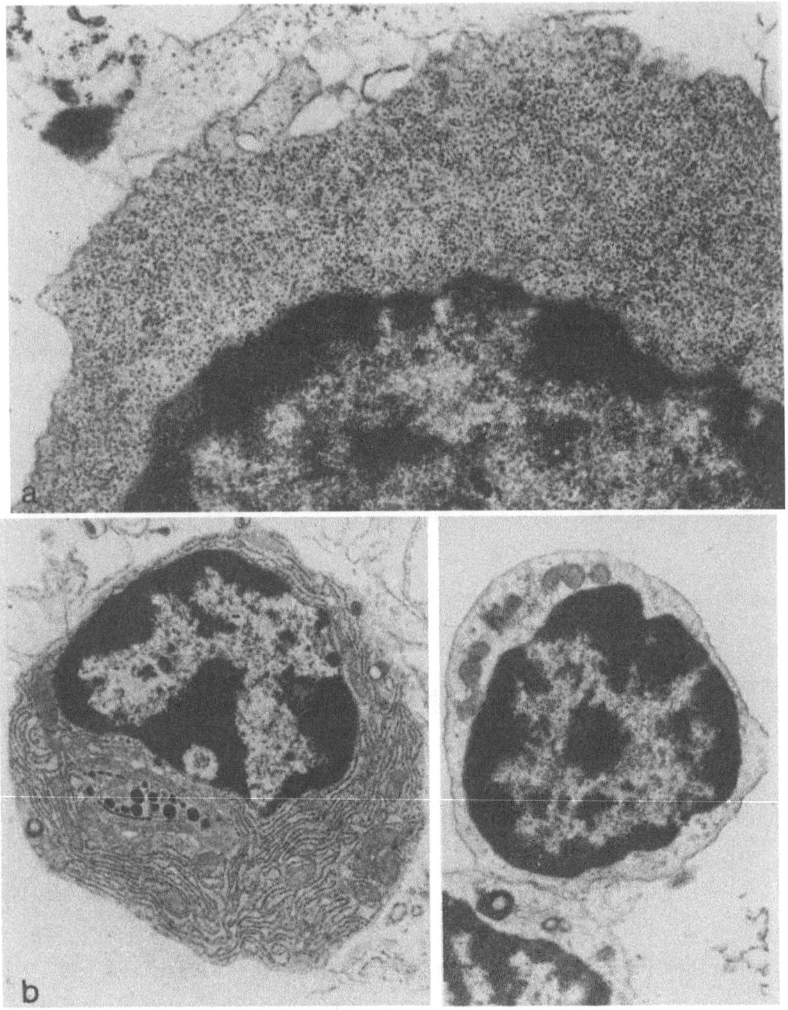

Fig. 2.13. Electron-microscopical aspect of immunoblast (upper) (many free polyribosomes), compared to plasma cell (below left) (see ergastoplasm), and to lymphocyte (below right)

factor; leucocyte inhibiting factor: LIF), and on various cells (lymphotoxins, interferon etc.).

When the antigen is a normal or tumoral constituent of a cell membrane, T-lymphocytes differentiate into cells endowed with a specific cytotoxic activity towards cells carrying the same antigenic determinants (CEROTTINI *et al.*, 1970).

The activity does not depend upon the presence of antibodies and complement, but results form the direct contact between the *cytolytic T-lymphocyte* (CTL) and the target cell via specific receptors on the surface of CTL, complementary to antigenic determinants on the surface of target cells (GOLDSTEIN *et al.*, 1971). Formation of CTL is observed in vivo during host-versus-graft and graft-versus-host reactions, tumour immunity and, in vitro, when culturing normal T-lymphocytes with allogeneic lymphoid cells (see CEROTTINI and BRUNNER, 1974).

It appears from recent studies that two types of T-lymphocytes co-operate in the development of certain cell-mediated immune responses. Termed amplifier and effector T-cells by Cantor, they interact synergistically in the development of graft-versus-host reaction (CANTOR and ASOFSKY, 1972) and in the induction of CTL in vitro (WAGNER, 1973; COHEN and HOWE, 1973). They can be differentiated on the basis of their anatomical origin, life span, and recirculating properties.

T-lymphocytes are also implicated in humoral immunity. Antibody formation against most naturally occurring antigens (thymus-dependent antigens) requires co-operative interaction between T- and B-lymphocytes (CLAMAN *et al.*, 1966; MILLER and MITCHELL, 1968; TAYLOR, 1969). The nature of this *T-cell helper function* has not yet been fully clarified. T-cells may bind antigens via specific membrane receptors, providing collected antigen directly to B-cells (MITCHISON *et al.*, 1970). A second hypothesis postulates that B-cells must receive two separate signals to be triggered: signal 1 is delivered from its own immunoglobulin-receptors after binding of antigenic determinants, and signal 2 is provided by activated T-cells (BRETSCHER and COHN, 1970). Evidence has now been accumulated to support the idea that signal 2 is a diffusable chemical mediator. T-cell factors, which can replace T-cells in the induction of an antibody response, are numerous. Some of them are antigen-specific (FELDMANN and BASTEN, 1972; GISLER *et al.*, 1973; YU and GORDON, 1973; TAUSSIG, 1974) and are described as secreted T-cell receptors which resemble monomeric IgM (FELDMANN *et al.*, 1973) and which bind antigen at the surface of macrophages (FELDMANN, 1972a, 1972b) or as a product of immune response genes of the I region of the major histocompatibility complex (TUASSIG and MUNRO, 1975). Other factors are produced by T-cells after antigenic or mitogenic stimulation and have a non-specific stimulatory effect on B-cells (DUTTON *et al.*, 1971; SCHIMPL and WECKER, 1972; DORIA *et al.*, 1972; GORCZYNSKI *et al.*, 1972; ARMERDING and KATZ, 1974).

During the last five years it has become obvious that T-lymphocytes can also exert an active suppressing influence on immune responses. Specific T-cell suppression has been reported in a variety of in vivo situations: graft-versus-host reaction (GERSHON *et al.*, 1972; HARDIN *et al.*, 1973; LAPP *et al.*, 1975), contact hypersensitivity (ZEMBALA and ASHERSON, 1973), allotypic suppression (HERZENBERG *et al.*,

1973), tolerance induction (GERSHON and KONDO, 1971a); HA and WAKSMAN, 1973; HUCHET and FELDMANN, 1974; PHILLIPS-QUAGLIATA, 1975), and antibody formation against a hapten-carrier conjugate (TADA *et al.*, 1972; TADA and TAKEMORI, 1974).

In vitro systems make it possible to demonstrate that T-cell suppression can be mediated via a T-cell product which appears to be similar to the cooperative factor involved in B-cell triggering, and which probably acts when combined with the antigen (TAUSSIG *et al.*, 1974; FELDMANN, 1974; ZEMBALA and ASHERSON, 1974). A specific suppressor factor has also been extracted from the thymus of hyperimmunized animals and it may be identical to T-cell receptors (OKOMURA and TADA, 1974).

Specific T-suppressor cells, can with the co-operation of macrophages, exert a nonspecific inhibitory effect in the presence of the inductive antigen (ASHERSON and ZEMBALA, 1974).

Nonspecific T-suppressor cells are considered as responsible for the phenomenon of antigenic competition (GERSHON and KONDO, 1971b); SJOBERG and BRITTON, 1972) and for the regulation of antibody responses against thymus-independent antigens (KERBEL and EIDINGER, 1972; ROTTER and TRAININ, 1974; BAKER *et al.*, 1974). They have also been induced in vitro by mitogens (DUTTON, 1973; RICH and PIERCE, 1973; FOLCH *et al.*, 1973, 1974).

Recent findings strongly suggest that these different T-cell functions are exerted by different subsets of T-cells, which can be distinguished by cell size and antigen binding properties (ELLIOTT and HASKILL, 1974), by the presence of different LY antigens at the cell surface (SHIKU *et al.*, 1975) and by their life span (KAPPLER *et al.*, 1974; JANEWAY *et al.*, 1975).

2.2.4. B-Lymphocyte Functions

B -lymphocytes are caused to proliferate by antigenic stimuli which lead to antibody formation. They differentiate into plasma cells which are highly specialized in the synthesis and excretion of immunoglobulins (LANGEVOORT, 1963; HUMMELER *et al.*, 1966), and are morphologically characterized by a conspicuous endoplasmic reticulum responsible for the excretion phenomenon (Figs. 2.7 and 2.8). Intermediate stages between lymphocytes and plasma cells are observed during such immunological reactions (HUMMELER *et al.*, 1966; LEDUC *et al.*, 1968; GUDAT *et al.*, 1970). *B immunoblasts* are morphologically identical to T immunoblasts at light microscopy and conventional electron microscopy. However, they can be distinguished by scanning EM (DANTCHEV, 1976) (Fig. 2.11). In prolonged stimulation, immunoblasts exhibit a development of rough endoplasmic reticulum and progressively differentiate into plasmoblasts and then into plasma cells.

Not all B-immunoblasts become plasma cells: some revert to lymphocytes which exert memory functions, just as some T-immunoblasts become *memory-T-lymphocytes.* this was shown in humoral responses by cell transfer experiments (MILLER and SPRENT, 1971). These *B-memory cells* could derive from an asymetric differentiation of B-lymphocytes (BYERS and SERCARZ, 1968), or by a return of antibody-producing cells to an undifferentiated state (NOSSAL and LEWIS, 1971).

It has already been shown that B-lymphocytes require the help of T-lymphocytes before differentiating into antibody-forming cells in response to many antigens. Some antigens can trigger B-cells in the absence of T-cells, and are called *thymus-independent antigens.* Most of them are polymers with a rigid backbone structure (polymerized flagellin, polyvinyl-pyrrolidone, pneumoccocal polysaccharides, bacterial lipopolysaccharides), and their capacity to bind to multiple B-cell receptors may account for their capacity to trigger these lymphocytes directly (FELDMANN and BASTEN, 1971). Other properties have been held responsible for the B-cell activating potential: mitogenicity for B-cells (COUTINHO and MÖLLER, 1973), ability to activate the third component of complement (PRYJMA *et al.*, 1974; PEPYS, 1974; MÖLLER and COUTINHO, 1975).

2.2.5. Macrophages Functions

Since the dawning of modern immunology, evidence has accumulated which supports the view that macrophages play an important role as nonspecific accessory cells at different stages in the complex of events that lead to an immune response.

Primarily, interaction of many antigens with the macrophages appears to be an important step in the normal induction of the immune response (UNANUE, 1972). Antigen taken up poorly by macrophages can induce tolerance, whereas immunity can be obtained with the appropriate form of antigens, which are easily concentrated by macrophages (DRESSER and MITCHISON, 1968).

The role of macrophages in the inductive phase of immunity is still unclear. They may simply take up and concentrate antigen in close proximity to immunologically reactive cells in the lymphoid tissues (NOSSAL and ADA, 1971; ASKONAS and ROELANTS, 1974). They may process antigen to a more immunogenic form by its digestive modification (CRUCHAUD and UNANUE, 1971; PEARSON and RAFFEL, 1971), or by formation of an antigen - RNA complex or "superantigen" (ASKONAS and RHODES, 1965; FISHMAN and PALMER,1967; GOTTLIEB, 1968). The contribution of macrophage-cell surface membrane may be crucial in respect of antigenic determinant presentation in the process of triggering T-lymphocytes (KATZ and UNANUE, 1973; ROSENTHAL and SHEVACH, 1973) or B-lymphocytes (FELDMANN, 1972a). Moreover, macrophages and antigens have been shown to be necessary for the induction of T cell helper function, in vitro (ERB and FELDMANN, 1975).

On the other hand, it is now well established that, in delayed hypersensitivity and in graft rejection processes, the interaction of sensitized lymphocytes and specific antigen changes the behaviour of bystander macrophages (MACKANESS, 1970). These cells undergo a variety of morphological and physiological alterations (NELSON, 1974; DAVID, 1974), and such *activated macrophages* are thought to be the effector cells in cell-mediated immunity. They express an increased antimicrobial activity in a non-specific form (MACKANESS, 1962; RUSKIN and REMINGTON, 1969) and can exert potent tumoricidal activity (EVANS and ALEXANDER, 1970; YASPHE, 1971; KELLER and HESS, 1972; HIBBS *et al.*, 1972a). This latter action will be the object of a detailed analysis below.

Recent studies have shown that macrophages become activated when cultured in the presence of a cell-free supernatant containing T-lymphocyte soluble mediators (NATHAN *et al.*, 1971). The factor responsible for macrophage activation (MAF) appears to be similar or identical to that of migration-inhibitory factor (MIF) (NATHAN *et al.*, 1971, 1973).

2.2.6. K-Cell Function

In addition to cytolytic T-lymphocytes and to macrophages, which can also display cytotoxic activity, evidence has recently been obtained that a third type of cell is involved in cell-mediated cytotoxicity, in vitro.

Designated as *K-cells* (K from "killer"), they are defined by their function: the ability to destroy target cells coated with antibodies directed against the cell-surface antigens, the function of which is referred to as *antibody dependent cell-mediated cytotoxicity* (ADCMC).

Cytotoxicity of K-cells is induced by IgG antibodies (MacLENNAN *et al.*, 1970; PERLMANN and PERLMANN, 1970), and its induction requires the presence of an intact Fc fragment (MÖLLER and SVEHAG, 1972; LARSSON and PERLMANN, 1972). Minute amounts of antibodies are needed; lysis of target cells coated with as few as a hundred antibody molecules has been obtained (PERLMANN *et al.*, 1972a). Cytotoxicity is independent of complement.

It must be emphasized that the specificity of the cytotoxic reaction is due to the antibody which must be specifically directed against the target cell. K-cells may come from non-immune animals.

Considerable controversy surrounds the nature of these cells. In fact, ADCMC is the property of a number of different cells which are all characterized by the presence of surface receptors for the Fc portion of IgG. Thus effector cells have been identified as: a) subset of B-lymphocytes bearing receptors for the third component of complement (VAN BOXEL *et al.*, 1973; PERLMANN *et al.*, 1975); b) a subset of B-lymphocytes with few detectable immunoglobulins and distinct from antibody-forming cell precursors (FORMAN and MÖLLER, 1973; PROSS *et al.*, 1974); c) cells of the monocyte series which can be subdivided into phagocytic and non-phagocytic effector cells (HOLM and HAMMARSTRÖM, 1973; GREENBERG *et al.*, 1973); d) null cells (GREENBERG *et al.*, 1975; WISLOFF and FROLAND, 1973; ALLISON, 1972; WYSS, 1976). It is now well demonstrated that T-lymphocytes are not involved in ADCMC (HARDING *et al.*, 1971; VAN BOXEL *et al.*, 1972).

It appears from the experimental data that the type of K-cell which is detected in an assay is dependent upon the nature of the target cells which are used. Thus, human Rh-positive red blood cells coated with appropriate isoantibody are exclusively killed by monocytes. Chicken red cells sensitized with specific heteroantibody are lysed by monocytes, B-cells, and "null" cells. These latter cells seem to be the only effector cells implicated in the lysis of mouse L1210 leukaemic cells coated with antimouse antibody.
At present there is a widespread agreement that the term "K cell" be used exclusively to designate the "null cells" endowed with cytotoxic activity in ADCMC.

2.3. The Immune Machinery and Cancer Cells

The large amount of data demonstrating that most neoplasms possess tumour-associated antigens has already been mentioned. Consequently, the tumour-bearing host should be capable of recognizing its growing neoplasm as antigenically foreign and of developing an immune response against it. Thus the immune machinery should be able to prevent, or at least, to limit, tumour growth. Therefore, the different immunological mechanisms which can lead to tumour cell elimination will be described.
As many tumours grow and kill their hosts, one can assume that other mechanisms exist which enable tumours to escape immunological destruction.

2.3.1. Mechanisms Involved in Tumour Cell Rejection

2.3.1.1. In vivo Evidence for Tumour Immunity

Some clinical and experimental observations provide evidence of tumour immunity operating in vivo.

The presence of lymphocyte in certain tumours, such as the so-called medullary breast carcinoma (Fig. 2.14), and their abundance in certain other types of neoplasias such as the lymphoid predominance type (Fig. 2.15) of Hodgkin's disease, are indications of good prognosis, while their rarity in the lymphoid depletion type (Fig. 2.15) of this tumour is a factor of unfavourable evolution.

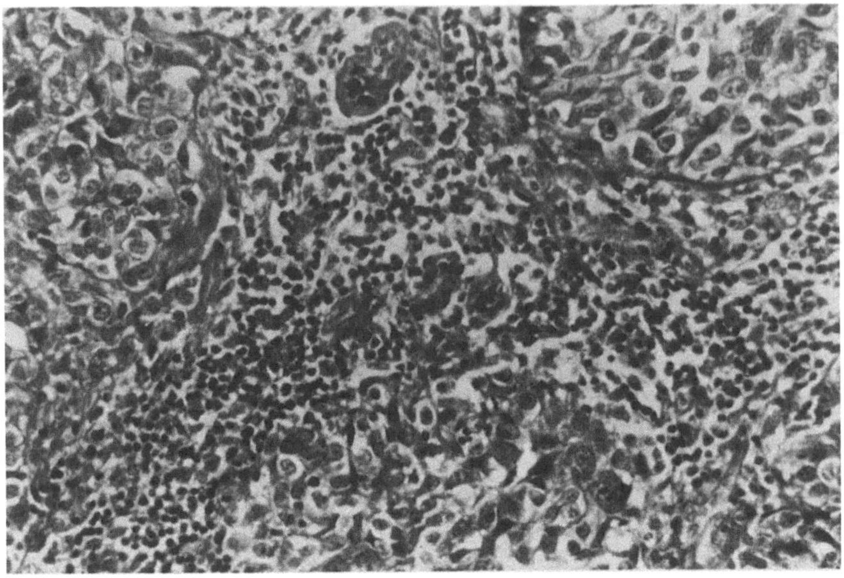

Fig. 2.14. Medullary breast carcinoma: intense infiltration by lymphocytes and plasma cells (Kindly prepared by G. Contesso)

A tumour bearing animal if often capable of rejecting a small graft of its own growing tumour. This phenomenon is referred to as *concomitant tumour immunity* (GERSHON *et al.*, 1967). In humans, spontaneous regressions of clinically and histologically diagnosed malignant tumours have been reported (EVERSON, 1964). Delayed hypersensitivity reactions have been elicited in cancer patient with the inoculation of soluble antigens prepared from the autologous tumour (HUGHES and LYTTON, 1964; STEWART and ORIZAGA, 1971; HOLLINSHEAD *et al.*, 1972). Correlation between delayed hypersensitivity to tumour antigens and the clinical course of the disease has been described in patients with Burkitt's lymphosarcoma (FASS *et al.*, 1970; BLUMING *et al.*, 1971), and acute lymphoid leukaemia (CHAR *et al.*, 1973; GUTTERMAN *et al.*, 1973b), although it should be accepted with caution.

A number of observations support the notion that tumour immunity, like allograft immunity, is essentially cell-mediated. Tumour growth can be inhibited if tumour cells are inoculated together with immune lymphoid cells into isogeneic recipients sublethally irradiated; such *neutralization tests*, originally introduced by WINN (1961), have been performed successfully in mice with different chemically or virus-induced tumours (OLD *et al.*, 1962; SLETTENMARK and KLEIN, 1962; FEFER *et al.*, 1967). Tumour specific immunity can be adoptively transferred by spleen, lymph node, or peritoneal cells from immune animals, and (a) they protect the recipients against a subsequent challenge with tumour cells (OLD *et al.*, 1962; FISCHER and HAMMOND, 1966;

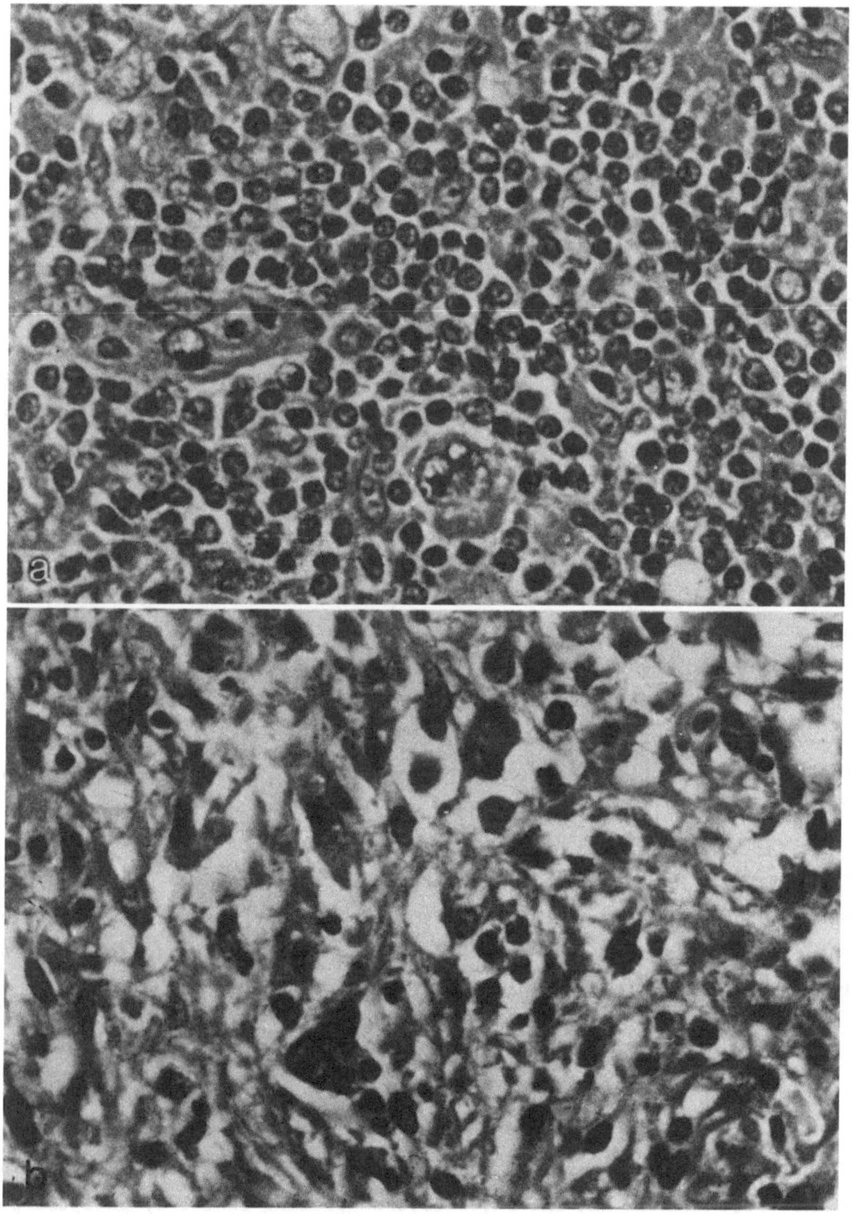

Fig. 2.15. Hodgkin's disease. Above: lymphocyte predominant type. Below: lymphocyte depletion type

WEPSIC *et al.*, 1970), or (b) they inhibit the growth of an already established tumour (DELORME and ALEXANDER, 1964; BORBERG *et al.*, 1972). In contrast, the transfer of the serum alone does not appreciably affect the tumour growth (OLD *et al.*, 1962; HELLSTRÖM and HELLSTRÖM, 1969a).

During the past decade, a variety of in vitro techniques have been developed to detect cell-mediated immunity to tumour-associated antigens and, although their value is debatable (LEVY and LECLERC, 1975), they have been useful in the understanding of the effector mechanisms involved in tumour immunity.

2.3.1.2. In vitro Assays for Cell Mediated Immunity

Three tests were the basis for the concepts presented below:

(a) *Chromium release test,* chromium 51 labelled tumour cells are incubated with the specifically sensitized lymphoid cells for a few hours and the amount of chromium released from damaged cells is determined. This assay measures only the cytotoxic effect of the effector cells on the target cells (BRUNNER *et al.,* 1968).

(b) and (c) *Colony inhibition and microcytotoxicity assays,* the target tumour cells and the effector cells are cultured together for 48-72 h. after which time the tests are scored by counting the number of colonies growing out of the plated tumour cells for the former assay, and the number of surviving tumour cells for the latter assay, (HELLSTRÖM and HELLSTRÖM, 1971; TAKASUGI and KLEIN, 1970; HELLSTRÖM, 1967). These two assays measure more than just cytotoxicity; they also detect cytostatic effects on tumour cells as well as any effects which result in loss of adherence of the plated cells.

These methods have given rise to extensive research which suggested that the mechanism of tumour cell-mediated immunity involves different types of effector cells. But the question arises as to whether the in vitro tests are true reflections of the state of immunity in cancer subjects.

2.3.1.3. Role of T-Lymphocytes in Tumour Rejection

Neonatal thymectomy increases the frequency of experimentally induced primary tumours (LAW, 1966; ALLISON and TAYLOR, 1967), suggesting that T-lymphocytes play an important part in the "defence" mechanisms against tumours. Transfer of adoptive tumour immunity can be abolished if immune lymphoid cells are previously treated by anti-θ serum and complement in order to eliminate the T-lymphocytes (ALLISON, 1972b; ROUSE *et al.*, 1972; GORCZYNSKI, 1974a).

There is strong evidence that tumour-bearing host lymphocytes exert in vitro cytolytic and cytostatic effects on isogeneic or autochthonous tumour cells (PERLMANN and HOLM, 1969; HELLSTRÖM and HELLSTRÖM, 1974b; CEROTTINI and BRUNNER, 1974; HERBERMAN, 1974).

T-lymphocytes are the most significant and possibly the most exclusive cytolytic effector cells in the chromium release test. Treatment with anti θ serum and complement was shown to inhibit the cytotoxic activity of lymphoid cells in different animal tumour systems (LECLERC *et al.*, 1973; HERBERMAN *et al.*, 1973; PLATA *et al.*, 1974; SCOLLARD, 1975). The cytotoxic cells are not retained on immunoglobulin-anti-immunoglobulin coated glass beads (PLATA *et al.*, 1974), and are not inhibited by preincubation with anti-immunoglobulin serum and complement (LECLERC *et al.*, 1973), two procedures which eliminate B-cells.

By using one or other of the cell purification techniques, it was shown that T-lymphocytes are either partially (PERLMANN *et al.*, 1972b; LAMON *et al.*, 1972, 1973; PLATA *et al.*, 1974; SCOLLARD, 1975) or exclusively (WHITNEY *et al.*, 1975) implicated in effector mechanisms responsible for tumour cell damage in the colony-inhibition or microcytotoxicity assays (Table 2.IV).

Table 2.IV. Nature of the effector cells, present in the spleens of MSV-induced sarcoma bearing mice, detected in the chromium release test (CRT) and the microcytotoxicity assay (MA)

Tumour state	Assay system	Nature of effector cells	Relative activity
During growth	CRT	T	+
	MA	T and non T (K-cell-macrophage)?	+ +
Maximal size	CRT	T	+ +
	MA		-
Regression	CRT	T	+
	MA	T and non T	+ +
After regression	CRT		-
	MA	non T	+ +

MSV = murine sarcoma virus CRT = chromium release test MA = microcytotoxicity assay

Cytotoxic lymphocytes have been demonstrated mainly in strongly immunogenic tumour systems. The Moloney sarcoma virus (MSV) induced tumour system, which is characterized by possible regressions, has been the most extensively studied. In mice inoculated with MSV, cytotoxic T-lymphocytes have been detected in the spleen by the chromium release test when the tumour became visible. The cytotoxic activity reached its maximum when the tumour began to regress and slowly declined thereafter (LECLERC *et al.*, 1973; PLATA *et al.*, 1974). T-cell activity, measured in the microcytotoxicity assay, was shown to follow distinct kinetics; it appeared just prior to tumour development, disappeared at peak tumour size, and was again evident when tumour regression took place (LAMON *et al.*, 1973; PLATA *et al.*, 1974) (Table 2.IV). These findings suggest that two subpopulations of T-effector cells are involved in the two different in vitro tests.

Clinical studies have generally been performed with unpurified lymphocytes, so that direct evidence of the participation of T-cells in the assay cannot be obtained. Lymphocytes from patients with acute leukaemia were found, by the chromium release test, to be cytotoxic against autologous blast cells and not against autologous remission leucocytes (LEVENTHAL *et al.*, 1972; HERBERMAN *et al.*, 1973). Cytotoxic reactivity was also found against allogeneic neoplastic cells in leukaemic patients, members of their family, and unrelated normal individuals (ROSENBERG *et al.*, 1972; HERBERMAN *et al.*, 1973). A particularly interesting finding was recently reported by JONDAL *et al.*, (1975) who have isolated T-cells from a Burkitt lymphosarcoma biopsy, which were able to kill the autologous tumour cells and EBV-positive cell lines.

The in vivo role of cytotoxic T-lymphocytes has not yet been established. The observation that cytotoxic T-lymphocytes, which are obtained by in vitro immunisation against isogeneic tumour cells, transferred specific immunity against this tumour in a neutralizing test (ROLLINGHOFF and WAGNER, 1973), favours the hypothesis that cytotoxic T-lymphocytes would be responsible for the early elimination of strongly immunogenic tumour cells.

T-lymphocytes, recognizing the tumour-associated antigens as foreign, may develop into cytotoxic effector cells, but they may also cooperate with B-cells for antibody production or they may release mediators of cell-mediated immunity that can stimulate nonspecific effector cell functions, and particularly macrophage functions.

2.3.1.4. *Role of Antibody in Tumour Cell Rejection*

Humoral antibodies to tumour-associated antigens have been detected in many tumour systems. They are cytotoxic in vitro for neoplastic cells carrying the same antigenic determinants in the presence of complement (OLD *et al.*, 1968; TAMERIUS and HELLSTRÖM, 1974).

Complement-dependant cytotoxicity assays are relatively insensitive. Two or more IgG antibody molecules, in close proximity on the cell surface, are required to activate complement and cause a hole in the cell membrane, and this implies that about 1000 IgG molecules would attach at random to the cell surface (HUMPHREY and DOURMASHKIN, 1969).

Moreover, an appropriate source of complement may be difficult to obtain for use in such cytotoxicity assays. Guinea pig or human sera may be ineffective for the lysis of human nucleated cells, and the presence of xeno-antibodies could interfere with the cytotoxic reaction.

Whether cytotoxic antibodies play any complement-dependant role in in vivo tumour rejection is difficult to assess. When an antibody response is detected in an experimental tumour system, it is usually found after the appearance of cytotoxic T-lymphocytes and sometimes only following surgical removal of the tumour (Table 2.V) (BANSAL and SJOGREN, 1973). Cytotoxic antibodies were detected in the sera of 30% of the patients with melanoma, and their presence seems to be related to the extent of the disease (LEWIS *et al.*, 1969). Antibodies directed against EBV-induced membrane antigens are present, at a relatively steady level, in patients in long term remission of Burkitt's lymphoma, whereas the concentration falls sharply some months prior to clinically diagnosed recurrence (KLEIN, 1971).

Two in vivo methods can assess humoral antibody-mediated tumour immunity; the use of a neutralization test and the technique of passive immunity. In the first method, tumour cells are incubated with antibody molecules before they are transplanted; neutralization of lymphosarcoma cells by antibodies and complement has been observed (WINN, 1960). Passive transfer of antibodies to tumour-bearing animals is a less sensitive technique, but it resembles more closely the conditions in which antibodies and cell surfaces interact in spontaneous neoplasias. Few reports can be found in which antibodies reacting against tumour-associated antigens have defavourized the tumour growth (BUBENIK *et al.*, 1965; FEFER, 1969). The rarity of such positive results may be explained by the fact that, in this type of in vivo experiment, the life span of the antibody is short, and its concentration, as well as that of complement components in different parts of the tumour, may be limiting factors (WINN, 1972).

Antibody bound specifically to target cells can induce cytotoxicity by activating non-immune effector cells, the K-cells.

2.3.1.5. Role of K-Cells

Recent studies of cell-mediated immunity against tumour-associated antigens have revealed that T-cells and non-T-cells are involved as effector cells in the in vitro

Table 2.V. Evolution of anti-tumour immunity in the rat hepatoma model (from BALDWIN *et al.*, 1973a; BANSAL and SJOGREN, 1973)

	Cell-mediated cytotoxicity [a]	Cytotoxic antibodies	Blocking serum factors	Unblocking serum factors
Progressive tumour growth	+	-	+	-
After complete tumour excision	+	+	-	+
After tumour recurrence	+	?	+	-
After immunization	+	+	±	±

[a] Microcytotoxicity assay

cellular cytotoxicity assays. K-cells appear as fully capable effector cells in the microcytotoxicity assay and probably in the colony-inhibition test. In such systems, where the lymphoid cells from tumour-bearing animals are cultivated with tumour cells, the cytotoxic reaction could be based on the presence of immune B-cells secreting antibody which binds specifically to the tumour cells. The antibody-coated target cells are then lysed by the non-T-cells carrying Fc receptors (K-cells) (LAMON *et al.*, 1974).

In the MSV-induced tumour system, K-cell activity, as well as cytotoxic T-lymphocyte activity, both measured in the microcytotoxicity assay, are detectable just prior to tumour development, are depressed at peak tumour size and emerge just as the tumour begins to regress. The K-cell activity becomes the only component of the cytotoxic reaction after complete tumour regression (LAMON *et al.*, 1973; PLATA *et al.*, 1974). The development of effector cells in cell-mediated cytotoxicity against MSV-associated tumour antigens is summarized in Table 2.IV.

In patients with bladder carcinoma, therapy such as surgery or local radiotherapy, results in loss of T-cell independent cytotoxicity, measured by the microcytotoxicity assay (O'TOOLE *et al.*, 1972). Reappearance of this activity long after therapy usually correlates with recurrence of the tumour (O'TOOLE *et al.*, 1973).

The microcytotoxicity assay and the colony inhibition test have been widely used for detecting tumour immunity in animal and human tumour systems. Unfortunately, except in the more recent experiments described above, these studies have been performed without characterization of the effector cells and, moreover, it was generally postulated that T-lymphocytes were involved. Lymphocytes from cancer patients were shown to be able to specifically destroy neoplastic cells of the same histological type as the patients own tumour, irrespective of whether or not the patient himself had symptoms of a growing tumour. Similar studies were performed in animals and, in most of them, lymphocytes from tumour-bearing and from tumour-free animals were equally effective in tumour cell killing. Some investigators have established a correlation between cytotoxic activity and tumour volume. No lymphocyte reactivity was found in animals with large tumours, whereas it was detectable early after induction or transplantation of the tumour or after tumour removal or regression (BARSKI *et al.*, 1974).

The phenomenon of antibody-dependent cell-mediated cytotoxicity (ADCMC) can provide a highly sensitive test for the detection of specific antitumour antibodies in the sera of cancer patients (ZIGHELBOIM *et al.*, 1974a).

It has been shown that certain sera from tumour-bearing animals (POLLACK *et al.*, 1972; SKURZAK *et al.*, 1972; HARADA *et al.*, 1973), or from patients during the period of tumour remission (HELLSTRÖM *et al.*, 1971a), potentiate the cytotoxic activity of lymphocytes reactive to their respective type of tumour target cells in the microcytotoxicity assay. These sera can also "arm" lymphoid cells from nonsensitized donors, so that they become specifically cytotoxic (POLLACK, 1973; HELLSTRÖM *et al.*, 1973a). The in vivo role of the arming phenomenon, and more generally that of K-cells in tumour immunity, has still to be established. The ADCMC phenomenon may account for the protective effect of passively administered specific antibodies against transplanted isogeneic tumour cells (PEARSON *et al.*, 1973; ZIGHELBOIM *et al.*, 1974b).

Another approach to this problem was to determine if cancer patients could manifest some defect in the population of cells responsible for ADCMC. Decreased or absent ADCMC activity has been reported in leukaemic children treated with chemotherapy and radiotherapy (CAMPBELL *et al.*, 1973), in patients with chronic lymphoid leukaemia (GALE *et al.*, 1975), in a patient with macroglobulinaemia (CAMPBELL *et al.*, 1972), and in 44 patients with various types of tumours (TING and TERASAKI, 1974).

2.3.1.6. Role of Macrophages

Many points of evidence provide compelling reasons for considering macrophage reactions as important in tumour rejection since the observation, made by GORER in 1956, that transplanted tumours elicit strong macrophage accumulation in the draining lymph node and at the site of the graft (LEVY and WHEELOCK, 1974).

Clinical observations support the idea that macrophages play a significant role in the defence against neoplasias. Histiocytic infiltration of lymph nodes draining mammary carcinoma has been correlated with a favourable clinical course (BLACK *et al.*, 1955; BERG, 1959) (Fig. 2.16). The infiltration of Burkitt's (African) lymphosarcoma by many macrophages (Fig. 2.17) compared to their absence in other diffuse (lymphoblastic or immunoblastic) lymphosarcomas is correlated with the high therapeutic sensitivity and excellent prognosis of the African tumour, contrasting with the much poorer sensitivity and prognosis of the other (MATHÉ *et al.*, 1975a; MATHÉ and RAPPAPORT, 1976). Depressed macrophage phagocytic or bactericidal activity has been observed in patients with Hodgkin's disease (GHOSH *et al.*, 1973), myelomonocytic leukaemia and lymphosarcomas (CLINE, 1973), but correlation between depressed macrophage function and defective antineoplastic "defence" has not been clearly established.

Macrophage content of experimental tumours was shown to vary from one neoplasia to another. Tumours with many resident macrophages are less likely to metastasize than those with few (BIRKBECK and CARTER, 1972; ECCLES and ALEXANDER, 1974).

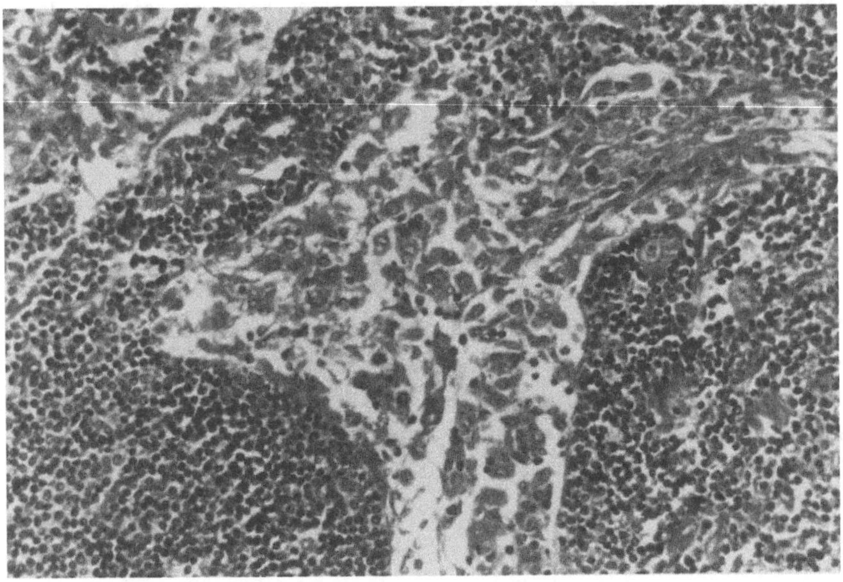

Fig. 2.16. Sinusal reticulohistiocytosis in lymph node draining mammary carcinoma (Kindly prepared by G. Contesso)

It has been shown that specific immunity against tumours could be transferred more efficiently by peritoneal cells than by the lymph node or spleen cells (OLD *et al.*, 1963; TSOI and WEISER, 1968; DULLENS and DEN OTTER, 1973), and that macrophages alone were effective (BENNETT, 1965). On the other hand, ZARLING and TEVETHIA (1973) showed that the efficiency of immune spleen cells in a neutralizing test was decreased if the recipients were pretreated with silica which poisoned macrophages.

Macrophages from animals immunized against allogeneic or isogeneic tumour cells were able to kill these tumour cells in vitro, specifically, by a process involving cell to cell contact, but not phagocytosis (GRANGER and WEISER, 1964; EVANS and ALEXANDER, 1970; LOHMANN-MATTHES *et al.*, 1972).

Non-immune macrophages could be rendered specifically cytotoxic to tumour cells after incubation with immune lymphoid cells or after being put in a cell-free supernatant from the culture of immune lymphoid cells with the specific tumour cells (EVANS and ALEXANDER, 1972; VAN LOVEREN and DEN OTTER, 1974). This arming phenomenon could be mediated (a) by a specific *macrophage-arming factor* (SMAF) which may be a T-cell product (EVANS *et al.*, 1972; LOHMANN-MATTHES *et al.*, 1973), or (b) by cytophilic antibodies (GRANGER and WEISER, 1964) and, in this latter case, *armed macrophages* closely resembled K-cells (SHIN *et al.*, 1972).

Armed macrophages, after interaction with the specific tumour cells, acquired the ability to kill antigenetically unrelated tumour cells and are then referred to as *activated macrophages* (ALEXANDER *et al.*, 1972) because they manifest their function in an immunologically nonspecific manner.

Besides the possibility that macrophages may be concerned in an immunologically specific way with the rejection of tumour cells, mononuclear inflammatory reactions, associated with delayed hypersensitivity to unrelated antigens, appeared to amplify the natural "defences" against neoplasias.

Mice or rats chronically infected with a variety of intracellular parasites, have increased resistance to autochthonous (WEATHERLY, 1970; HIBBS *et al.*, 1971a) or transplanted isogeneic tumours (KELLER *et al.*, 1971; HIBBS *et al.*, 1971b). The peritoneal macrophages from the infected animals transferred tumour resistance when injected into isogeneic animals (KELLER and JONES, 1971). Resistance to neoplasia could also be induced by the administration of agents such as adjuvants of immunity, which provoke intensive stimulation of the mononuclear-phagocyte cell system (HALPERN, *et al.*, 1966; YASPHE, 1971; HIBBS *et al.*, 1972b ; KELLER and HESS , 1972). Macrophages taken from the peritoneal cavity of infected or adjuvant-treated animals are activated, since they have acquired the ability to destroy neoplastic cells and all cells with

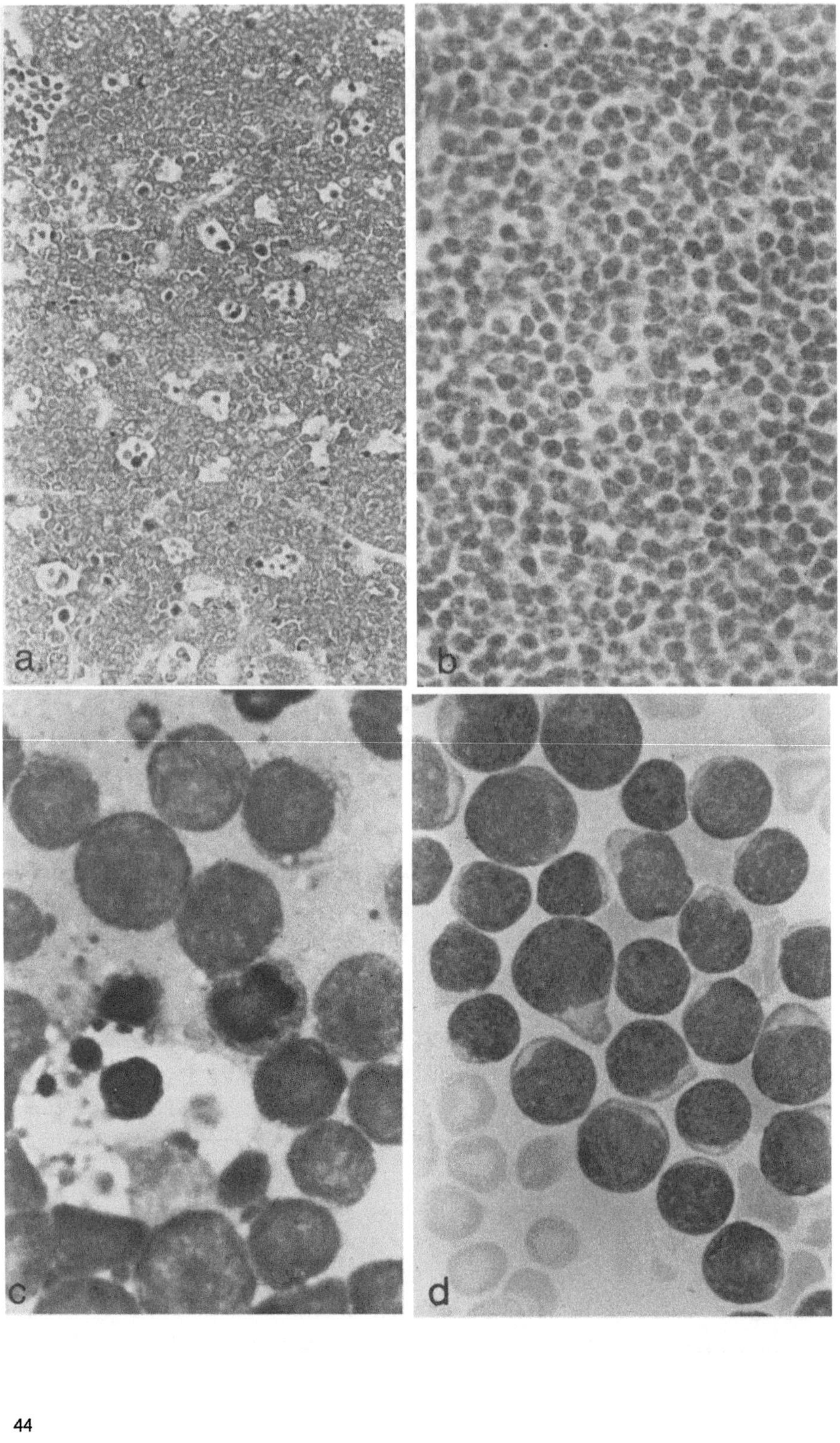
a
b
c
d

abnormal growth properties, in vitro, in a nonspecific way (HIBBS *et al.*, 1972a); KELLER, 1973; GHAFFAR *et al.*, 1974; CLEVELAND *et al.*, 1974; MATHÉ *et al.*, 1975b).

Mediators of cell-mediated immunity, acting on macrophages, have been shown to exert antitumoural activity. Intradermal injection of MIF in guinea pigs induces an inflammatory reaction and results in complete suppression of an isogeneic tumour graft at the reaction site (BERNSTEIN *et al.*, 1971).

Regression and eradication of skin tumours has been obtained by the induction, at the tumour site, of delayed hypersensitivity reactions either to haptens, to BCG alone, or to this mycobacteria mixed with tumour cells (ZBAR *et al.*, 1970; KLEIN and HOLTERMANN, 1972; GUTTERMANN *et al.*, 1973c); during this process the main histologically visible phenomenon was an intense infiltration by macrophages (HANNA, 1974; LIKHITE, 1974a, 1974b).

2.3.2. Escape Mechanisms to Tumour Destruction

The immune machinery is capable of destroying tumour cells by various mechanisms involving T-lymphocytes, B-lymphocytes, K-cells and macrophages, although their relative importance in vivo, in tumour rejection is still unknown. Tumour development is therefore dependent upon alteration of these mechanisms, occurring either prior to tumour establishment or accompanying tumour growth.

2.3.2.1 Escape Mechanisms Prior to Tumour Establishment

A theory based on immune alteration in the form of immunological tolerance was much favoured at one time. Such a phenomenon has been suspected by SJÖGREN (1964) in an experimental tumour system where an oncogenic virus (polyoma) was introduced into immunologically incompetent newborn hosts. But it was then shown that, in fact, newborn inoculated animals could mount an immune response against virus-induced tumour antigens, as tumour apparition could be prevented by the inoculation of a second dose of virus or by the injection of irradiated tumour cells, towards the end of the oncogenic latency period (EDDY *et al.*, 1964; GOLDNER *et al.*, 1964).

DORÉ *et al.* (1970b), have shown that, in AKR mice, where the RNA oncogenic virus is transmitted vertically from parents to offspring, (GROSS, 1970), a graft of isogenic

◀ *Fig. 2.17. Histological (upper) and cytological (below) aspect of African Burkitt's lymphosarcoma (left) (see the macrophages) and of diffuse common lymphosarcoma (right) no macrophages)*
(Description to Fig. 2.17, page 43)

leukaemic cells could be rejected after specific immunization or immunostimulation, and antibodies with G specificity could be detected (Fig.2.18. Table 2.VI). ALLISON (1970) and OLDSTONE *et al.* (1972) confirmed this observation.

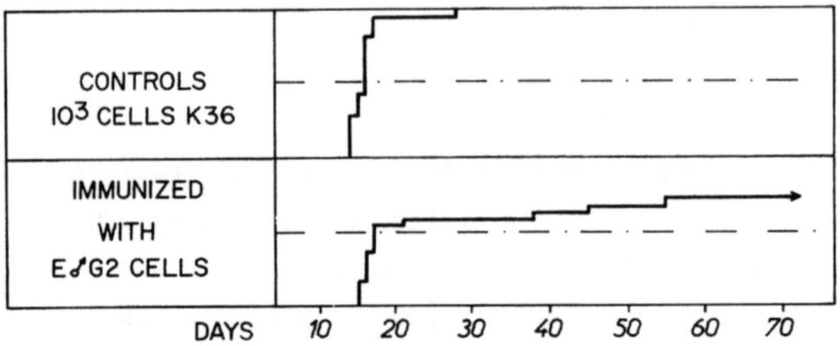

Fig. 2.18. Cumulative mortality of 2-month-old normal AkR mice immunized with E♂ G2 and then grafted with 10^3 K36 cells. (DORÉ et al., 1970b)

Table 2.VI. Antibody detected in AkR mice after specific immunization with leukaemic cells absorption of an AkR anti-K36 (AkR) serum with tumour cells of known phenotype

Tumour cells	Strain	Induction	Phenotype	Cytotoxicity index
Control (unabsorbed)	-	-	-	0,71
Leukaemia	AkR	Spontaneous	G+[a]	0,19
K36	AkR	Gross	G+ L+[b]	0,20
E♀K1	C_{57}Bl/6	Gross	G+	0,07
E (AkR)	C_{57}Bl/6	Gross	G+	0,53
E ♂ G2	C_{57}Bl/6	Gross	G+ L-	0,08
EL 4	C_{57}Bl/6	DMBA	G- L+	0,60
BP 8	C_3H	Benzpyrene	G-	0,58
5563	C_3H	Plasmocytoma		0,36
RC 19	BALB/C	Rauscher	G-	0,54
L 1210	DBA/2	DMBA	G-	0,22

[a] G (Gross) cell surface antigen

[b] L (L, C_3H) antigen

Conclusions. These data indicate G-specificity and no L-specificity.

At the present time, immunosuppression seems a more acceptable hypothesis than tolerance in explaining how the immune machinery may be incapable of eradicating the first tumour cells.

Numerous experimental works have reported increased tumour incidence after immunosuppressive treatments, including neonatal thymectomy (ALLISON and TAYLOR, 1967), antilymphocyte serum (ALS) administration (ALLISON and LAW, 1968; GAUJAS *et al.*, 1973; BALDWIN, 1973), X-ray irradiation (LAW and DAWE, 1960) etc.

Cancers are abnormally frequent in patients suffering from congenital immunological deficiency diseases (Table 2.VII) (WALDMANN, 1972; GATTI and GOOD, 1971), and that, for most types of cancer, the curve of tumour incidence rises logarithmically with age (Fig. 2.19) (Nat. Cancer Inst. Monogr., 1971), while there is

Table 2.VII. Neoplasms in patients with immunodeficiency diseases (WALDMANN, 1972)

Immunodeficiency Diseases and Cancers Complicating their Evolution	Number of Cases	
Combined immunodeficiency	5	
Lymphosarcoma		3
Hodgkin's disease		1
Leukaemia		1
Bruton-type agammaglobulinaemia	5	
Leukaemia		4
Haematosarcoma		1
Common variable immunodeficiency	24	
Lymphoreticular malignancy		11
Lymphosarcoma and lymphoma		4
Hodgkin's disease		1
Reticulosarcoma		1
Nonlymphoreticular malignancy		2
Cancer of bladder		1
Cancer of breast		1
Gastrointestinal carcinoma		3
IgA deficiency	4	
Haematosarcoma		1
Leukaemia		1
Gastrointestinal carcinoma		1
Pulmonary carcinoma		1

Table 2.VII. (Continuation)

Immunodeficiency Disease	Number of Cases	
Ataxia-telangiectasia	42	
Lymphosarcoma		18
Reticulosarcoma		7
Hodgkin's disease		3
Leukaemia		8
Ovarian dysgerminoma		1
Mixed cystic glioma		1
Cerebellar meduloblastoma		1
Gastric adenocarcinoma		2
Basal cell carcinoma		1
Wiskott-Aldrich syndrome	13	
Malignant reticuloendotheliosis		5
Lymphosarcoma, reticulosarcoma		5
Cerebellar astrocytoma		1
Myelogenous leukaemia		1
Leiomyosarcoma (multiple)		1

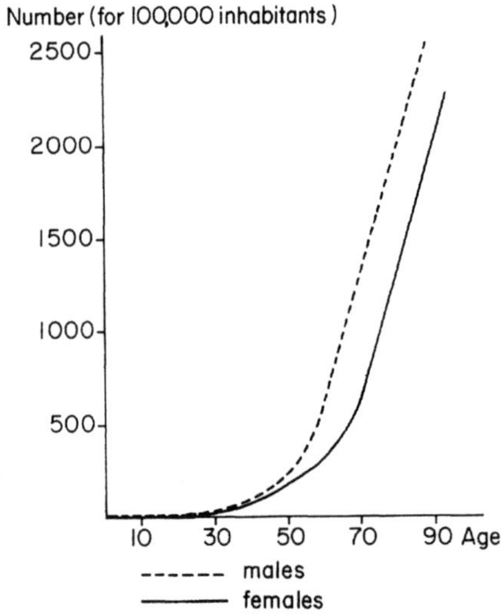

Fig. 2.19. Increase of incidence of cancer with age (Nat. Cancer Inst. Monogr. 1971)

a progressive weakening of immunological responses (WIGZELL and STJERNSWARD, 1966). We also know that the risk of a second cancer in Hodgkin's disease patients submitted to intensive radiochemotherapy is about 25 times that of the normal population (CANELLOS *et al.*, 1974, BELPOMME *et al.*, 1974), and that the incidence of malignant neoplasias is abnormally frequent in subjects receiving immunosuppressive therapy for organ transplantation (MacKHANN, 1969; PENN, 1975) (Table 2.VIII).

Table 2.VIII. Neoplasias in organ transplanted subjects (PENN, 1975)

231 *Kidney transplantation*		
2 *Heart transplantations*		
1 *Liver transplantation*		
Cancers of the skin and lips	95	(39%)
Haematosarcomas	66	(27%)
Carcinomas of the cervix	18	(8%)
Carcinomas of the lung	12	(5%)
Miscellaneous	50	(21%)

If those tumours are due to, or favourized by a permanent or known long-term immunodepression, the question arises as to the possible role of a transitory unknown one. As the doubling time of human tumours is long, the event responsible for transitory immunodepression is probably forgotten by the time the neoplasia is detected.

The immunosuppressive agent in the environment may be the carcinogen itself. This has been shown in the case of cigarette smokers (THOMAS *et al.*, 1973). Animals exposed to chemical carcinogens, or infected with leukaemogenic viruses, undergo a profound depression in their immune responsiveness to a variety of antigens (STJERNSWARD, 1966; SIEGEL and MORTON, 1966; SZAKAL and HANNA, 1972). With DORÉ *et al.* (1970b), the author has shown that cell-mediated immunity is severely impaired in two-month-old AkR mice, that is to say, in the preleukaemic phase.

2.3.2.2. Escape Mechanisms after Tumour Establishment

It has already been seen that tumour immunity is a multifacetted process involving different effector mechanisms. If some of these immune reactions lead to tumour cell destruction, the others result in tumour protection against the immune aggression mediated by the former. One important step has been the discovery that humoral factors may interfere with cell-mediated immunity. These factors were called blocking factors because they could specifically stop immune lymphocytes from destroying tumour cells in vitro. More recent findings suggest that cells with immunosuppressive functions could be activated during tumour development.

2.3.2.2.1. Role of Serum Blocking Factors

In 1958, KALISS described, under the name of *immunological enhancement,* the progressive growth of normally rejected tumour allografts in recipients pretreated with hyperimmune-alloantisera directed against tumour cells, demonstrating that humoral factors can protect tumours from immunological destruction. Enhancement of isogeneic tumour growth has also been observed after treatment with hyperimmune sera directed against tumour-associated antigens (MÖLLER, 1964; BUBENIK and KOLDOVSKY, 1965; BATCHELOR, 1968). Nevertheless, this work did not receive much attention until there was growing evidence to show that tumour-bearing hosts have factors in their serum which specifically inhibit cellular immunity against tumour cells in vitro; this argues for the role of enhancement in the development of autochthonous tumours (HELLSTRÖM and HELLSTRÖM, 1974b).

In 1969, HELLSTRÖM and his co-workers first claimed that sera from mice with growing MSV-induced sarcoma abrogate the destruction of sarcoma cells by immune lymphocytes in an immunologically specific way (Fig. 2.20). Blocking factors have then been said to be detected in the sera of animals with many different types of tumours HELLSTRÖM and HELLSTRÖM, 1970a) and of human patients with a large variety of neoplasms (HELLSTRÖM *et al.*, 1971b).

Enhancement of tumour growth, observed after transfer of hyperimmune sera, has been said to be mediated by specific antibodies (CHARD, 1968; TAKASUGI and HILDEMANN,

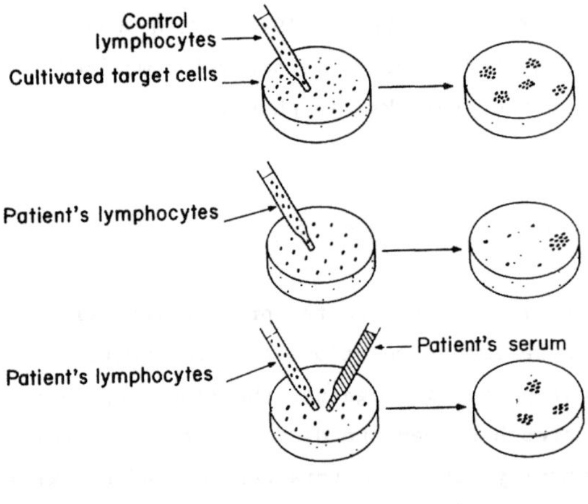

Fig. 2.20. Hellström's test demonstrating what were considered to be blocking factors in sera of cancer patients

1969). Hyperimmune sera against tumour alloantigens or tumour-associated antigens could block, in vitro, lymphocyte-mediated cytotoxicity after incubation with the specific target cells, but not with the effector lymphocytes (efferent blocking) (MOLLER, 1965; BALDWIN and BARKER, 1967; BRUNNER *et al.*, 1968).

The blocking effect of cellular cytotoxicity observed in vitro with sera from tumour-bearing hosts rapidly appeared as a more complex phenomenon as far as the nature of the blocking factors and their mode of action were concerned.

The observation that blocking activity in the serum of tumour-bearing animals could be abrogated by precipitation with anti-immunoglobulin sera demonstrated that tumour specific antibodies were involved (HELLSTRÖM and HELLSTRÖM, 1969b). Nevertheless, difficulties in detecting specific antitumour antibodies by complement dependent cytotoxic activity or immunofluorescent staining of tumour cell-surface antigens (BALDWIN *et al.*, 1973, 1973), strongly suggested that free antibodies were not resuming the blocking process.

The presence of free circulating tumour-specific antigens were detected in blocking sera from rats with chemically-induced hepatoma (BALDWIN *et al.*, 1973a) or with established sarcomata (THOMSON *et al.*, 1973).

A number of observations support the idea that serum blocking factors in tumour-bearing hosts are complexes of antigen(s) released from the tumour and of antibodies formed by the host. Serum blocking factors from MSV-induced sarcoma bearing mice or from hepatoma-bearing rats could be separated by ultrafiltration into high and low molecular weight fractions, after exposition at low pH (SJÖGREN *et al.*, 1971; BALDWIN *et al.*, 1973a). Neither fraction blocked when added to the target cells, but blocking activity was restored when the two fractions were mixed. The low molecular weight fraction alone blocked only when it was added to the effector lymphocytes. It contained tumour specific antigens, demonstrated by its capacity to neutralize specific antibodies, whereas the high molecular weight fraction contained specific antibodies (THOMSON *et al.*, 1973; BALDWIN *et al.*, 1973a).

Blocking factors have been eluted from animal and human tumours, and injections of the eluates could lead to tumour growth enhancement in vivo (BANSAL *et al.*, 1972; RAN and WITZ, 1972).

Immunoglobulins belonging to the IgG_2 class and directed against tumour antigens have been detected in the tumour eluates dissociated at low pH (RAN and WITZ, 1970). Ultrafiltration procedures separate tumour eluates into two fractions identical to those obtained from tumour-bearing host sera (SJÖGREN and BANSAL, 1972).

Blocking complexes have been artificially prepared by mixing nonblocking hyperimmune sera (free of soluble tumour antigens) with papain solubilized tumour antigens. These complexes were blocking after incubation with the target cells (BALDWIN *et al.*, 1972).

In all these studies, it was assumed that T-lymphocytes were the cytotoxic effector cells and, consequently, it was postulated that blocking affected this type of cell. However, as the experiments were carried out with the microcytotoxicity or the colony-inhibition assays, blocking of K-cells must also be considered: a) antigen-antibody complexes could inhibit ADCMC through interaction with the Fc receptor of the effector cells, inducing competition with the antibody-coated target cells; b) excess of soluble antigens could cause inhibition by saturation of the antigen-binding sites of the antibody (MacLENNAN, 1972). Blocking of cytotoxic T-lymphocytes, measured in the chromium release assay, has rarely been studied, and has generally been done in allogeneic situations. In these assays, blocking of the effector cells, by the use of solubilized antigens, has been unsuccessful. This contrasts strongly with the observations of BALDWIN *et al.*, 1973b, 1973c) when they found that lymphocyte cytotoxicity could be inhibited in the microcytotoxicity assay by incubation of the effector cells with papain-solubilized tumour antigens. Antigen-antibody complexes could block cytotoxic T-lymphocytes, the antibody moiety probably acting by carrying the antigen to the receptor sites of the lymphocyte, which was consequently "desensitized". T-lymphocyte cytotoxicity could also be blocked by previous interaction of the target-cells with specific antibody (BONAVIDA, 1974).

Definite conclusions about the mechanisms underlying blocking in tumour immunity will not be reached until further studies are performed with non-criticizable in vitro assays (LEVY and LECLERC, 1975).

The crucial question is whether the blocking factors detected in vitro can affect tumour growth in vivo. Blocking activity in the serum of rats with transplanted isogeneic polyoma virus-induced sarcoma became detectable simultaneously with the appearance of cytotoxic lymphocytes, increased with tumour growth and widespread metastasis, but disappeared shortly after tumour removal; (BANSAL and SJOGREN, 1973). Similar observations were reported in rats with transplanted hepatoma and sarcoma by BALDWIN *et al.* (1973d) and by THOMSON *et al.* (1973). It appeared from these studies that the relative proportion of free and immune complexed tumour antigens on the one hand and specific antibodies on the other, in the serum of the animals, varied at different stages of tumour growth (Table 2.V.).

In three different mouse tumour systems, BARSKI and YOUN (1969), Le FRANCOIS *et al.* (1971) and BELEHRADEK *et al.* (1972) have described, under the name of *"immunological eclipse"*, the disappearance of cell-mediated immunity which was concomitant with the appearance of blocking activity in the serum. BANSAL *et al.* (1972) succeeded in enhancing the growth of transplanted sarcoma by the administration of blocking sera in sufficient quantities so that blocking activity could be detected in the serum of the recipients.

In humans, sera from most patients with growing neoplasms of various histological types were shown to contain blocking factors in contrast to the sera from patients who became clinically tumour-free after therapy (HELLSTRÖM *et al.*, 1971c, 1973). Blocking activity seemed to reappear, in many cases, before clinically overt tumour recurrency (HELLSTRÖM *et al.*, 1973a).

CURRIE and BASHAM (1972) showed that lymphocytes from patients with widespread metastatic melanoma, could recover cytotoxic activity against melanoma cells in vitro after repeated washings. This observation strongly suggested that tumour-associated antigens, either free or complexed to specific antibody, could inhibit lymphocyte cytotoxicity directly and that this mechanism may be operative in vivo.

Correlation between tumour evolution and the presence of circulating blocking factors has not always been so evident. Blocking activity was found in the sera of tumour-free animals (HEPPNER, 1969; ANKERST, 1971; BALDWIN, 1973) and in clinically tumour-free periods in patients with melanoma (HELLSTRÖM *et al.*, 1973a). Similarly, blocking factors were not always detected in patients with growing neoplasms (HELLSTRÖM *et al.*, 1971a; BYRNE *et al.*, 1973).

Moreover, these studies lead to the discovery of a new phenomenon called *unblocking:* nonblocking sera from tumour-free animals or patients were able to neutralize the blocking activity of sera from tumour bearing hosts. Unblocking activity was first demonstrated in the sera of mice in which MSV-induced sarcoma had spontaneously regressed (HELLSTRÖM and HELLSTRÖM, 1970b). It was also found in the sera of rats after surgical removal of polyoma-induced transplanted sarcoma (BANSAL and SJOGREN, 1971). In man HELLSTRÖM *et al.* (1971b) observed the disappearance of the blocking activity in the serum of melanoma patients in remission and the concomitant appearance of factors able to cancel the blocking activity of sera from patients with growing neoplasm.
Injections of unblocking sera immediately after a polyoma-induced sarcoma isograft in rats, resulted in complete tumour regression after a short period of tumour growth. Blocking activity did not appear in the sera of these animals. Strong cell-mediated immunity was found, as well as the presence of cytotoxic antibodies. These effects were less marked when the unblocking sera were injected into animals with already developed tumours. Only the slowing down of tumour growth was generally observed (SJOGREN and BANSAL, 1972; BANSAL and SJOGREN, 1974).

The nature and mechanism of action of unblocking factors is still unknown. The unblocking activity was shown to be associated with immunoglobulins; moreover it could be absorbed by tumour cells from the same histological type of neoplasia (SJOGREN and BANSAL, 1972). It may be due to free antibody which could act by removing tumour antigen from receptor sites of lymphocytes (SJOGREN, 1974).

2.3.2.2.2. Role of Suppressor Cells

The immunosuppressive effect of lymphoid cells in anti-tumour immunity could be suspected in PREHN's observation (1972): injection of a small number of immune spleen cells mixed with tumour cells into isogeneic recipients, in a neutralizing test, resulted in enhanced tumour growth.

Suppressor cells may also account for tumour enhancement sometimes observed after adoptive transfer of tumour immunity. UMIEL and TRAININ (1974) have reported acceleration of Lewis carcinoma growth and increase of metastatic incidence in mice receiving T-lymphocytes from Lewis tumour-bearing animals, and, to a lesser extent, in mice receiving non-immune lymphoid cells. Similarly, mice deprived of short-lived T - lymphocytes after adult thymectomy manifested a striking reduction in Lewis tumour development.

More direct evidence of suppressor cells in tumour immunity has been shown in the MSV-induced tumour system. KIRCHNER *et al.* (1974) and GORCZYNSKI (1974b) demonstrated that suppressor cells were present in the spleen and lymph nodes of mice bearing growing MSV-induced sarcoma, but not in mice in which tumours had regressed.

In the experiments of KIRCHNER and his co-workers, suppressor cells were made evident by their capacity to inhibit the proliferative response to phytohaemagglutinin (PHA) of lymphocytes from normal mice, and their presence was accounted for by the marked impairment of this response in tumour-bearing animals. Only lymphocyte functions involving a proliferative step were depressed regardless of whether tumour immunity was involved (KIRCHNER *et al.*, 1975). Suppressor activity appeared to be mediated by radio-resistant, adherent, phagocytic cells, not sensitive to treatment with anti-θ serum and complement. This data strongly favoured the hypothesis that suppressor cells belong to the monocyte-macrophage series.

GORCZYNSKI demonstrated the presence of suppressor cells by their capacity to inhibit the PHA response and tumour-antigen-specific responses of lymphocytes from mice in which the tumour had regressed. These suppressor cells were isolated by velocity sedimentation and appeared as θ negative-immunoglobulin positive cells (KILBURN *et al.*, 1974). When incubated in vitro, they released soluble factors which inhibited in vitro tumour-specific and nonspecific immune responses, and which appeared to be similar to antigen-antibody complexes present in the sera of the tumour bearing animals and which blocked specific cell-mediated immunity in the MSV tumour system (GORCZYNSKI *et al.*, 1975). It was suggested that nonspecific suppression of T-cells could occur by means of these immune complexes which bind to the Fc receptors as recently demonstrated at the surface of activated T-cells (YOSHIDA and ANDERSSON, 1972).

Nonspecific suppressor cells have also been detected in the spleens of mice with progressively growing chemically-induced tumours. POUPON (1975) identified the cells as adherent, θ negative cells which can bear immunoglobulins probably absorbed at the cell surface. The suppressive activity was demonstrated as nonspecific and affecting both T- and B-lymphocyte responses in vitro. It disappeared shortly after surgical removal of the tumour.

EGGERS and WUNDERLICH (1975) have also reported the presence of adherent, θ negative suppressor cells in the spleens of mice with progressively growing methylcholanthrene-induced tumours. These cells were shown to inhibit the generation of cytotoxic T-lymphocytes during in vitro immunization of normal lymphocytes against histocompatibility antigens, but not the expression of the cytotoxic activity, which is a proliferation - independent process.

Hence, the exact nature of the suppressor cells demonstrated in tumour immunity systems is not yet determined, and little is known of their role in the in vivo growth of tumour.

2.3.2.2.3. Role of the Tumour Itself

The tumour may contribute to its own promotion in different ways including, mainly, the expression of its cell surface antigens, the excretion of factors which can interfere with immune effector mechanisms, and its localization in the organism.

All the immunological mechanisms described above, and which can lead either to tumour rejection or to tumour growth enhancement, require that tumour cells bear antigens which are recognized as foreign by the host. There are some examples of tumours which do not appear noticeably antigeneic in transplantation tests. This was shown in a number of chemically induced rat tumours and spontaneous carcinomas and sarcomas (PREHN, 1969; BALDWIN, 1973).

If, as is generally the case, tumour cells possess tumour-associated antigens, loss of antigens would lead to the selective overgrowth of cells with a lower antigenicity. It is not known to what extent tumour-associated antigens may be lost. It has been reported that metastases of virus-induced sarcomas in hamsters are less immunogenic than the primary tumours (DEICHMAN, 1969). Moloney virus-induced lymphoma sublines, exhibiting decreased immunosensitivity to specific cytotoxic antibodies, have been isolated following successive exposure to cytotoxic antiserum plus complement in vitro and reinoculations to preimmunized recipients (FENYÖ *et al.*,1968).

Antigenic loss may be due to *antigenic modulation.* In vivo or in vitro exposure of cells to antibodies directed against cell surface constituents may lead to the specific disappearance of these antigenic determinants. This phenomenon has been first described for TL-antigens which are expressed on thymocytes and on a number of mouse leukaemic cells (OLD *et al.*, 1968b), and was also observed for H-2 antigens (SCHLESINGER and CHAOUAT, 1972).

Antigen modulation seems to be related to the mobility of cell surface antigens. After interaction with specific antibodies, antigens migrate and gather at one pole of the cell surface to form caps, and immune antigen-antibody complexes are then rapidly pinocytosed into the cell. Cap formation and pinocytosis have been observed for lymphocyte surface immunoglobulins, θ antigen and histocompatibility antigens (TAYLOR *et al.*,1971; TAKAHASHI, 1971).

There is no evidence that capping and antigenic modulation phenomena may occur for tumour-associated antigens in vivo, but they could be responsible for perturbations in the circulation of antigen-induced signals to the effector cells of the immune machinery.

It has been shown that many tumour cells are surrounded by a sialomucin coat called *glycocalix* (Fig. 2.21) which may protect them from immunological attack (CURRIE and BAGSHAWE, 1968; WEISS, 1973). Neuraminidase treatment which removed sialic acid (Fig. 2.5) was shown to increase tumour cell susceptibility to cytotoxic antibody and complement (SANFORD, 1967; RAY *et al.*, 1970; DORÉ *et al.*, 1974). Increased immunogenicity of neuraminidase-treated cells was demonstrated by their ability to induce good antitumour immunity when injected into isogeneic animals (SIMMONS *et al.*, 1971).

Tumour cells may release factors which can interfere with the immune "defence" of the host. Emphasis has already been placed on circulating soluble tumour antigens, either free or complexed to specific antibodies, as mediators of the blocking phenomenon, in vitro. Excess of soluble antigen was taken as being responsible for some phenomena of immune paralysis in the draining lymph nodes of a growing neoplasm. ALEXANDER *et al.* (1969) observed that these lymph nodes are often prevented from releasing cytotoxic immunoblasts in the circulation (Fig. 2.22). Lymph node cells from patients with sarcomas were not stimulated in vitro by tumour-associated antigens, whereas peripheral blood lymphocytes from the same patients gave positive reactions. Presence of tumour antigens at the surface of the lymph node cells could be suspected, as these cells were able to stimulate autologous peripheral lymphocytes in vitro (STJERNSWARD and VANKY, 1972).

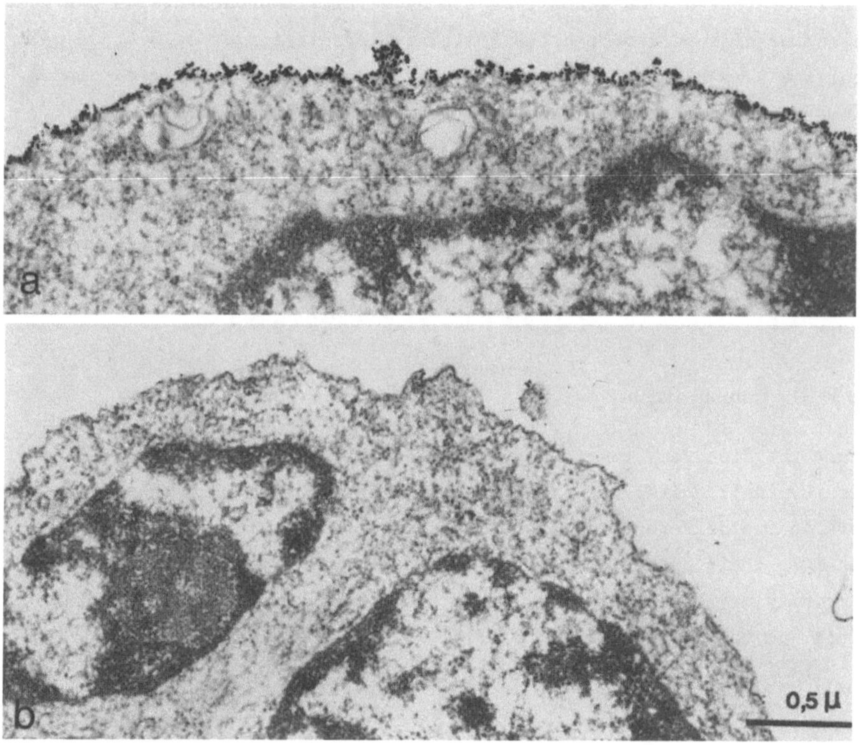

Fig. 2.21. Above: Glycocalix coating the cell. Below: Effect of neuraminidase on glycocalix (kindly provided by M. PAINTRAND)

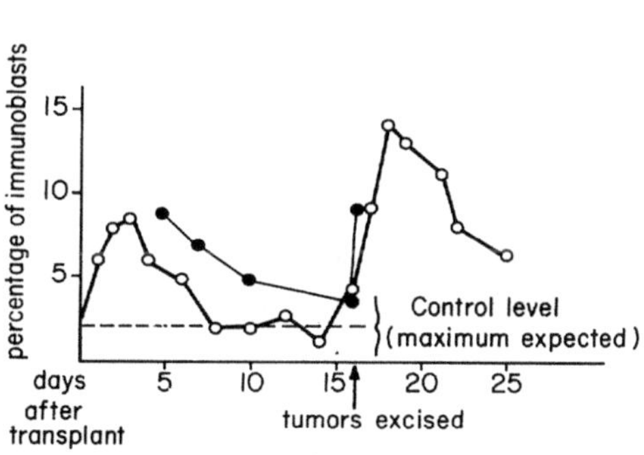

Fig. 2.22. Increase of percentage of immunoblasts in thoracic duct of rats after graft of syngeneic carcinoma. Once tumour growth is maximal, the number of the immunoblasts goes down to the normal level and then goes up (taken from ALEXANDER et al., 1969)

On the other hand, the tumour could release nonspecific factors. FAUVE *et al.* (1974) reported that mouse teratocarcinoma, melanoma, and Lewis carcinoma tumour cells excreted a factor with anti-inflammatory properties, and therefore could prevent infiltration of the tumour with macrophages. HUCHET and LHÉRITIER (1976) showed, in the laboratory of the author, that nonspecific immunodepression can be induced by injection of some tumour cells killed by irradiation or of a crude cell-free extract from the same tumours (Table 2.IX). In the case of the Lewis carcinoma, this effect seems to be due to a non-oncogenic contaminant virus.

Tumours may be localized in immunologically privileged sites so that the immune machinery fails to destroy primary tumours or metastases, when the tumour load is still small. For example, the brain may be inaccessible to the immunological effector cells (ALBRINK and GREENE, 1953). In CBA mice, the bone-marrow, where leukaemic cells proliferate, is poor in T-lymphocytes (RAFF, 1971).

2.3.3. The Immune Balance

Since the discovery that tumours possess antigens which could be recognized as foreign by their hosts, there has been growing evidence that the immune machinery may work in two different ways: one leading to the destruction of the tumour cells, the other to their protection against immunological attack. Consequently, the immune machinery may be conceived as an immune balance with two plates: the anti-cancer plate and the pro-cancer plate (Fig. 2.23).

Until recently, the anti-cancer plate (Fig. 2.24) was thought to carry cytotoxic T-lymphocytes and cytotoxic antibodies, and it was opposed to the pro-cancer one carrying factors with blocking activity. The anti-cancer plate grew rich with the addition of the macrophages acting in specific (armed macrophages) or nonspecific ways (activated macrophages) and of the K-cells which act via specific antibodies. The pro-cancer plate was loaded with the suppressor cells which may belong to different subsets of lymphoid or monocytoid cells. Moreover, blocking factors were shown to be formed more readily by antigen-antibody complexes or free soluble antigen(s) than by free antibodies.

Table 2. IX. Effect of i.p. injection of a crude cell-free extract from Lewis tumour on the immune status of (DBA/2 x C57 Bl/6) F1 mice (HUCHET and LHÉRITIER, 1976)

PHA responsiveness	Antitumour immunity to L1210 (induced by immunization with irradiated L1210 cells)	Skin allograft rejection	Humoral response to a thymus-independent antigen (TNP-polymerized flagellin)	Humoral response to a thymus-dependent antigen (TNP-hemocyanin)
Abolished	Abolished	Delayed	Normal	Abolished

Cytolytic T-lymphocytes (S)

Armed or "immune" macrophages (S)

Activated macrophages (NS)

B lymphocytes

Plasma cells

Antibodies

Suppressor cells (S, NS)

(T lymphocytes, B lymphocytes, macrophages ?)

Cytotoxic antibodies (S)
(plus complement)

Soluble antigen ?
Antigen-antibody complexes ?
Antibodies ?
= Blocking factors (S)

IgG Antibodies

= K cells (B lymphocytes, "null" cells, monocytes ?)
Antibody-dependent-cell-mediated-cytotoxicity
(S for the tumour cells; NS for the effector cells)

Unblocking factors. Antibodies ? (S)

S = Specific, NS = Non Specific

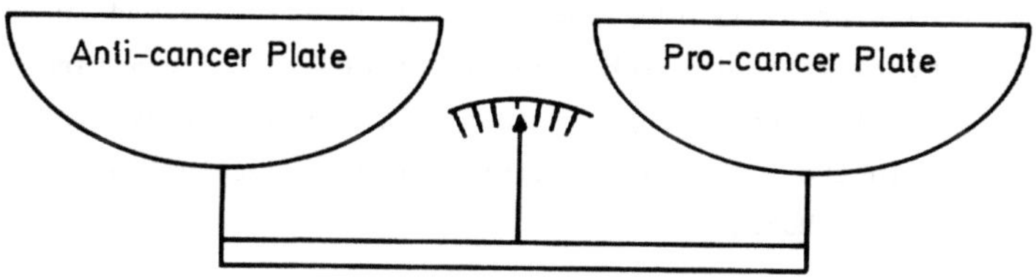

Fig. 2.23. Immune balance and cancer

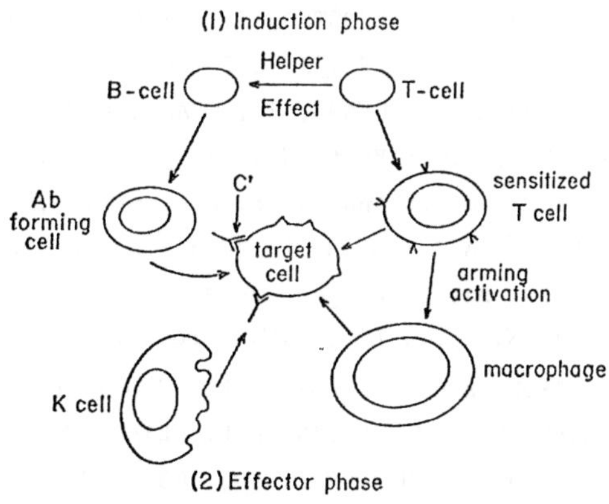

Fig. 2.24. Scheme of the possible mechanisms of cancer cell attack by immune machinery

The immune balance concept, which is schematically represented in Fig. 2.23, results essentially from in vitro observations. It has already been shown that (a) the in vitro assays are not always well defined concerning the nature of the effector cells, and (b) the correlation to mechanisms occurring in vivo is often lacking or questionable. Taking, for example, the blocking phenomenon, it was demonstrated with the microcytotoxicity assay, for which it was postulated in these studies that T-lymphocytes are the effector cells. Other investigators similarly showed that K-cells are predominantly involved. Until now, there has been no evidence to support the theory that K-cells, which are easily blocked (in contrast to T-cells, studied with the chromium release test), play any role in tumour rejection in vivo. Consequently, the blocking phenomenon, which was considered as the most important escape mechanism in vivo, has lost its popularity, perhaps momentarily (LEVY and LECLERC, 1975). Moreover, the tumour-associated antigens are numerous and it is, therefore, difficult to demonstrate that these tumour antigens detected in one in vitro assay are the same as the tumour-associated transplantation antigens implicated in tumour rejection. This could explain the lack of specificity which is increasingly reported in the in vitro assays.

In conclusion, there are probably many phenomena which are involved in cancer establishment and development, most of which occur in other life circumstances, pathological (auto-immunization) or physiological (mother tolerance of embryo): they are considered as defects in the first two conditions and as survival adaptive mechanisms in the third (JACOBS and UPHOFF, 1974).

References

ABDOU, N.I., ABDOU, N.L.: Science 175, 46 (1972)

ALBRINK, W.S., GREENE, H.S.N.: Cancer Res. 13, 64 (1953).

ALEXANDER, P., BENSTED, J., DELORME, E.J., HALL, J.G., HODGETT, J.: Proc. Royal Soc. London. B 174, 237 (1969).

ALEXANDER, P., EVANS, R., GRANT, C.K.: Ann. Inst. Pasteur 122, 645 (1972).

ALLISON, A.C.: Brit. Med. J. 4, 419 (1970).

ALLISON, A.C.: Ann. Inst. Pasteur 123, 585 (1972a).

ALLISON, A.C.: p.23 in "Frontiers of Radiation Therapy and Oncology" (J.M. Vaeth, ed.) Basel: Karger 1972b.

ALLISON, A.C., LAW, L.W.: Proc. Soc. Exp. Biol. Med. 127, 207 (1968).

ALLISON, A.C., TAYLOR, R.B.: Cancer Res. 27, 703 (1967).

ANDERSON, P.N., KLEIN, D.L., BIAS, W.B., MULLINS, G.M., BURKE, P.J., SANTOS, G.W.: Israel J. Med. Sci. 10, 1033 (1974).

ANKERST, J.: Cancer Res. 31, 997 (1971).

ANKERST, J., STEELE, G., SJOGREN, H.O.: Cancer Res. 34, 1794 (1974).

ARMERDING, D., KATZ, D.H.: J. Exp. Med. 140, 19 (1974).

ASHERSON, G.L., ZEMBALA, M.: Europ. J. Immunol. 4, 804 (1974).

ASKONAS, B.A., ROELANTS, G.E.: Europ. J. Immunol. 4, 1 (1974).

ASKONAS, B.A., RHODES, J.M.: Nature 205, 470 (1965).

BACH, M.L., BACH, F.H., JOO, P.: Science 166, 1520 (1969).

BAKER, P.J., STASHAK, P.W., AMSBAUGH, D.F., PRESCOTT, B.: J. Immunol. 112, 404 (1974).

BAKER, M.A., TAUB, R.N.: Nature, New Biol. 241, 93 (1973).

BALDWIN, R.W.: Adv. Cancer Res. 18, 1 (1973).

BALDWIN, R.W., BARKER, C.R.: Brit. J. Cancer 21, 793 (1967).

BALDWIN, R.W., BOWEN, J.G., PRICE, M.R.: Brit. J. Cancer 28, 16 (1973a).

BALDWIN, R.W., EMBLETON, M.J., PRICE, M.R.: Int. J. Cancer 12, 84 (1973c).

BALDWIN, R.W., EMBLETON, M.J., ROBINS, R.A.: Int. J. Cancer 11, 1 (1973d).

BALDWIN, R.W., PRICE, M.R., ROBINS, R.A.: Nature, New Biol. 238, 185 (1972).

BALDWIN, R.W., PRICE, M.R., ROBINS, R.A.: Int. J. Cancer 11, 527 (1973b).

BANSAL, S.C., HARGREAVES, R., SJOGREN, H.O.: Int. J. Cancer 9, 97 (1972).

BANSAL, S.C., SJOGREN, H.O.: Nature 233, 76 (1971).

BANSAL, S.C., SJOGREN, H.O.: Fed. Proc. 32, 165 (1973).

BANSAL, S.C., SJOGREN, H.O.: Israel J. Med. Sci. 10, 939 (1974).

BARBIERI, D., BELEHRADEK, J. jr., BARSKI, G.: Int. J. Cancer 7, 364 (1971).

BARSKI, G., YOUN, J.K.: J. Nat. Cancer Inst. 43, 111 (1969).

BARSKI, G., YOUN, J.K., LE FRANCOIS, D., BELEHRADEK, J. jr.: Israel J. Med. Sci. 10, 913 (1974).

BARTLETT, G.L.: Nat. Cancer Inst. Monogr. 35, 27 (1972).

BASTEN, A., MILLER, J.F.A.P., SPRENT, J., PYE, J.: J. Exp. Med. 135, 610 (1972a).

BASTEN, A., SPRENT, J., MILLER, J.F.A.P.: Nature New Biol. 235, 178(1972b).

BATCHELOR, J.R.: Cancer Res. 28, 1410 (1968).

BAUER, H.: Adv. Cancer Res. 20, 275 (1974).

BELEHRADEK, J., BARSKI, G., THONIER, M.: Int. J. Cancer 9, 461 (1972).

BELLANTI, J.A.: Immunology. Philadelphia: Saunders (1971).

BELPOMME, D., CARDE, P., OLDHAM, R.K., MATHÉ, G., JACQUILLAT, C., CHELLOUL, N., WEIL, M., AUCLERC, G., WEISBERGER, G., TANZER, T., BERNARD, J.: P. 115 in "Complications of Cancer Chemotherapy" (G. Mathé, R.K. Oldham, eds.). Vol. 49 R.R.C.R., Heidelberg: Springer-Verlag (1974).

BELPOMME, D., LE BORGNE DE KAOUEL, C., AJURIA, E., JASMIN, C., DORE, J.F.: Nature 222, 890 (1969).

BELPOMME, D., JOSEPH, R., LELARGE, N.: Cancer Immunol. Immunoth. 1, 113 (1976).

BENNETT, B.: J. Immunol. 95, 656 (1965).

BERG, J.W.: Cancer 12, 714 (1959).

BERNSTEIN, I.D., THOR, D.E., ZBAR, B., RAPP, H.J.: Science 172, 729 (1971).

BIANCO, C., PATRICK, R., NUSSENZWEIG, V.: J. Exp. Med. 132, 702 (1970).

BIAS, W.B., SANTOS, G.W., BURKE, J.P., MULLINS, P.J., HUMPHREY, R.L.: Science 178, 304 (1972).

BIGLEY, N.J.: Immunological fundamentals. Chicago: Year Book Med. Publ. 1975.

BINET, J.L., MATHÉ, G.: New York Acad. Sci. 99, 426 (1962).
BIRBECK, M.S.C., CARTER, R.L.: Int. J. Cancer 9, 249 (1972).
BLACK, M.M., OPLER, S.R., SPEER, F.D.: Surg. Gynec. Obstet. 100, 543 (1955).
BLOMGREN, H., ANDERSSON, B.: Exp. Cell Res. 57, 185 (1969).
BLUMING,A.Z., ZIEGLER, J.L., FASS, L., HERBERMAN, R.B.: Clin. Immunol. 9, 713 (1971).
BONAVIDA, B.: J. Immunol. 112, 1308 (1974).
BORBERG, H., OETTGEN, H.F., CHOUDRY, K., BEATTIE, E.J.: Brit. J. Cancer 10, 539 (1972).
BOYSE, E.A., ITAKURA, K., STOCKERT, E., IRITANI, C.A., MIURA, M.: Transplant. 11, 351 (1971).
BOYSE, E.A., MIYAZAWA, M., AOKI, T., OLD, L.J.: Proc. Royal Soc. B. 170, 175 (1968b).
BOYSE, E.A., OLD, L.J., STOCKERT, E.: p. 23 in "Fourth International Symposium on Immunopathology" (P. Grabar, P.A. Miescher, eds.). Basel: Schwabe & Co 1965.
BOYSE, A.B., OLD, L., STOCKERT, E., SHIGENO, N.: Cancer Res. 28, 1280 (1968a).
BRETSCHER, P., COHN, M.: Science 169, 1042 (1970).
BRUNNER, K.T., MAUEL, J., CEROTTINI, J.C., CHAPUIS, B.: Immunology 14, 181 (1968a).
BRUTON, O.C.: Pediatrics 9, 722 (1952).
BUBENIK, J., IVANYU, J., KOLDOVSKI, P.: Folia Biol. (Praha) 11, 426 (1965).
BUBENIK, J., KOLDOVSKY, P.: Folia Biol.(Praha) 11, 258 (1965).
BUBENIK, J., PERLMANN, P., HASEK, M.: Transplantation 10, 290 (1970).
BUBENIK, J., PERLMANN, P., HELMSTEIN, K., MOBERGER, G.: Int. J. Cancer 5, 310 (1970b).
BUCKLEY, R.H.: p. 1061 in "Progress in Immunology" (B. Amos, ed.). New-York: Academic Press 1971.
BURKITT, D.P., WRIGHT, D.H.: Burkitt's Lymphoma. Edinburgh: Livingstone 1970.
BURNET, F.M.: Immunological surveillance. 1 vol., Oxford: Pergamon Press 1970.
BYERS, V.S., SERCARZ, E.E.: J. Exp. Med. 127, 307 (1968).
BYRNE, M., HEPPNER, G., STOLBACH, L., CUMMINGS, F., McDONOUGH, E., CALABRESI, P.: Nat. Cancer Inst. Monogr. 37, 3 (1973).
CAMPBELL, D.H., GARVEY, J.S., CREMER, N.E., SUSSDORF, D.H.: Methods in Immunology. 2nd ed., New-York: W.A. Benjamin 1970.
CAMPBELL, A.C., HERSEY, P., MacLENNAN, I.C.M., KAY, H.E.M., PIRE, M.C.: Brit. Med. J. 2, 385 (1973).
CAMPBELL, A.C., MacLENNAN, I.C.M., SNAITH, M.L., BARNETT, I.G.: Clin. Exp. Immunol., 12, 1 (1972).
CANELLOS, G.P., DE VITA, V.T., ARSENEAU, J.C., JOHNSON, R.C.: p. 108 in "Complications of Cancer Chemotherapy" (G. Mathé, R.K., Oldham, eds.). vol 49 R.R.C.R. Berlin-Heidelberg-New York: Springer-Verlag 1974.
CANTOR, H., ASOFSKY, R.: J. Exp. Med. 135, 764 (1972).
CEROTTINI, J.C., BRUNNER, K.T.: Adv. Immunol. 18, 67 (1974).
CEROTTINI, J.C., NORDIN, A.A., BRUNNER, K.R.: Nature 228, 1308 (1970).
CHALMERS, D.G., COOPER, E.H., COULSON, A.S., IMMAN, D.R., TOPPING, N.E.: Int. Arch. Allergy 30, 83 (1966).

CHAR, D.H., LEPOURLHIET, A., LEVENTHAL, R.G., HERBERMAN, R.B.: Int. J. Cancer 12, 409 (1973).
CHARD, J.: Immunology 14, 583 (1968).
CIKES, M.: Transplant. Proc. 3, 1161 (1971).
CLAMAN, H.N., CHAPERON, E.A., TRIPLETT, R.F.: Proc. Soc. Exp. Biol (N.Y.) 122, 1167 (1966).
CLEVELAND, R.P., MELTZER, M.S., ZBAR, B.: J. Nat. Cancer Inst. 52, 1887 (1974).
CLINE, M.J.: J. Clin. Invest. 52, 2185 (1973).
COHEN, J.J., FISCHBACH, M., CLAMAN, H.N.: J. Immunol. 105, 1146 (1970).
COHEN, L., HOWE, M.L.: Proc. Nat. Acad. Sci. 70, 2707 (1973).
Cold Spring Harbor Symposia of Quantitative Biology. Antibodies. Vol. 32, New-York: Cold Spring Harbor 1967.
COOPER, M.D., PETERSON, R.D.A., SOUTH, M.A., GOOD, R.A.: J. Exp. Med. 123, 75 (1966).
COUTINHO, A., MOLLER, G.: Nature New Biol. 245, 12 (1973).
CRUCHAUD, A., UNANUE, E.R.: J. Immunol. 107, 1329 (1971).
CULLEN, P.R., MASON, D.Y.: Clin. Exp. Immunol. 571, 1974 (1974).
CURRIE, G.A., BAGSHAWE, K.D.: Brit. J. Cancer 22, 843 (1968).
CURRIE, G.A., BASHAM, C.: Brit. J. Cancer 26, 427 (1972).
DAVIES, D., OLD, L., STOCKERT, E.: J. Exp. Med. 125,549 (1966).
DAMESHEK, W.: Blood 21, 243 (1963).
DANTCHEV, D.: Europ.J.C. in press (1976).
DAVID, J.R.: Karolinska Symposia on Research Methods in Reproductive Endocrinology. P.1. Copenhagen: Forum Printers 1974.
DAVIES, A.J.S., LEUCHARS, E., WALLIS, V., KOLLER, P.C., Transplant. 4, 438 (1966).
DEICHMAN, D.I.: Adv. Cancer Res. 12, 101 (1969).
DELORME, E.J., ALEXANDER, P.: Lancet 2, 117 (1964).
DiGEORGE, A.M.: p. 116 in "Immunologic deficiency diseases of man" (D. Bergsma, R.A. Good, eds.). New-York: National Foundation 1968.
DORE, J.F., AJURIA, E., MATHÉ, G.: Europ. J. Clin. Biol. Res. 15, 81 (1970b).
DORE, J.F., GUIBOUT, C.: 1st International Workshop on Human Leukaemia associated antigens. Duke University, June 1974, unpublished data.
DORE, J.F., HADJIYANNAKIS, M.J., COUDERT, A., GUIBOUT, C., MARHOLEV, L., IMAI, Y.: p. 387 in "Investigation and Stimulation of Immunity in Cancer Patients" (G. Mathé, R. Weiner, eds.). Vol. 47 R.R.C.R. Berlin-Heidelberg-New York: Springer-Verlag 1974.
DORE, J.F., KOURILSKY, F.M.: p. 59 in "Present Problems in Haematology" (J. Libansky, L. Donner, eds.). 1 vol., Amsterdam, Prague: Excerpta Medica and Avicenum Czechoslovak Medical Press 1974.
DORE, J.F., MARHOLEV, L., AJURIA, E., BELPOMME, D., MATHÉ, G.: p. 276 in "Abstract of the XIIITH International Congress of Haematology". Munich 1970a.

DORE, J.F., MARHOLEV, L., COLAS DE LA NOUE, H., VASSAL, F. DE, MOTTA, R., HRSAK, I., SEMAN, G., MATHÉ, G.: Lancet 2, 1496 (1967).
DORIA, G., AGAROSSI, G., DI PIETRO, S.: J. Immunol. 108, 268 (1972).
DRESSER, D.W., MITCHISON, N.A.: Adv. Immunol. 8, 129 (1968).
DULLENS, H.F., DEN OTTER, W.: Experientia 29, 479 (1973).
DUTTON, R.W.: J. Exp. Med. 138, 1496 (1973).
DUTTON, R.W., FALKOFF, R., HIRST, J.A., HOFFMAN, M., KAPPLER, J.W., KETTMAN, J.R., LESLEY, J.F., VANN, D.: p. 355 in "Progress in Immunology"(B. Amos, ed.). New-York: Academic Press 1971.
ECCLES, S.A., ALEXANDER, P.: Nature 250, 667 (1974).
EDDY, B.E., GRUBBS, G.E., YOUNG, R.D.: Proc. Soc. Exp. Biol. Med. 117, 575 (1964).
EDELMAN, G.M., POULIK, M.D.: J. Exp. Med. 113, 861 (1961).
EDWARDS, G.E., MILLER, R.G., PHILLIPS, R.A.: J. Immunol. 105, 719 (1970).
EGGERS. A.E., WUNDERLICH, J.R., J. Immunol. 114, 1554 (1975).
EILBER, F.R., MORTON, D.L.: Nature 225, 1137 (1970a).
EILBER, F.R., MORTON, D.L.: J. Nat. Cancer Inst. 44, 651 (1970b).
ELLIOTT, B.E., HASKILL, J.S.: Nature 252, 607 (1974).
EPSTEIN, S.S.: in "10th Annual Meeting of the American Society Clinical Oncology and 65th Annual Meeting of the American Association for Cancer Research",Houston, March 27-30th 1974.
EPSTEIN, M.A., HENLE, G., ACHONG, B.G., BARR, Y.M.: J. Exp. Med. 121, 761 (1965).
ERB, P., FELDMANN, M.: Nature, 254, 352 (1975).
EVANS, R., ALEXANDER, P.: Nature, 228,620 (1970).
EVANS, R., ALEXANDER, P.: Nature, 236, 168 (1972).
EVANS, R., GRANT, C.K., COX, H., STEELE, K., ALEXANDER, P.: J. Exp. Med. 136, 1318 (1972).
EVERSON, T.C.: Ann. N.Y. Acad. Sci., 114, 721 (1964).
FASS, L., HERBERMAN, R.B., ZIEGLER, J.L.: New England J. Med. 282, 776 (1970).
FAUVE, R.M., HEVIN, B., JACOB, H., GAILLARD, J.A., JACOB, F.: Proc. Nat. Acad. Sci. USA 71, 4052 (1974).
FEFER, A.: Cancer Res. 29, 2177 (1969).
FEFER, A., McCOY, J.L., GLYNN, J.P.: Cancer Res. 27, 2207 (1967).
FELDMANN, M.: J. Exp. Med. 135, 1049 (1972a).
FELDMANN, M.: J. Exp. Med. 136, 737 (1972b).
FELDMANN, M.: Europ. J. Immunol. 4, 667 (1974).
FELDMANN, M., BASTEN, A.: J. Exp. Med. 134, 103 (1971).
FELDMANN, M., BASTEN, A.: Nature New Biol. 237, 13 (1972).
FELDMANN, M., CONE, R.E., MARCHALONIS, J.P.: Cell. Immunol. 9, 1 (1973).
FENYO, E.M., KLEIN, E., KLEIN, G., SWIECH, K.: J. Nat. Cancer Inst. 40, 69 (1968).
FISHMAN, M., PALMER, J.: Cold Spring Harbor Symp. 32, 343 (1967).
FISCHER, A., HAMMOND, W.G.: Surg. Forum 17, 102 (1966).

FOLCH, H., WAKSMAN, B.M.: J. Immunol. 113, 140 (1974).

FOLCH, H., YOSHINAGA, M., WAKSMAN, B.M.: J. Immunol. 110, 835 (1973).

FOLEY, E.J.: Cancer Res. 13, 835 (1953).

FORD, C.E., HAMERTON, J.L., BARNES, D.W.H., LOUTIT, J.F., Nature, 177, 452 (1956).

FORD, C.E., MICKLEM, H.S., EVANS, E.P., GRAY, J.G., OGDEN, D.A.: Ann. N.Y. Acad. Sci. 129, 283 (1966).

FORMAN, J., MOLLER, G.: Transplant. Rev. 17 (1973).

FOSSATI, G., GOLNACHI, H., DELLA PORTA, G., CASCINELLI, N., VERONESI, U.: Int. J. Cancer 8, 344 (1971).

FRIDMAN, W.H., KOURILSKY, F.M.: Nature 224, 277 (1969).

FRIEND, C.: J. Exp. Med. 105, 30 (1957).

FROLAND, S.S., NATVIG, J.B.: p. 107 in "Progress in Immunology" (B. Amos, ed.). New-York: Academic Press 1971.

FROLAND, S.S., NATVIG, J.B.: Transplant. Rev. 16, 114 (1973).

FROLAND, S.S., NATVIG, J.B., WISLOFF, F.: J. Immunol. 2, 83 (1973).

GALE, R.P., ZIGHELBOIM, J., OSSORIO, R.C., FAHEY, J.L.: Clin. Immunol. Immunopath. 3, 377 (1975).

GATTI, R.A., GOOD, R.A.: Cancer 28, 89 (1971).

GAUJAS, J.M., ALLISON, A.C., CHESTERMAN, F.C., REES, R.J.W., HIRSCH, M.S.: Brit. J. Cancer 27, 10 (1973).

GEFFARD,M., ORBACH-ARBOUYS, S.: Cancer Immunol. Immunoth. 1, 41(1976).

GERSHON, R.K., CARTER, R.L., KONDO, K.: Nature 213, 674 (1967).

GERSHON, R.K., COHEN, P., HENCIN, R., LIEBHABER, S.A.: J. Immunol. 108, 586 (1972).

GERSHON, R.K., KONDO, K.: Immunology 21, 903 (1971a).

GERSHON, R.K., KONDO, K.: J. Immunol. 106, 1524 (1971b).

GHAFFAR, A., CULLEN, R.T., DUNBAR, N., WOODRUFF, M.F.A.: Brit. J. Cancer 29, 199 (1974).

GHOSH, M.L., HUDSON, G., BLACKBURN, E.K.: Brit. J. Haematol. 25, 293 (1973).

GIRALDO, C., BETH , E., HIRSHAUT, Y., AOKI, T., OLD, L.J., BOYSE, E.A., CHOPRA, H.C.: J. Exp. Med. 133, 454 (1971).

GISLER, R.H., STABER, F., RUDE, E., DUKOR, P.: Europ. J. Immunol. 3, 650 (1973).

GLANZMANN, E., RINIKER, P.: Ann. Paediat. 175, 1 (1950).

GLYNN, L.E., HOLBOROW, E.J.: Auto-immunity and disease. Oxford: Blackwell 1964.

GOLD, P., FREEDMAN, S.O.: J. Exp. Med. 122, 467 (1965).

GOLDNER, H., GIRARDI, A.J., LARSON, V.M., HILLEMAN, M.R.: Proc. Soc. Exp. Biol. Med. 117, 851 (1964).

GOLDSTEIN, P., SVEDMYR, E.A., WIGZELL, H.: J. Exp. Med. 134, 1385 (1971).

GOLDSTEIN, I.J., IYER, R.N.: Biophys. Acta 121, 197 (1966).

GORCZYNSKI, R.M.: J. Immunol. 112, 533 (1974a).

GORCZYNSKI, R.M.: J. Immunol. 112, 1826 (1974b).

GORCZYNSKI, R.M., KILBURN, D.G., KNIGHT, R.A., NORBURY, C., PARKER, D.C.: Nature 254, 141 (1975).

GORCZYNSKI, R.M., MILLER, R.G., PHILLIPS, R.A.: J. Immunol. 108, 547 (1972).
GORER, P.A.: Adv. Cancer Res. 4, 149 (1956).
GOTTLIEB, A.A.: p. 479 in "Nucleic Acids in Immunity" (O.J. Plescia, W. Braun, eds.). Heidelberg: Springer-Verlag 1968.
GOWANS, J.L., McGREGOR, D.D.: p. 89 in "Third International Symposium on Immunopathology" (P. Grabar, P.A. Miescher, eds.). Basel: Schwabe 1963.
GOWANS, J.L., McGREGOR, D.D.: Progr. Allergy 9, 1 (1965).
GOWANS, J.L., McGREGOR, D.D., COWEN, D.M.: p. 20 in "The immunologically competent cells: Ciba Found. Study Group no. 10" (G.E.W. Wolstenholme, J. Knight, eds.). London: Churchill 1963.
GOWANS, J.L., UHR, J.W.: J. Exp. Med. 124, 1017 (1966).
GRANGER, G.A., WEISER, R.S.: Science 145, 1427 (1964).
GREENBERG, A.H., HUDSON, L., SCHEN, L., ROITT, I.M.: Nature, New Biol. 242, 111 (1973).
GREENBERG, A.H., SHEN, L., WALKER, L., ARNAIZ-VILLENA, A., ROITT, I.M.: Eur. J. Immunol. 5, 457 (1975).
GREENSPAN, I., BROWN, E.R., SCHWARTZ, S.O.: Blood 21, 717 (1963).
GROSS, L.: Oncogenic Viruses. 2nd ed. Oxford: Pergamon Press 1970.
GUDAT, F.G., HARRIS, T.N., HARRIS, S., HUMMELER, K.: J. Exp. Med. 132, 448 (1970).
GUTTERMAN, J.U., HERSH, E.M., McCREDIE, K.B., BODEY, G.P., RODRIGUEZ, V., FREIREICH, E.J.: Cancer Res. 32, 2524 (1972).
GUTTERMAN, J.U., HERSH, E.M., FREIREICH, E.J., ROSSEN, R.D., BUTLER, W.T., McCREDIE, K.B., BODEY, G.P., RODRIGUEZ, V., MAVLIGIT, G.M.: Nat. Cancer Inst. Monogr. 37, 153 (1973b).
GUTTERMAN, J.U., McBRIDE, C.M., FREIREICH, E.J., MAVLIGIT, G., FREI, E. III, HERSH, E.M.: Lancet 1, 1208 (1973c).
GUTTERMAN, J.U., ROSSEN, R.D., BUTLER, W.T., McCREDIE, K.B., BODEY, G.P., FREIREICH, E.J., HERSH, E.M.: New Engl. J. Med. 288, 169 (1973a).
HA, T.Y., WAKSMAN, B.H.: J. Immunol. 110, 1290 (1973).
HALPERN, B.N., BIOZZI, G., STIFFEL, C., MOUTON, D.: Nature 212, 853 (1966).
HAN, T., WANG, J.: Clin. Exp. Immunol. 12, 171 (1972).
HANNA, M.G. jr.: Seminars in Oncology 1, 319 (1974).
HARADA, M., PEARSON, G., PETTIGREW, H., REDMON, L., ORR, T.: Cancer Res. 33, 3886 (1973).
HARDIN, J.A., CHUSED, T.M., STEINBERG, A.D.: J. Immunol. 111, 650 (1973).
HARDING, B., PUDIFIN, D.J., GOTCH, F., MacLENNAN, I.C.M.: Nature, New Biol. 232, 80 (1971).
HARRIS, R.: Nature 241, 95 (1973).
HARRIS, R., VIZA, D., TODD, R., PHILLIPS, J., SUGAR, R., JENNISON, R.F., MARRIOTT, G., GLEESON, M.H.: Nature 233,556 (1971).
HAYWOOD, G., McKHANN, C.F.: Fed. Proc. 29, 371 (1970).
HELLSTROM, I.: Int. J. Cancer 2, 65 (1967).

HELLSTROM, I., HELLSTROM, K.: Int. J. Cancer 4, 587 (1969b).
HELLSTROM, K.E., HELLSTROM, I.: Adv. Cancer Res. 12, 167 (1969a).
HELLSTROM, I., HELLSTROM, K.E.: Int. J. Cancer 5, 195 (1970b).
HELLSTROM, K., HELLSTROM, I.: Ann. Rev. Microbiol. 24, 373 (1970a).
HELLSTROM, K.E., HELLSTROM, I.: Ann. Rev. Med. 23, 19 (1972).
HELLSTROM, I., HELLSTROM, K.E.: p. 409 in "In vitro methods in cell-mediated immunity" (B.R. Bloom, P.R. Glade, eds.). New-York: Academic Press 1971.
HELLSTROM, K.E., HELLSTROM, I.: Adv. Immunol. 18, 209 (1974b).
HELLSTROM, I., HELLSTROM, K.E.: p. 37 in "The Role of Immunological Factors in Viral and Oncogenic Processes" (R.F. Beers, jr., R.C. Tilghman, E.G. Bassett, eds.). Baltimore: Johns Hopkins University Press 1974a.
HELLSTROM, I., HELLSTROM, K.E., BILL, A.H., PIERCE, C.E., YANG, J.P.S.: Int. J. Cancer 6, 172 (1970).
HELLSTROM, I., HELLSTROM, K., EVANS, E., HEPPNER, A., PIERCE, G.H., YANG, J.P.S.: Proc. Nat. Acad. Sci. 62, 362 (1969).
HELLSTROM, I., HELLSTROM, K.E., SJOGREN, H.O., WARNER, G.A.: Int. J. Cancer 8, 185 (1971a).
HELLSTROM, I., HELLSTROM, K.E., SJOGREN, H.O., WARNER, G.A.: Int. J. Cancer 7, 1 (1971c).
HELLSTROM, I., HELLSTROM, K.E., WARNER, G.A.: Int. J. Cancer 12, 348 (1973a).
HELLSTROM, I., SJOGREN, H.O., WARNER, G., HELLSTROM, K.: Int. J. Cancer 7, 226(1971b).
HELLSTROM, I., WARNER, G.A., HELLSTROM, K.E., SJOGREN, H.O.: Int. J. Cancer 11, 280 (1973a).
HENLE, G., HENLE, W.: J. Bact. 91, 1248 (1966).
HENLE, W., HENLE, G., BURTIN, P., CACHIN, Y., CLIFFORD, P., SCHRYVER, A. De, THE, G. De, DIEHL, V., HO, H.C., KLEIN, G.: J. Nat. Cancer Inst. 44, 225 (1970)
HEPPNER, G.H.: Int. J. Cancer 4, 608 (1969).
HERBERMAN, R.B.: Nat. Cancer Inst. Monogr. 37, 1 (1973).
HERBERMAN, R.B.: Adv. Cancer Res. 19, 207 (1974).
HERBERMAN, R.B., ROSENBERG, E.B., HALTERMAN, R.H., McCOY, J.L., LEVENTHAL, B.G.: Nat. Cancer Inst. Monogr. 35, 259 (1973).
HERZENBERG, L.A., CHAN, E.L., RAVITCH, M.M., RIBLET, R.J., HERZENBERG, L.A.: J. Exp. Med. 137, 1311 (1973).
HIBBS, J.B., LAMBERT, L.H., REMINGTON, J.S.: J. Clin. Invest. 50, 45 (1971a).
HIBBS, J.B., LAMBERT, L.H., REMINGTON, J.S.: J. Infect. Dis. 124, 587 (1971b).
HIBBS, J.B., LAMBERT, L.H., REMINGTON, J.S.: Proc. Soc. Exp. Biol. Med. 139, 1053 (1972b).
HIBBS, J.B., LAMBERT, L.H., REMINGTON, J.S.: Nature, New Biol. 235, 48 (1972a).
HOLLINSHEAD, A., GLEW, D., BUNNAG, B., GOLD, P., HERBERMAN, R.B.: Lancet 1, 1191 (1970).
HOLLINSHEAD, A.C., MacWRIGHT, C.G., ALFORD, T.C., GLEW, D.H., GOLD, P., HERBERMAN, R.B.: Science 177, 887 (1972).

HOLM, G., HAMMARSTROM, S.: Clin. Exp. Immunol. 13, 29 (1973).
HUBER, H., DOUGLAS, S.D., FUDENBERG, H.H.: Immunology 17, 7 (1969).
HUBER, H., POLLEY, M.S., LINSCOTT, W.D., FUDENBERG, H.H., MULLER-EBERHARD, H.J.: Science 162, 1281 (1968).
HUCHET, R., FELDMANN, M.: Europ. J. Immunol. 4, 768 (1974).
HUCHET, R., LHERITIER, J.: in "Lymphocytes, macrophages and cancer" (G. Mathé, I. Florentin, M.C. Simmler, eds), Heidelberg: Springer-Verlag 1976.
HUGHES, L.E., LYTTON, B.: Brit. Med. J. 1, 209 (1964).
HUMMELER, K., HARRIS, T.N., TOMASSINI, N., HECHTEL, M., FARBER, M.B.: J. Exp. Med. 124, 255 (1966).
HUMPHREY, J.H., DOURMASHKIN, R.R.: Adv. Immunol. 11, 75 (1969).
IMMAN, D.R., COOPER, E.H.: J. Cell. Biol. 19, 441 (1963).
INBAR, M., SACHS, L.: Proc. Nat. Acad. Sci. 63, 1418 (1969).
JACOBS, B.B., UPHOFF, D.E.: Science 185, 582 (1974).
JANEWAY, C.A. jr., SHARROW, S.O., SIMPSON, E.: Nature, 253, 544 (1975).
JANOSSY, G., GREAVES, M.F.: Clin. Exp. Immunol 10, 525 (1972).
JANOSSY, G., GREAVES, M.F.: Transplant. Rev. 24, 177 (1975).
Joint report of the Fourth International Histocompatibility Workshop. Los Angeles, 1970. Histocompatibility Testing, Copenhagen: Munksgaard 1970.
Joint report of the Fifth International Histocompatibility Workshop. Evian, 1972, Histocompatibility Testing, Copenhagen: Munksgaard 1972.
JONDAL, M., HOLM, G., WIGZELL, H.: J.Exp.Med., 136, 207 (1972).
JONDAL, M., SVEDMYR, E., KLEIN, E., SINGH, S.: Nature 255, 405 (1975).
JOSEPH, R., BELPOMME, D., MATHÉ, G.: Brit. J. Cancer in press (1976)
KALISS, N.: Cancer Res. 18, 992 (1958).
KAPPLER, J.W., HUNTER, P.C., JACOBS, D., LORD, E.: J. Immunol. 113, 27 (1974).
KATZ, D.H., UNANUE, E.R.: J. Exp. Med. 137, 967 (1973).
KELLER, R.: J. Exp. Med. 138, 625 (1973).
KELLER, R., HESS, M.W.: Brit. J. Exp. Path. 53, 570 (1972).
KELLER, R., JONES, V.E.: Lancet 2, 847 (1971).
KELLER, R., OGILVIE, B.M., SIMPSON, E.: Lancet 1, 678 (1971).
KERBEL, R.S., EIDINGER, D.: Europ. J. Immunol. 2, 114 (1972).
KHALIL, A., BOURUT, C., HALLE-PANNENKO, O., MATHÉ, G., RAPPAPORT, H.: Biomedicine 22, 112 (1975).
KILBURN, D.G., SMITH, J.B., GORCZYNSKI, R.M.: Europ. J. Immunl. 4, 784 (1974).
KIRCHNER, H., CHUSED, T.M., HERBERMAN, R.B., HOLDEN, H.T., LAVRIN, D.H.: J. Exp. Med. 139, 1473 (1974).
KIRCHNER, H., MUCHMORE, A.V., CHUSED, T.M., HOLDEN, H.T., HERBERMAN, R.B.: J. Immunol. 114, 206 (1975).
KLEIN, G.: Israel J. Med. 2, 135 (1966).
KLEIN, G.: Adv. Immunol. 14, 187 (1971).

KLEIN, G.: p. 15 in "The Role of Immunological Factors in Viral and Oncogenic Processes" (R.F. Beers jr., R.C. Tilghman, E.G. Bassett, eds.). Baltimore: John Hopkins University Press 1974.

KLEIN, E., HOLTERMANN, O.A.: Nat. Cancer Inst. Monogr. 35, 379 (1972).

KLEIN, E., KLEIN, G.: J. Nat. Cancer Inst. 32, 547 (1964).

KLEIN, G., SJOGREN, H.O., KLEIN, E., HELLSTROM, K.E.: Cancer Res. 20, 1561 (1960).

KLEIST, S. von, BURTIN, P.: C.R. Acad. Sci., Paris 263, 1543 (1966).

KLEIST, S. von, CHAVANEL, O., BURTIN, P.: Proc. Nat. Acad. Sci. 69, 2492 (1972).

KOLDOVSKY, P.: Folia Biol. (Praha) 7, 170 (1961).

KOLDOVSKY, P.: Tumor specific transplantation antigen. 1 vol., Heidelberg: Springer-Verlag 1969.

KONDA, S., STOCKERT, E., SMITH, R.T.: Cell. Immunol. 7, 275 (1973).

LAFLEUR, L., MILLER, R.G., PHILLIPS, R.A.: J. Exp. Med. 135, 1363 (1972).

LAFLEUR, L., MILLER, R.G., PHILLIPS, R.A.: J. Exp. Med. 137, 954 (1973).

LAMON, E.W., ANDERSSON, B., WIGZELL, H., FENYO, E.M., KLEIN, E.: Int. J. Cancer 13, 91 (1974).

LAMON, E.W., SKURZAK, H.M., KLEIN, E., WIGZELL, H.: J. Exp. Med. 136, 1072 (1972).

LAMON, E.W., WIGZELL, H., KLEIN, E., ANDERSSON, B., SKURZAK, H.M.: J. Exp. Med. 137, 1472 (1973).

LANCE, E.M., TAUB, R.H.: Nature 221, 841 (1969).

LANDSTEINER, K., CHASE, M.W.: Proc. Soc. Exp. Biol. Med. 49, 688 (1942).

LANGEVOORT, H.L.: Invest. 12, 106 (1963).

LAPP, W.S., ELIE, R.E., PARTHENAIS, E.: p. 127 in "Suppressor Cells in Immunity", (Singhal, S.K., Sinclair, N.R.C. eds.), London, The University of Western Ontario 1975.

LAPPE, A.: Nat. Cancer Inst. Monogr. 35, 49 (1972).

LARSSON, A., PERLMANN, P.: Int. Arch. Allergy Appl. Immunol. 43, 80 (1972).

LAW, L.W.: Cancer Res. 26, 551 (1966).

LAW, L.W.: Transplant. Proc. 2, 117 (1970).

LAW, L.W., DAWE, C.J.: Proc. Soc. Exp. Biol. Med. 105, 414 (1960).

LAWRENCE, H.S., LANDY, M.: Mediators of Cellular Immunity. New-York: Academic Press 1969.

LAY, W.H., NUSSENZWEIG, V.: J. Exp. Med. 128, 991 (1968).

LECKBAND, E., BOYSE, E.A.: Science 172, 1258 (1971).

LECLERC, J.C., GOMARD, E., PLATA, F., LEVY, J.P.: Int. J. Cancer 11, 426 (1973).

LEDUC, E.M., AVRAMEAS, S., BOUTEILLE, M.: J.Exp.Med. 127, 109 (1968).

LEE, A.: Proc. Soc. Exp. Biol. (N.Y.) 128, 891 (1968).

LE FRANCOIS, D., YOUN, J.K., BELEHRADEK, J. jr., BARSKI, G.: J. Nat. Cancer Inst. 46, 981 (1971).

LEVENTHAL, R.G., HALTERMAN, R.H., HERBERMAN, R.B.: Proc. Amer. Assoc. Cancer Res. 12, 51 (1971).

LEVENTHAL, B.G., HALTERMAN, R.H., ROSENBERG, E.B., HERBERMAN, R.B.: Cancer Res. 32, 1820 (1972).

LEVY, J.P.: 249 in "Progress in Immunology (II)" (L. Brent, J. Holborow, eds.). Vol. 3, Amsterdam: North Holland 1974.

LEVY, N.L. AMOS, D.B., SOLOVIEFF, G.V., DOS REIS, A.P.: Nat. Cancer Inst. Monogr. 35, 5 (1972).

LEVY, J.P., LECLERC, J.C.: Biomedicine 23, 249 (1975).

LEVY, M.H., WHEELOCK, E.F.: Adv. Cancer Res. 20, 131 (1974).

LEWIS, M.G.: Lancet 2, 921 (1967).

LEWIS, M.G., IKONOPISOV, R.L., NAIRM, R.C., PHILLIPS, T.M., FAIRLEY, G.H., BODENHEM, D.C., ALEXANDER, P.: Brit. Med. J. 3, 547 (1969).

LIKHITE, V.V.: Cancer Res. 34, 1027 (1974a).

LIKHITE, V.V.: Cancer Res. 34, 2790 (1974b).

LOHMANN-MATTHES, M.L., SCHIPPER, H., FISHER, H.: Europ. J. Immunol. 2, 45 (1972)

LOHMANN-MATTHES, M.L., ZIEGLER, F.G., FISHER, H.: Europ. J. Immunol. 3, 56 (1973).

MacGREGOR, D.D., GOWANS, J.L.: J. Exp. Med. 117, 303 (1963).

MacKANESS, G.B.: J. Exp. Med. 116, 381 (1962).

MacKANESS, G.B.: p. 62 in "Infections Agents and Host Reactions" (S. Mudd, ed.). Philadelphia: Saunders 1970.

MacKHANN, C.F.: Transplantation 8, 209 (1969).

MacLENNAN, I.C.: Clin. Exp. Immunol. 10, 275 (1972).

MacLENNAN, I.C., LOEWI, G., HARDING, B.: Immunology 18, 397 (1970).

MANN, D.L., LEVENTHAL, B., HALTERMAN, R.: J. Nat. Cancer Inst. 54, 345 (1975).

MANN, D.L., ROGENTINE, G.N., HERBERMAN, R.: Science 174, 1136 (1971).

MARTIN, F., KNOBEL, S., MARTIN, M., BORDES, M.: Cancer Res. 35, 333 (1975).

MARTIN, F., MARTIN, M.S.: Int. J. Cancer 6, 352 (1970).

MARTYRE, M.C., HALLE- PANNENKO, O., JOLLES, P.: p. 401 in "Investigation and Stimulation of Immunity in Cancer Patients" (G. Mathé, R. Weiner, eds.). Vol. 47 R.R.C.R., Berlin-Heidelberg-New York: Springer-Verlag 1974.

MARUYAMA, K., DMOCHOWSKI, L., BOWEN, J.M., HALES, R.L.: Texas Rep. Biol. 26, 545 (1968).

MATHÉ, G.: C.R. Acad. Sci., Paris 264, 2702 (1967a).

MATHÉ, G.: Rev. Fr. Et. Clin. Biol. 12, 380 (1967b).

MATHÉ, G., BELPOMME, D., DANTCHEV, D., POUILLART, P.: Biomedicine 23, 177 (1975a).

MATHÉ, G., HALLE-PANNENKO, O., FLORENTIN, I., BRULEY-ROSSET, M., KAMEL, M., HIU, I.J.,BOURUT, C.: Europ. J. Cancer 11, 8ol (1975b).

MATHÉ, G., RAPPAPORT, H.: Histocytological typing of the neoplastic diseases of the haematopoitic and lymphoid tissues. 1 Vol., Geneva, World Health Organization 1976.

METZGAR, R.S., MOHANAKUMAR, T., MILLER, D.S.: Science 178, 986 (1972).

MILLER, J.F.A.P.: Proc. Royal Soc. B. 156, 315 (1962).

MILLER, J.F.A.P., MITCHELL, G.F.: J. Exp. Med. 128, 801 (1968).

MILLER, J.F.A.P., SPRENT, J.: J. Exp. Med. 134, 66 (1971).

MITCHINSON, N.A., RAJEWSKY, K., TAYLOR, R.B.: p. 547 in "Developmental aspects of antibody formation and structure" (J. Sterzl, H. Riha). New-York: Academic Press 1970.
MOLLER, G.: Nature 204, 846 (1964).
MOLLER, G.: J. Exp. Med. 122, 11 (1965).
MOLLER, G., COUTINHO: J. Exp. Med. 141, 647 (1975).
MOLLER, G., SVEHAG, S.E.: Cell. Immunol. 4, 1 (1972).
MORTON, D.L., HOLMES, E.C., EILBER, F., WOOD, W.: Ann. Intern. Med. 74, 587 (1971).
MORTON, D.L., MALMGREN, R.A.: Science 162, 1279 (1968).
MOTTA, R.: Etude des antigènes de surface des cellules normales et leucémiques de souris et de leurs relations avec le système immunologique de l'hôte. Paris VII: Thèse de Doctorat d'Etat es-Science Naturelles 1974.
MUNA, N.M., MARCUS, S., SMART, C.: Cancer 23, 88 (1969).
NATHAN, C.F., KARNOVSKY, M.L., DAVID, J.R.: J. Exp. Med. 133, 1356 (1971).
NATHAN, C.F., REMOLD, H.G., DAVID, J.R.: J. Exp. Med. 137, 275 (1973).
NELSON, D.S.: Transplant. Rev. 19, 226 (1974).
NORDIN, A.A.: Nat. Cancer Inst. Monogr. 35, 21 (1972).
NOSSAL, G.J.V., ADA, G.L.: Antigens, lymphoid cells and the immune response. New-York: Academic Press 1971.
NOSSAL. G.J.V., LEWIS, H.: Immunology 20, 739 (1971).
NOSSAL, G.J.V., MAKELA, O.: J. Exp. Med. (1962).
NOSSAL. G.J.V., PIKE, B.L.: Immunology 25, 33 (1973).
NOWELL, P.C., HIRSCH, B.E., FOX, D.H., WILSON, D.B.: J. Cellular physiol., 75, 151, (1970).
OKOMURA, K., TADA T.: J. Immunol. 112, 783 (1974).
OLD, L.J., BOYSE, E.A., BENNETT, B., LILLY, F.: p. 89 in "Cell-bound Antibodies" Philadelphia: Wistar Institute Press 1963.
OLD. L.J., BOYSE, E.A., CLARKE, D.A., CARSWELL, E.: Ann. N.Y. Acad. Sci. 101, 80 (1962).
OLD, L.J., BOYSE, E.A., GEERING, G., OETTGEN, H.F.: Cancer Res. 28, 1288 (1968a).
OLD, L.J., STOCKERT, E., BOYSE, E.A., KIMM, J.H.: J. Exp. Med. 127, 523 (1968b).
OLDSTONE, M.B.A., AOKI, T., DIXON, F.J.: Proc, Nat. Acad. Sci. Wash. 69, 134 (1972).
ORDER, S.E., WAKSMAN, B.H.: Transplant. 8, 783 (1969).
OREN, M.E., HERBERMAN, R.B.: Clin. Exp. Immunol. 9, 45 (1971).
OSMOND, D.G., NOSSAL, G.J.V.: Cellular Immunol. 13, 132 (1974).
O'TOOLE, C., PERLMANN, B., UNSGAARD, B., MOBERGER, G., EDSMYR, F.: Int. J. Cancer 10, 77 (1972).
O'TOOLE, C., PERLMANN, P., WIGZELL, H., UNSGAARD, B., ZETTERLUND, C.G.: Lancet 1, 1085 (1973).
PARROTT, D.M., DESOUSA, M.A.B., ESAT, J.: J. Exp. Med., 123, 191 (1966).
Patterns in cancer mortality in the United States 1950-1967. Nat. Cancer Inst. Monogr. 33 (1971).
PEACOCKE, I., AMOS, B., LAZLO, J.: Blood 28, 665 (1966).

PEARSON, G., KLEIN, G., HENLE, G., HENLE, W., CLIFFORD, P.: J. Exp. Med. 129, 707 (1969).
PEARSON, M.N., RAFFEL, S.: J. Exp. Med. 133, 494 (1971).
PEARSON, G.R., REDMON, L.W., BASS, L.: Cancer Res. 33, 171 (1973).
PEGRUM, G.D.: Brit. J. Haemat. 24, 1 (1973).
PEGRUM, G.D., EVANS, C.A., MIDDLETON, V.L., BALFOUR, I.C.: Clin. Exp. Immunol. 10, 251 (1972).
PENN, I.: Transplant. Proc. 7, 323 (1975).
PEPYS, M.B.: J. Exp. Med. 140, 126 (1974).
PERLMANN, P., HOLM, G.: Adv. Immunol. 11, 117 (1969).
PERLMANN, P., O'TOOLE, C., UNSGAARD, B.: Nat. Cancer Inst. Monogr. 35, 223 (1972b).
PERLMANN, P., PERLMANN, H.: Cell. Immunol. 1, 300 (1970).
PERLMANN, P., PERLMANN, H., BIBERFELD, P.: J. Immunol. 108, 558 (1972a).
PERLMANN, P., PERLMANN, H., MULLER-EBERHARD, H.J.: J. Exp. Med. 141, 287 (1975).
PHILLIPS-QUAGLIATA, J.M.: Amer. Zool. 15, 199 (1975).
PLATA, F., GOMARD, E., LECLERC, J.C., LEVY, J.P.: J. Immunol. 112, 1477 (1974).
POLLACK, S.: Int. J. Cancer 11, 138 (1973).
POLLACK, S., HEPPNER, G., BRAWN, R.J., NELSON, K.: Int. J. Cancer 9, 316 (1972).
POLLIACK, A., LAMPEN, N., CLARKSON, B.D., DE HARVEN, E., BENTWICH, Z., SIEGAL, F.P., HUNKEL, H.G.: J. Exp. Med. 138, 607 (1973).
POPP, R.A.: Proc. Soc. Exp. Biol. Med. 108, 561 (1961).
PORTER, R.R.: Biochem. J. 73, 119 (1959).
PORTER, R.R.: p. 177 in "Basis Problems in Neoplastic Disease" (A. Gelhorn, E. Hirschberg, eds.). New-York: Columbia Universita Press 1962.
POUPON, M.F.: Cellules suppressives responsables de la suppression de la stimulation des lymphocytes par des mitogènes dans deux modèles expérimentaux: 1. Souris porteuses de tumeurs; 2. Souris traitées par l'endoxan. Thèse, Université de Paris-Sud, Faculté de Médecine du Kremlin-Bicêtre, 1975.
POWLES, R.L., BALCHIN, L.A., FAIRLEY, G.H., ALEXANDER, P.: Brit. Med. J. 1, 486 (1971).
PREHN, R.T.: Cancer Res. 20, 1614 (1960).
PREHN, R.T.: Ann. N.Y. Acad. Sci. 164, 149 (1969).
PREHN, R.T.: Science 176, 170 (1972).
PREHN, R.T., MAIN, J.M.: J. Nat. Cancer Inst. 18, 769 (1957).
PROSS, H.F., TRACEY, D.E., WIGZELL, H., SCHIRRMACHER, V.: Scand. J. Immunol. 3, 769 (1974).
PRYJMA, J., HUMPHREY, J.H., KLAUS, G.G.B.: Nature 252, 505 (1974).
RABELLINO, E., COLON, S., GREY, H.M., UNANUE, E.R.: J. Exp. Med. 133, 156 (1971).
RAFF, M.C.: Nature 224, 378 (1969).
RAFF, M.C.: Immunology 19, 637 (1970).
RAFF, M.C.: Transplant. Rev. 6, 52 (1971).
RAFF, M.C., CANTOR, H.: p. 83 in "Progress in Immunology". (B. Amos, ed.). New-York: Academic Press 1971.

RAN, M., WITZ, I.P.: Int. J. Cancer 6, 361 (1970).

RAN, M., WITZ, I.P.: Int. J. Cancer 9, 242 (1972).

RAY, P.K., GEWURZ, H., SIMMONS, R.L.: Fed. Proc. 29, 573 (1970).

REEDMAN, B.M., KLEIN, G.: Int. J. Cancer 11, 449 (1973).

REIF, A.E., ALLEN, J.M.V.: Nature 203, 886 (1964).

RICH, R.R., PIERCE, C.W.: J. Exp. Med. 137, 649 (1973).

ROITT, I.M.: Essential Immunology. 2nd ed., Oxford: Blackwell 1974.

ROLLINGHOFF, M., WAGNER, H.: Europ. J. Immunol. 3, 471 (1973).

ROSENBERG, E.B., HERBERMAN, R.B., LEVINE, P.H., HALTERMAN, R.H., McCOY, J.L., WUNDERLICH, J.R.: Int. J. Cancer 9, 648 (1972).

ROSENTHAL. A.S., SHEVACH, E.M.: J. Exp. Med. 138, 1194 (1973).

ROTTER, V., TRAININ, N.: Cell Immunol. 13, 76 (1974).

ROUSE, B.T., ROLLINGHOFF, M., WARNER, N.L.: Nature New Biol. 238, 116 (1972).

RUSKIN, J., REMINGTON, J.S.: Antimicrob. Agents Chemoth. 474 (1969).

RYSER, J.E., VASSALI, P.: J. Immunol. 113, 719 (1974).

SANFORD, B.H.: Transplantation 5, 1273 (1967).

SCHIMPL, A., WECKER, E.: Nature New Biol. 237, 15 (1972).

SCHLESINGER, M.: Progr. Allergy 16, 214 (1972).

SCHLESINGER, M., CHAOUAT, M.: Tissue Antigens 2, 427 (1972).

SCHLESINGER, M., GOLAKAI, V.K.: Science 155, 1114 (1967).

SCHWEITZER, M., MELIEF, C.J.M., EIJSVOOGEL, V.P.: Int. J. Cancer 11, 11 (1973).

SCOLLARD, D.: Transplantation 19, 87 (1975).

SHIKU, H., KISIELOW, P., BEAN, M.A., TAKAHASHI, T., BOYSE, E.A., OETTGEN, H.F., OLD, L.J.: J. Exp. Med. 141, 227 (1975).

SHIN, H.S., KALISS, N., BORENSTEIN, D., GATELY, M.: J. Exp. Med. 180, 375 (1972).

SIEGEL, B.V., MORTON, J.L.: Immunology 10, 559 (1966).

SIMON-REUSS, I., COOK, C.M.W., SEAMAN, C.V.F., HEARD, D.: Cancer Res. 24, 2038 (1964).

SIMMONS, R.L., RIOS, A.: Cancer 34, 1541 (1974a).

SIMMONS, R.L., RIOS, A.: p. 75 in "The Role of immunological Factors in Viral and Oncogenic Processes" (R.F. Beers, jr., R.C. Tilghman, E.G. Bassett, eds.). Baltimore: John Hopkins University press 1974b

SIMMONS, R.L., RIOS, A., LUNDGREN, G., RAY, P.K., McKHANN, C.F., HAYWOOD, G.R.: Surgery 70, 38 (1971).

SJOBERG, O., BRITTON, S.: Europ. J. Immunol. 2, 282 (1972).

SJOGREN, H.O.: J. Nat. Cancer Inst. 32, 661 (1964).

SJOGREN, H.O.: p. 51 in "The Role of Immunological Factors in Viral and Oncogenic Processes" (R.F. Beers. Jr., R.C. Tilghman, E.G. Bassett, eds.). Baltimore: John Hopkins University press 1974.

SJOGREN, H.O., BANSAL, S.C.: Nat. Cancer Inst. Monogr. 35, 231 (1972).

SJOGREN, H.O., HELLSTROM, I., BANSAL, S.C., HELLSTROM, K.: Proc. Nat. Acad. Sci., USA 63, 1372 (1971).

SJOGREN, H.O., HELLSTROM, I., BANSAL, S.C., WARNER, G.A., HELLSTROM, K.E.: Int. J. Cancer 9, 274 (1972).

SJOGREN, H., HELLSTROM, I., KLEIN, G.: Exp. Cell Res. 23, 204 (1961).

SKURZAK, H.M., KLEIN, E., YOSHIDA, T.O., LAMON, E.W.: J. Exp. Med. 135, 997 (1972).

SLETTENMARK, B., KLEIN, E.: Cancer Res. 22, 947 (1962).

STEELE, G. jr., SJOGREN, H.O.: Cancer Res. 34, 1801 (1974a).

STEELE, G., jr., SJOGREN, H.O.: Int. J. Cancer 13, 435 (1974b).

STEWART, T.H.M., ORIZAGA, M.: Cancer 28, 1472 (1971).

STJERNSWARD, J.: J. Nat. Cancer Inst. 36, 1189 (1966).

STJERNSWARD, J.: J. Nat. Cancer Inst. 40, 13 (1968).

STJERNSWARD, J., VANKY, F.: Nat. Cancer Inst. Monogr. 35, 237 (1972).

SZAKAL, A.K., HANNA, M.G. jr.: Nat. Cancer Inst. Monogr. 35, 173 (1972).

TADA, T., OKOMURA, K., TANIGUCHI, M., J. Immunol. 108, 1535 (1972).

TADA, T., TAKEMORI, T.: J. Exp. Med. 140, 239 (1974).

TAKAHASHI, T.: Transplant. Proc. 3, 1217 (1971).

TAKASUGI, M., HILDEMANN, W.H.: J. Nat. Cancer Inst. 43, 843 (1969).

TAKASUGI, M., KLEIN, E.: Transplantation 9, 219 (1970).

TAMERIUS, J.D., HELLSTROM, I.: J. Immunol. 112, 1987 (1974).

TAUSSIG, M.J.: Nature 248, 234 (1974).

TAUSSIG, M.J., MOZES, E., ISAC, R.: J. Exp. Med. 140, 301 (1974).

TAUSSIG, M.J., MUNRO, A.J.: p. 293 in "Membrane receptors of lymphocytes" (M. Seligmann. J.L. Preud'homme, F.M., Kourilsky, eds). Amsterdam: North Holland publ.1975.

TAYLOR, R.B.: Transplant. Rev. 1, 114 (1969).

TAYLOR, R.B., DUFFUS, P.H., RAFF, M.C., DE PETRIS, S.: Nature, New Biol. 233, 225 (1971).

THOMAS, W., HOLT, P.G., KEAST, D.: Nature 243, 240 (1973).

THOMSON, D.M.R., STEELE, K., ALEXANDER, P.: Brit. J. Cancer 27, 27 (1973).

TING, A., TERASAKI, P.I.: Cancer Res. 34, 2694 (1974).

TISELIUS, A., KABAT, E.A.: J. Exp. Med. 69, 119 (1939).

TRAININ, N.: Physiol. Rev., 54, 272 (1974).

TRENTIN, J., WOLF, N., CHENG, V., FAHLBERG, W., WEISS, D., BONHAG, R.: J. Immunol. 98, 1326 (1967).

TROUILLAS, P.: Lancet 2, 552 (1971).

TSOI, M.S., WEISER, R.S.: J. Nat. Cancer Inst. 40, 23 (1968).

TURK, J.L.: Immunology in Clinical Medicine. 1 Vol., London: Heinemann 1969.

TURK, J.L., OORT, J.: in "Germinal Centers in Immune Responses" (H. Cottier, N. Odortchenko, R. Schindler, C.C. Congdon, eds.). New-York: Springer-Verlag 1967.

UMIEL, T., TRAININ, N.: Transplantation 18, 244 (1974).

UNANUE, E.R.: Advance Immunol. 15, 95 (1972).

URIEL, J., NECHAUD, B. de, STANISLAWSKI-BIRRENCWAJC, M., MASSEYEFF, R., LEBLANC, L., QUENUM, C., LOISILLIER, F., GRABAR, P.: C.R. Acad. Sci., Paris 265, 75 (1967).

VAN BOXEL, J.A., PAUL, W.E., FRANK, M.M., GREEN, I.:J. Immunol. 110, 1027 (1973).

VAN BOXEL, J.A., STOBO, J.D., PAUL, W.E., GREEN, I.: Science 175, 194 (1972).

VAN FURTH, R.: Mononuclear Phagocytes. Oxford: Blackwell 1970.

VAN LOVEREN, H., DEN OTTER, W.: J. Nat. Cancer Inst. 52, 1917 (1974).

VIZA, D., DAVIES, D.A.L., HARRIS, R.: Nature 227, 1248 (1970).

VRIES, M.J. De, RUMKE, P., BERNHEIM, J.L.: Int. J. Cancer 9, 567 (1972).
WAGNER, H.: J. Exp. Med. 138, 1379 (1973).
WALDMANN, T.A.: Ann. Intern. Med. 77, 605 (1972).
WARNER, N.L., SZENBERG, A., BURNET, F.M.: Austr. J. Exp. Biol. 40, 373 (1962).
WEATHERLY, N.F.: J. Parasit. 56, 748 (1970).
WEIR, D.M. (ed.): Handbook of Experimental Immunology. Vol. 2 Cellular Immunology. 2nd ed., Oxford: Blackwell 1973a.
WEIR, D.M. (ed.): Handbook of Experimental Immunology. Vol. 3 Application of Immunological Methods. 2nd ed., Oxford: Blackwell 1973b.
WEISS. L.: J. Cell Biol. 26, 735 (1965).
WEISS, D.W.: Nat. Cancer Inst. Monogr. 35, 89 (1972).
WEISS, L.: J. Nat. Cancer Inst. 50, 3 (1973).
WEISS, L., MAYHEW, E., ULRICH, K.: Lab. Invest. 15, 1304 (1966).
WEISS, D.W., SULITZEANU, A., YOUNG, L., ADELBERG, M., SEGEV, Y.: p. 187 in "Immunological Parameters of Host-Tumour Relationships" (D.W. Weiss, ed.). 1 Vol., New-York Academic Press 1971.
WEISSMAN, I.L.: J. Exp. Med. 126, 291 (1967).
WEISSMAN, I.L.: J. Exp. Med. 137, 504 (1973).
WEPSIC, H.T., ZBAR, B., RAPP, H.J., BORSOS, T.: J. Nat. Cancer Inst. 14, 955 (1970).
WHITNEY, R.B., LEVY, J.G., SMITH, A.C.: Brit. J. Cancer 31, 157 (1975).
WIGZELL, H., STJERNSWARD, J.: J. Nat. Cancer Inst. 37, 513 (1966).
WINN, H.J.: J. Immunol. 84, 530 (1960).
WINN, H.J.: J. Immunol. 86, 228 (1961).
WINN, H.J.: Nat. Cancer Inst. Monogr. 35, 13 (1972).
WISLOFF, F., FROLAND, S.S.: Scand. J. Immunol. 2, 151 (1973).
WU, A.M., TILL, J.E., SIMINOVITCH, L., McCULLOCH, E.A.: J. Exp. Med. 127, 455 (1968).
WYBRAN, J., CARR, M.C., FUDENBERG, H.H.: J. Clin. Invest. 51, 2537 (1972).
WYSS, P.: in preparation (1976).
WYSS, P.: in preparation (1975).
YASPHE, D.S.: p. 90 in "Immunological Parameters of Host-Tumour Relationship" (D.W. Weiss, ed.). New York: Academic Press 1971.
YOSHIDA, T.O., ANDERSSON, B.: Scand. J. Immunol. 1, 401 (1972).
YOSHIDA, T.O., IMAI, K.: Europ. J. Clin. Biol. Res. 15, 61 (1970).
YU, H., GORDON, J.: Nature New Biol. 244, 20 (1973).
ZARLING, J.M., TEVETHIA, S.S.: J. Nat. Cancer Inst. 50, 149 (1973).
ZBAR, B., BERNSTEIN, I., TANAKA, T., RAPP, H.J.: Science 170, 1217 (1970).
ZEMBALA, M., ASHERSON, G.L.: Nature 244, 227 (1973).
ZEMBALA, M., ASHERSON, G.L.: Europ. J. Immunol. 4, 799 (1974).
ZIGHELBOIM, J., BONAVIDA, B., FAHEY, J.L.: J. Nat. Cancer Inst. 52, 879 (1974b).
ZIGHELBOIM, J., GALE, R.P., OSSORIO, R.C., FAHEY, J.L.: Transplantation 18, 27 (1974a).

Chapter 3
Systemic Active Immunotherapy

When we started the first clinical controlled trial on active immunotherapy in 1963 (MATHÉ *et al.*, 1968, 1969a), the favourable tumour growth plate of the immune balance was not known. Nevertheless, it appeared reasonable to conduct experiments on active immunotherapy of experimental established tumours, and to apply them to man in the conditions and modalities which were efficient in animals. Later, the knowledge of this plate made the respect of these conditions in human trials more strict.

3.1. Experimental

Leukaemias and solid tumours will be considered separately and experiments will be restricted to those concerning the treatment of established tumours and those concerning immunomanipulation of animals before the tumour is established will be described in the chapter devoted to immunoprophylaxis.

3.1.1. Leukaemias

3.1.1.1. Active Immunotherapy on Limited Tumour Cell Populations

Figure 3.1 describes an experiment in which (DBA/2 x C_{57} Bl/6) F_1 mice were carrying L 1210 leukaemia, from which they had received 10^2-10^7 cells 24 h before active immunotherapy. The latter comprised either BCG (1 mg per mouse i.v. repeated every 4 days) or 10^7 irradiated leukaemic cells (one injection SC) (in vitro irradiation of 14000 rads) or a combination of both (MATHÉ, 1968). One can see that active immunotherapy, whatever its form, is able to delay mortality compared with controls, and is able to cure animals in groups where all controls die. But it is able to do so only in mice which were grafted with 10^5 leukaemic cells or less. No effect was obtained in those which had been inoculated with 10^6 or more neoplastic cells.

This probably explains why it is efficient only if it is applied 1-3 days after the transplantation of the leukaemia and inactive if administered on day 4 (MATHÉ *et al.* 1969b).

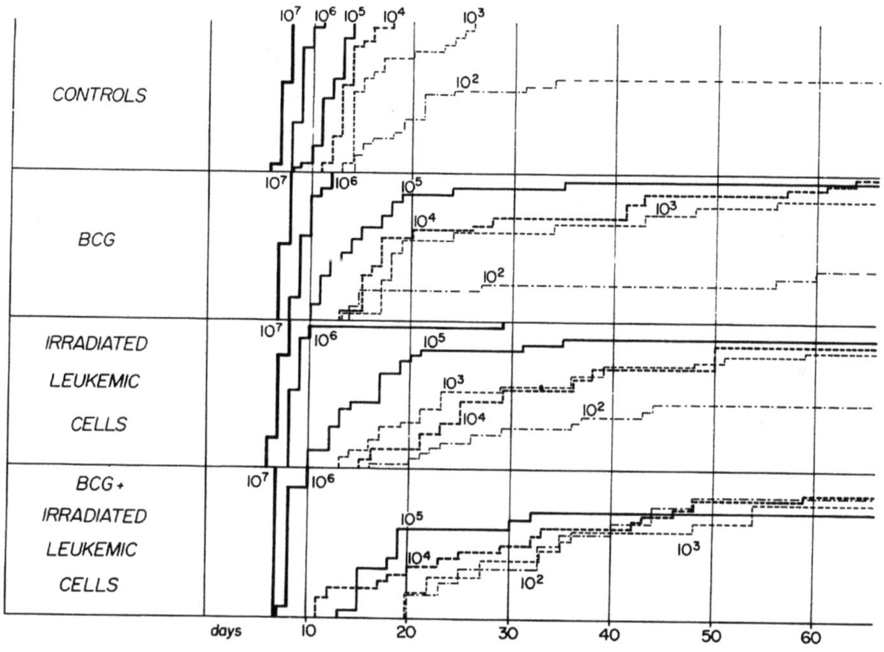

Fig. 3.1. Cumulative survival of mice grafted with 10^2 *to* 10^7 *L1210 leukaemia cells not treated or treated by BCG in the 24 h following the graft (repeated i.v. injections of 1 mg/mouse, each 4 days) or* 10^7 *irradiated leukaemic cells (1 s.c. injection), or a combination of both (MATHÉ, 1968)*

3.1.1.2. Type of Action: The Plateau of Tumour Growth Phenomenon

It seemed from the above mentioned experiment that immunotherapy obeys a "yes or no" rule, destroying the whole population of cells if the latter is $\leqslant 10^5$, presenting no effect at all if it is greater. The author's experiment (MATHÉ, 1968) does not eliminate the lysis of some neoplastic cells if their population is high: it only shows that it does not reduce it to a sufficiently small number to be followed by a prolongation of the cumulative survival of the animals.

On the other hand, measuring the volume of the tumour and observing its size after a successful active immunotherapy, one sees (Fig. 3.2) that the latter induces a regression only during the second phase of the gompertzian growth, the so-called saturation phase, while it exerts no effect on the exponential one. Finally, in the same group of identically treated animals, three types of result can be observed: absence of effect, regression, or prolongation of the plateau; the latter, which can be very long (60 days for L 1210 leukaemia), finally gave way to relapse (MATHÉ *et al.*,

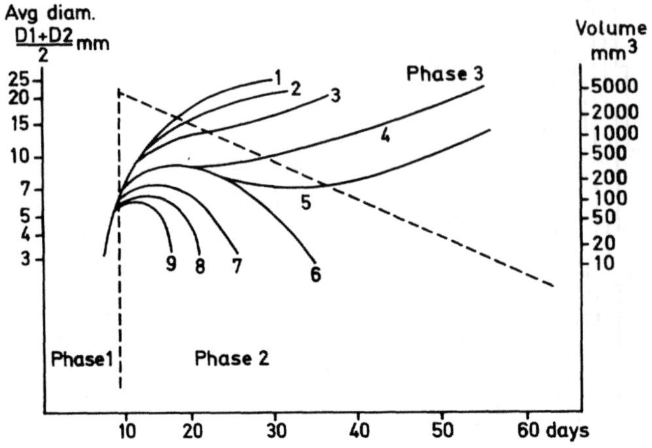

Fig. 3.2. Growth curve population of tumours obtained from mice treated with BCG and irradiated cells the day after injection with 10^4 L1210 cells (MATHÉ et al., 1969b; LHÉRITIER, not published)

1969b). This phenomenon suggests that, in some conditions, immunotherapy may, for a time, destroy as many cells as the tumour is producing. This has been demonstrated in our laboratory by LHÉRITIER (not publ.) using tritiated thymidine incorporation. As far as the final relapse after a long plateau is concerned, it suggests that either immunoresistance or another tumour protecting phenomenon may appear, or there may be a decrease of tumour destroying immunity. These possibilities will be discussed later.

3.1.1.3. Comparision of the Use of Sterilized Tumour Cells, a Systemic Immunity Adjuvant (BCG), or a Combination of Both

Simultaneous Combination Versus Cells Alone or the Adjuvant Alone. In the experiment (results of which are given in Fig. 3.1), it appeared that BCG is less efficient than the irradiated cells, and that the combination of both is more active than the two other modalities given alone (MATHÉ, 1968).

This observation was confirmed by another experiment in which the animals (same hybrid mice) received 10^3 leukaemic cells and, 24 hours later, were treated with either BCG (1 mg/mouse, i.v. injections) or 10^7 irradiated leukaemic cells administered subcutaneously once, or both (MATHÉ *et al.*, 1969b): one sees again in Fig. 3.3 that the cells are more efficient than BCG, and that the best result is obtained with a combination of both, which cures 40% of the animals.

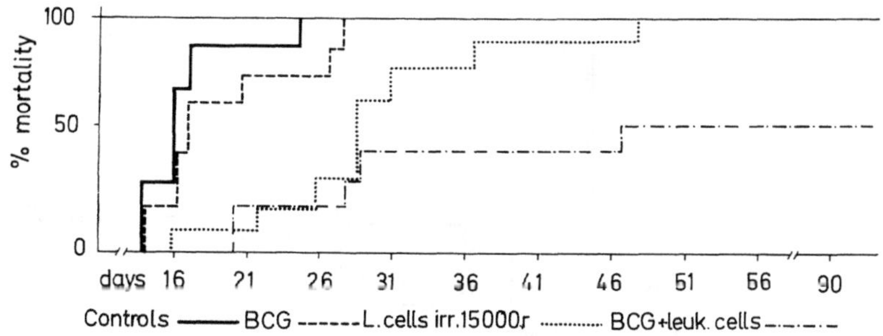

Fig. 3.3. Cumulative survival of mice grafted with 10^3 *L1210 leukaemia cells. Some mice were treated with BCG 24 h after graft (injections repeated every 4 days); others were treated with irradiated leukaemic cells (one injection 24 h after the graft); others with the combination of both (MATHE et al., 1969b)*

An even better result has been obtained with BCG and irradiated leukaemia cells on grafted Rauscher (RC 19) leukaemia (MATHÉ *et al.*, 1971a) (Fig. 3.4).

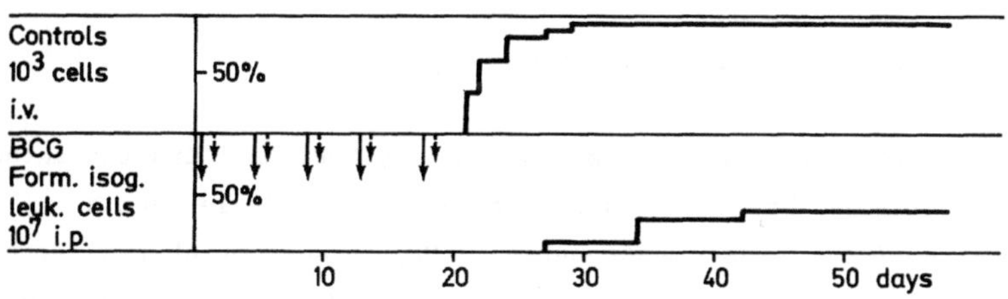

Fig. 3.4. Cumulative survival of mice carrying grafted RC 19 leukaemia, not treated, or treated with BCG plus 10^7 *formolized isogeneic leukaemia cells (MATHE et al., 1971a)*

This observation has been confirmed by Iris PARR (1972) on 51 78 Y grafted tumour.

Sequential Combination of the Adjuvant and Cells. An interesting result has been obtained by BANSAL and SJÖGREN (1972), who injected BCG and, 10-12 days later injected mitomycin-treated cells of a tumour into rats or rabbits: the serum of these animals, when injected into others carrying the same tumour growing intensively, retarded its growth (SJÖGREN, 1974). In the discussion on the mechanism of action of adjuvants below, the production, in treated animals, of the so-called "unblocking" factor will be discussed. In any case, it might be interesting to treat animals carrying a residual disease with this sequential combination instead of the simultaneous one.

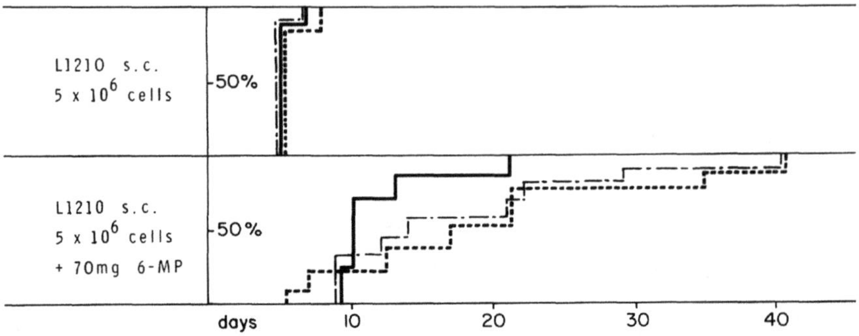

Fig. 3.5. Results of active immunotherapy on mice inoculated with 5 x 10^6 L1210 leukaemia cells and treated, 36 h later, with 70 mg of 6-mercaptopurine (6MP), and, one day after, with BCG or BCG and leukaemia cells. Whilst these two forms of immunotherapy are ineffective in animals where the number of leukaemic cells has not been reduced by chemotherapy, it is effective in those which have received a dose of 6MP able to reduce their number to 10^5. ——— no immunotherapy; ------ immunotherapy with BCG and irradiated cells (6MP was applied 36 h after SC tumour graft) (BCG + irradiated cells were applied 24 h after 6MP); ·-·-·- immunotherapy with BCG (MATHE, 1971)

3.1.1.4. Active Immunotherapy after Cell Reduction by Chemotherapy

A significant result has been obtained by POUILLART and the author (MATHÉ, 1971) in mice which had been inoculated with 5 x 10^6 L1210 leukaemia cells; 36 h later they received 70 mg of 6-mercaptopurine (6MP), a dose which reduces the number of the neoplastic cells to 10^5, according to SKIPPER *et al.* (1965), and then, one day later, BCG or BCG and irradiated cells (Fig. 3.5). However, the mice did not recover. This may have been due to the immunosuppressive effect of 6MP.

Another observation can be drawn from this experiment: BCG alone induces the same effect as BCG and cells given after chemotherapy: this may be due to the lysis of leukaemic cells after the effect of chemotherapy, which may, in some conditions, act in the same way as injection of irradiated cells.

Treating mice inoculated with 10^5 E ♂ G_2 leukaemia cells with procarbazin at a dosage which does not cure the animals, AMIEL and BERARDET (1969, 1970) cured all of them when they added either 10^7 allogeneic AkR leukaemic cells (which have the same tumour antigens as E♂G_2 cells), or 1 mg BCG on day 10.

Similarly, PEARSON *et al.* (1972, 1973), when treating LSTRA murine lymphoid leukaemia(induced by the graft of 10^4 cells) with BCNU (1,3-bis-2 chloroethyl)-1-nitrosourea) administered 7 days after the graft, prolonged the survival only in

12% of animals. When BCG was injected 19 days after BCNU, the proportion of long-term survivors was 68%.

Another remarkable result has been obtained by the author and co-workers (MATHÉ *et al.*, 1974a) with BCG alone, administered after cyclophosphamide (CPM), in treatment of L1210 leukaemia (Fig. 3.6 -3.7). Contrary to what has been observed with immunotherapy after the administration of 6MP, a significant number of mice were cured by the addition of BCG alone. This may be due to the fact that CPM is a B-lymphocyte

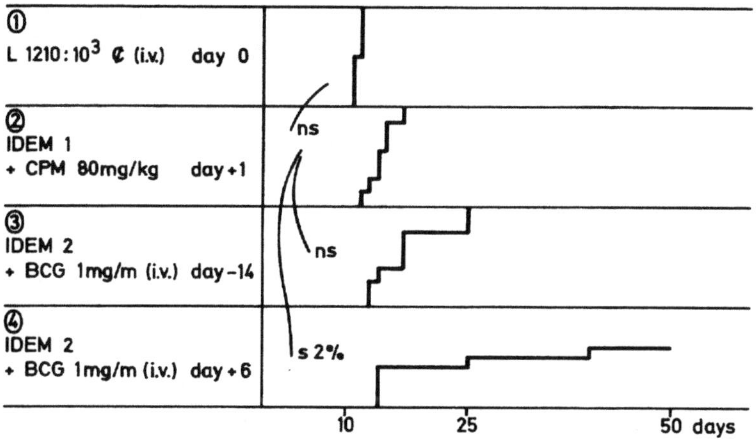

Fig. 3.6. Immunotherapeutic effect of BCG (1 mg/mouse i.v.) administered 5 days after injection of 80 mg/kg of cyclophosphamide (CPM) itself given one day after graft of 10^3 L1210 leukaemia cells (MATHÉ et al., 1974a)

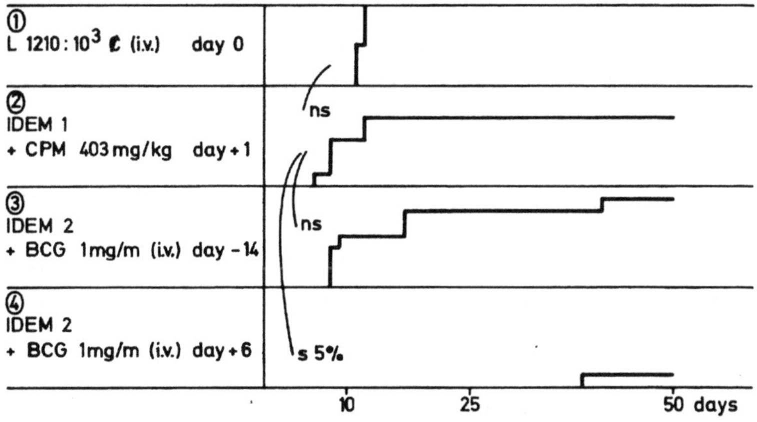

Fig. 3.7. Immunotherapeutic effect of BCG (1 mg/mouse i.v.) given 5 days after injection of 403 mg/kg of cyclophosphamide (CPM) and its protective effect against early bone-marrow (and lymphoid ?) toxicity. Mice were inoculated with 10^3 L1210 leukaemia cells one day before chemotherapy (MATHE et al., 1974a)

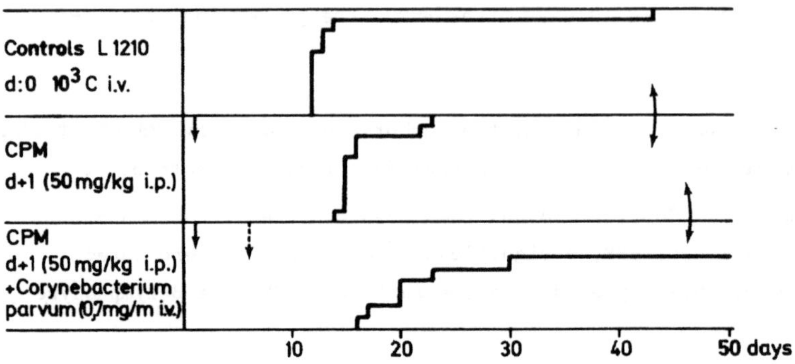

Fig. 3.8. Immunotherapeutic effect of C. parvum (0.7 mg/mouse) administered 5 days after injection of 50 mg/kg of cyclophosphamide (CPM) given 1 day after L 1210 10^3 cell inoculation in (DBA/2 x $C_{57}Bl/6$)F1. (MATHÉ, 1976)

immunity depressor and, therefore, favours the antitumour plate of the immune balance (TURK and POULTER, 1972; TURK *et al.* 1972).

In all these experiments, a minimal time gap has to be left between chemotherapy and immunotherapy: 6 days for L1210 leukaemia and 10 for E♂G$_2$ leukaemia.

The author obtained a similar result with an i.v. injection of Corynebacterium parvum (another immunity adjuvant) 5 days after injection of CPM, itself applied at 50 mg/kg 1 day after inoculation of L1210 10^3 cells (Fig. 3.8) MATHÉ, 1976).

3.1.1.5. Immunotherapy and immunological parameters

In a recent study conducted with OLSSON *et al.* (1976) on active immunotherapy (with BCG and irradiated tumour cells) of a C57 Bl/6 transplantable leukaemia originally induced by Gross virus (EAkR leukaemia), we have tried to establish some relationship between the growth rate and the cytological characteristics of the tumour (growing as an ascitis) and the tumour-specific immune responses of the host (Table 3.I).

BCG alone prolonged the survival time of most of the animals and cured 22% of them whereas BCG plus irradiated tumour cells cured only about 10% and irradiated cells alone had no curative effect. As we already observed with immunotherapy of the L1210 leukaemia, BCG effect on tumour growth became apparent in the second phase of the gompertzian curve.

Cytological examinations revealed that BCG not only delayed tumour cell dissemination but also changed the cell composition of the primary ascitic tumour: an inverse relationsship between tumour size and the number of lymphoid cells (lymphocytes and immunoblasts) in the ascitis was observed.

Table 3.I. Correlation between EAkR leukemia evolution and immunological parameters during active immunotherapy with BCG and irradiated tumour cells (OLSSON *et al.*, 1976)

Parameters		Experimental Groups					
	I	II		III		IV	V
		A	B	A	B		
Tumour-size	-	↓	→	↓	→	—	Controls
% TC in peritoneal fluid	-	↓	→	↓	→	—	Controls
% Lymphocytes at day 13 and 18	-	↑	→	↑	→	—	Controls
% Immunoblast-like cells at day 13 and 18	-	↑	→	↑	→	—	Controls
Complement-dependent antibody-mediated cytotoxicity (CDAC)	0	+ +	+ +	+ +	+ +	+(+)	0
Direct lymphocyte cytotoxicity (DLC) — Spleen	+	+ +	+ +	+ +	+ +	+	(+)
Direct lymphocyte cytotoxicity (DLC) — Lymph nodes	0	0	0	0	0	0	0
Antibody-dependent lymphocyte-mediated cytotoxicity (ADLMC) — Spleen	(+)	+ +	(+)	+ +	(+)	(+)	(+)
Antibody-dependent lymphocyte-mediated cytotoxicity (ADLMC) — Lymph nodes	0	+ +	0	+ +	0	0	0

Groups:

I. Normal controls

II. Mice receiving 10^5 viable tumour cells (T.C.)i.p. on day 0 and weekly i.v. injections of 1 mg BCG.

III. Mice receiving tumour cells i.p. plus weekly injections of BCG i.v. and of 10^7 irradiated tumour cells s.c.

IV. Mice receiving tumour cells i.p. plus weekly injections of irradiated tumour cells s.c.

V. Mice receiving 10^5 tumour cells i.p. on day 0.

Subgroups:

IIA and IIIA mice with tumour cell number far below mice from group V.

IIB and IIB mice with tumour cell number not significantly different from group V control mice.

Tumour-bearing mice were examined individually for the cytotoxic potential of antibodies and lymphoid cells to tumour cells in vitro. The amount of complement-dependent cytotoxic antibodies in the serum and the activity of spleen and lymph node cells in terms of direct T-lymphocyte cytotoxicity were not correlated with tumour size whatever the form of immunotherapy.

In contrast, antibody-dependent cell mediated cytotoxicity (test performed with serum and spleen or lymph node cells of the tumour-bearing animal) was closely related to tumour evolution since such activity was not detectable in untreated tumour-bearing hosts and was high only in treated mice with a tumour cell number below untreated controls. These results suggest that K-cells can play an important role in the BCG-induced tumour-reductive mechanisms.

3.1.1.6. *Active Immunotherapy of Perceptible Voluminous Leukaemia*

As far as grafted leukaemias are concerned, we are inclined to conclude that the rule of the small maximal number of cells (10^5) which immunotherapy is able to eradicate, is absolute and applicable to all models. We may be wrong if we extrapolate to spontaneous tumours, as the doubling time and the diffusion of the latter may be much slower that those of transplanted neoplasias.

We have treated the so-called spontaneous, but GROSS virus (1970) induced leukaemia in AkR mice when they were 6 month old. Fig. 3.9. shows that immunotherapy with BCG and allogenic leukaemia cells carrying the same tumour-associated antigens obtained a significant reduction of mortality at 11 months of age; more important, their survival curve presented as a plateau after this age, which is the statistical expression of cure (MATHÉ *et al.*, 1973a). This result confirms that, on the one hand, active immunotherapy is not efficient in animals carrying a great population of tumour cells, which was the case with most AkR mice at 6 months of age. On the other hand, it shows that active immunotherapy may be reasonably indicated for subjects in which the number of neoplastic elements cannot be evaluated exactly, but seems to be relatively small.

LEMONDE (1974), treating 275-350 day old AkR mice with one i.v. injection of 1 mg of BCG, noticeably reduced the incidence of spontaneous leukaemias in males (42% vs 76% in controls), while he noted no significant difference in females.

Administering levamisole [a synthetic phenyl imidothiazole, considered as a systemic immunity adjuvant (RENOUX and RENOUX,1972a], to mice with the same spontaneous leukaemias when adenomegalies were palpable, RENOUX *et al.* (1976) observed that the animals were divided into two categories, those whose tumour load was strikingly reduced, and those whose course of disease was not modified at all compared with controls. If this result is confirmed, and if levamisole is only an immunity adjuvant, this would prove that a leukaemia with an abundant (because perceptible) cell population may be sensitive to active immunotherapy.

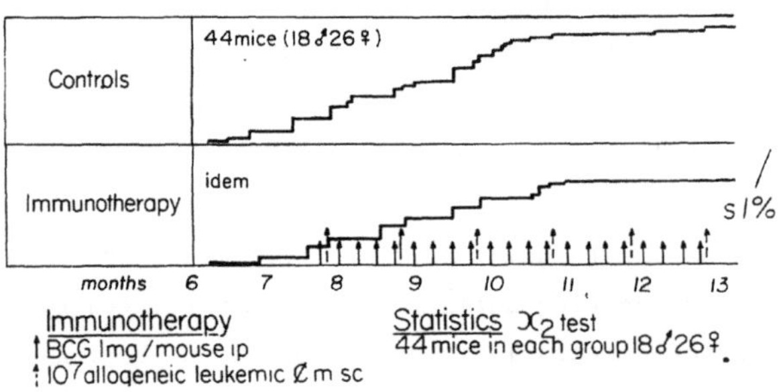

Fig. 3.9. Cumulative survival curve of spontaneous AkR leukaemia submitted to active immunotherapy at six months. Note that curve of treated animals reaches a plateau after 11 months (MATHÉ et al., 1973a)

3.1.2. Solid Tumours

3.1.2.1. Active Immunotherapy on Limited Tumour Cell Population

CURRIE and BAGSHAWE (1970) found that a single ID injection of 200 µg of C.parvum administered on day 6 after the graft of a murine fibrosarcoma, when the mean tumour diameter was about 2-3 mm, produced a significant depression in the rate of growth. The effect was still detectable, though it was not significant until day 18. However, it should be noted that no complete remission was obtained.

Similarly, SCOTT (1974) was able to prolong survival of mice carrying a mastocytoma, by treating them with C.parvum (700 µg) 2 days after tumour implantation.

WOODRUFF (1974), who used a preparation of the same organism against different established solid tumours, also underlined that the optimum time of injection of this adjuvant, for obtaining the maximum effect, is a few days after tumour inoculation.

MILAS *et al.* (1974), using C. granulosum at a dose of 0.25 mg given on days 3 and 7 after the inoculation of an isogeneic methylcholanthrene-induced fibrosarcoma, were able to obtain complete regression in about 50% of the mice. Treatment of animals on day 13 after tumour graft only slowed the growth.

As was the case with GREAGER *et al.* (1975), the author and his co-workers have not obtained a favourable effect on Lewis tumour carrying mice, with systemic BCG applied after the neoplasia graft, whatever the time of mycobacteria injection and whatever the number of inoculated tumour cells (unpublished). This negative result may be due to some immunosuppressive properties of the neoplasia, as shown in the author's laboratory by HUCHET and LHERITIER (1976).

Favourable results on different solid tumours have been obtained by WEISS (1972) with mycobacterial methanol extract residue (MER) (see p. 228), which is also an adjuvant, when it is applied just after neoplasia establishment.

Japanese workers (CHIHARA *et al.*, 1969, 1970; MAEDA and CHIHARA, 1971; MAEDA *et al.*, 1971) have published a report on the positive effect on sarcoma 180 of lentinan, another adjuvant, which is a polysaccharide extracted from Lentinus idodes. Results were obtained if its repeated administration (1 mg/kg in 10 daily doses) was started on the day after tumour inoculation.

3.1.2.2. Active Immunotherapy after Cell Reduction by Surgery and/or Radiotherapy and/or Chemo- or Hormonotherapy

SPARKS *et al.* (1973) have used immunotherapy following *surgical excision* of a grafted metastazing isogeneic, mammary carcinoma in rats: subcutaneously injected BCG or BCG combined with tumour cells considerably increased the survival, compared to controls only treated by surgery. This is the typical model for the application of immunotherapy to residual disease, compared to that described in leukemia.

LEVY *et al.* (1974) treated DBA/2j mice, transplanted with a methylcholanthrene-induced tumour, first with surgery and then with injection of tumour cell extracts alone or combined with BCG. They obtained a significant reduction of recurrences when they operated on the animals at the time the tumours were small (1 g), while they obtained only a slight prolongation if they operated when the tumours were large (2.5 g).

BALDWIN and PIMM (1973) treated a lung metastizing subcutaneously inoculated carcinoma by surgical removal followed by i.v. BCG, and they obtained a significant increase in survival (which was, however, not prolonged by combining irradiated tumour cells with BCG).

YRON *et al.* (1975) have also decreased the local recurrence of mammary carcinoma in mice with MER following surgical removal.

LACOUR *et al.* (1972, 1974) were able to increase moderately, but significantly, the survival of C_3H/H_e mice operated on for spontaneous mammary tumours, by administering to them i.v. 150 µg of poly A-poly U, another immunity adjuvant (see p. 228), once a week for one and a half months.

BARSKI and YOUN (1969), BARSKI *et al.* (1972) operated on mice during the immunological "eclipse" phenomenon which was observed when the tumours were large (see p. 52). Cell-mediated immunity reappeared after a certain time. This recovery was accelerated significantly by reinjection of x-irradiated and frozen cells.

Radiotherapy of the tumour is also an efficient means of reducing tumour volume. Though an intense and prolonged immunodepression is induced even by localized irradiation to a part of one member, as the author showed with MARTIN *et al.* (1975),

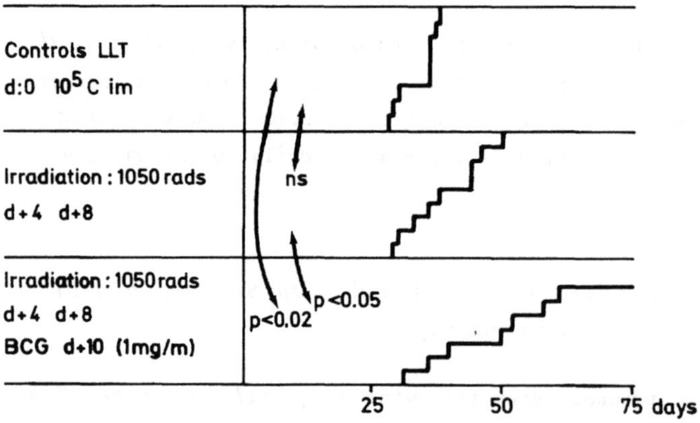

Fig. 3.10. Treatment of Lewis tumour in $C_{57}Bl/6$ mice by radiotherapy followed by active systemic immunotherapy (MARTIN et al., 1975)

immunotherapy with BCG applied after fractionated radiotherapy for the treatment of Lewis solid tumour, gave a significant result (Fig. 3.10). YRON *et al.*(1973) did not obtain a significant increase of the radiotherapy effect on a solid tumour by secondarily adding MER.

CURRIE and BAGSHAWE (1970) obtained a remarkable effect in mice carrying an isogeneic fibrosarcoma when they injected ID 200 µg of C parvum, 12 days after *chemotherapy* by cyclophosphamide (complete regression in 70% of the animals): this interval between chemotherapy and immunotherapy appeared to be very critical: when it was 6 or 16 days, no complete regression occurred (Fig. 3.11).

The reduction of the tumour mass may be obtained by *hormonomanipulation* in the case of hormonodependent neoplasias. PIESSENS *et al.* (1970, 1971) chose a dimethylbenz(a)anthracene-induced mammary carcinoma in rats as a model. They submitted the animals to ovariectomy, which induces a tumour regression followed by recurrence.

A further single subcutaneous injection of BCG delayed the relapse for three weeks. Repeated weekly injections delayed it for three months.

With THOMPSON and ALBEROLA (1972), the author and his co-workers treated mice carrying Lewis tumour with *surgery followed by chemotherapy* with cyclophosphamide which was then followed, 6 days later, by active immunotherapy with BCG and poly I-poly C, another immunity adjuvant (see p. 228). Active immunotherapy, applied after the sequence of surgery-chemotherapy, increased the number of cured animals in

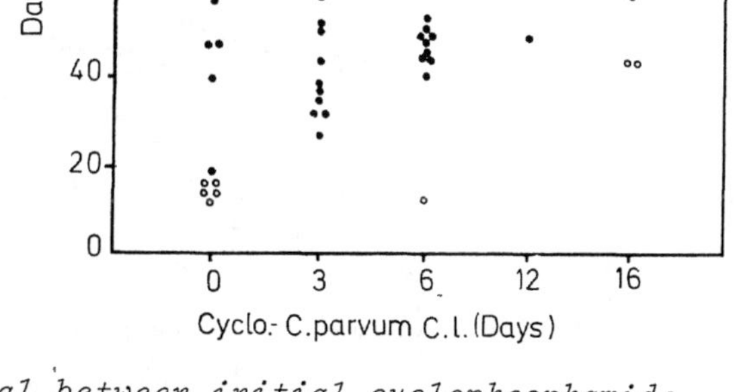

Fig. 3.11. Effect of time interval between initial cyclophosphamide, given on day 6, and a subsequent intradermal injection of C. parvum on their anti-tumour action on mouse fibrosarcoma. When the chemoimmunotherapy interval (CI) is zero, the combination is toxic, whereas a CI of 12 days leads to 70% lasting regressions. Black circles denote death from tumour. Open circles denote tumour-free animals (CURRIE and BAGSHAWE, 1970)

the group which had received 100 mg CPM/kg, but decreased it in the groups which had received 125 mg or 140 mg (Fig. 3.12). The reason for this may be that, for active immunotherapy to be effective, it must be separated from chemotherapy by intervals that vary according to dosages. In this model, surgery alone was not efficient, and chemotherapy used alone, whatever the dose, was not able to cure any of the animals.

3.1.2.3. *Active Systemic Immunotherapy of Perceptible Voluminous Solid Tumours*

Some authors have treated neoplasias of larger volume than in the above-mentioned experiments, and they sometimes obtained favourable results.

WISSLER *et al.* (1968) observed an inhibition in the growth of the well-established "minimum deviation" MORRIS (1960) hepatoma, and delayed mortality of animals by a combined i.v. and subcutaneous administration of tumour cells and pertussis vaccine, a classical immunity adjuvant (see p. 228).

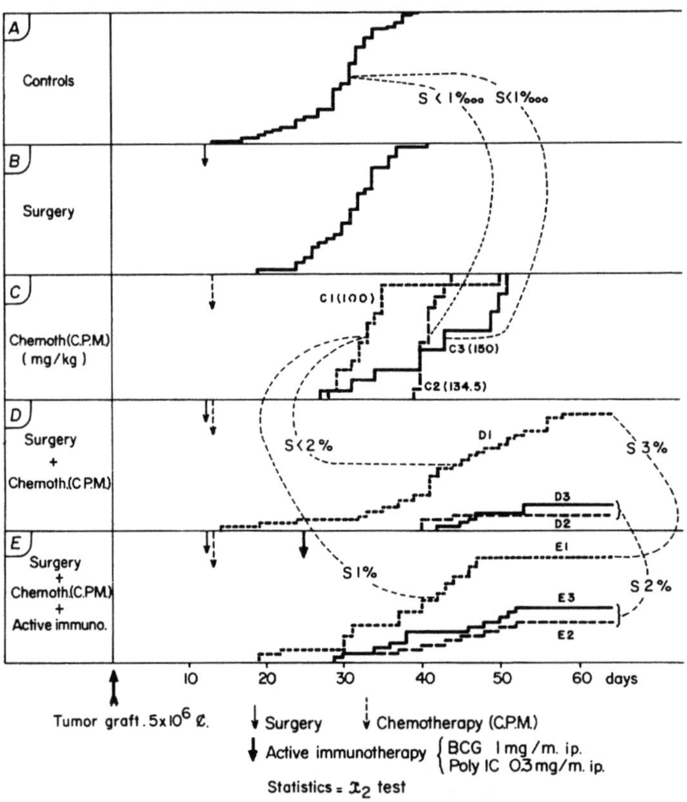

Fig. 3.12. Effects of surgery, chemotherapy, surgery + chemotherapy or surgery + chemotherapy + active immunotherapy, on Lewis tumour. The effect of immunotherapy depends on the dose of cyclophosphamide (THOMPSON et al., 1972)

Treating Lewis tumour with levamisole, RENOUX and RENOUX (1972b) obtained significant results on established solid neoplasias, while GREAGER *et al.* (1975) and ourselves (MATHÉ *et al.*, 1975a) obtained no results on such tumours with systemic BCG as will be illustrated later.

The most remarkable regressions of firmly established, isogeneic grafted (methylcholanthrene-induced) fibrosarcomas are those obtained by SIMMONS *et al.* (1971), who challenged the tumour-carrying mice with neoplastic cells treated in vitro by Vibrio cholerae neuraminidase (see p. 217). The effect of this enzyme on the cells used for active specific immunotherapy will be dealt with in detail in the chapter devoted to Developmental Pharmacology. However, it may be mentioned here that this regression is immunospecific and the effect can be augmented by simultaneous application of an adjuvant such as BCG (SIMMONS and RIOS 1971, 1972).

SEDLACEK *et al.* (1975) treated dogs carrying multiple spontaneous mammary carcinomas with autologous neuraminidase and mitomycine-treated cells after resection of most tumours; they observed some regressions of the neoplasias which remained.

3.1.2.4. Immunotherapy of Metastasis

An experiment by BALDWIN and PIMM (1973) suggests that active systemic immunotherapy may even be effective against metastasis of solid tumours. They developed an imperfect but simple model injecting dissociated sarcoma cells i.v. with which they obtained lung nodules. The development of the latter could be inhibited and the survival of the animals prolonged by one or more i.v. injections of BCG or by a subcutaneous inoculation of BCG-tumour cell mixture, if the tumour was antigenic enough. On the other hand, i.v. injection of BCG does not inhibit, and may even enhance, tumour growth in the lung when tumours lack immunogenicity.

The same authors (BALDWIN and PIMM, 1973) estimated the value of systemic active immunotherapy by BCG injected i.v. in the treatment of postsurgical pulmonary metastases of a rat transplanted carcinoma. They obtained partial control of lung heterotopias which was more significant than that observed on such localizations induced by subcutaneous injection of tumour cells in admixture with viable BCG. In another experiment, BALDWIN and PIMM (1974) treated rats presenting spontaneous pulmonary metastases from a 4-dimethylamino-benzene-induced hepatoma with significant success by i.v. BCG.

In one of their experiments, consisting of i.v. injection of BCG after surgical removal of the subcutaneous carcinoma, they observed that its favourable effect on survival and on the number of metastasis, was not improved by the combination of mycobacteria and irradiated tumour cells.

YRON *et al.* (1975) have decreased the metastatic spread of mammary carcinoma in mice by MER treatment.

Submitting mice to C. granulosum, MILAS *et al.* (1975) were able to protect them against artificially induced pulmonary metastases of spontaneous mammary carcinoma MC2.

In the experiment of RENOUX and RENOUX 1972b) quoted above, on Lewis tumour, the metastases were inhibited by levamisole in a significant percentage of animals.

An agent remarkable for its power in mobilizing lymphoytes (ORMAI and DE CLERCQ, 1969), polymethacrylic acid, has been shown by FRANCHI *et al.* (1973) to be able to decrease Lewis tumour lung metastasis when injected i.p. or i.v. early after inoculation, even at doses which are unable to affect the growth of the primary tumour. The same authors observed that this agent decreases the incidence of lymph node metastasis of osteosarcoma C22LR, while it was ineffective against lung heterotopias.

3.1.3. Problems Common to Leukaemias and Solid Tumours

3.1.3.1. The Myth of Immunological Tolerance in the Case of Tumours Associated with Vertically Transmitted Spontaneous Virus

The author's L1210 leukaemia model (MATHÉ, 1968; MATHÉ *et al.*, 1969b) was considered, at the time of his first clinical trial (1963), to be an unsuitable model for active immunotherapy in children suffering from acute lymphoid leukaemia (ALL). As a matter of fact, it was admitted, at that time, that the patients should be "immunologically tolerant" to their tumour-associated antigens, as AkR mice carrying spontaneous leukaemia were considered to be tolerant to their neoplasia-associated antigens. The reason for this last assumption was that the Gross virus, responsible for this tumour, is vertically transmitted (GROSS, 1970), while that historical "immunological immune tolerance" had been originally described for antigens administered at birth (BILLINGHAM *et al.*, 1956).

As mentioned above the author and his co-workers showed, with DORÉ and AJURIA (1970), that this "tolerance" was a myth; it was demonstrated by the rejection of isogeneic leukaemia cells by AkR mice that had been pretreated with allogeneic cells carrying the same tumour-associated antigen(s) (Fig. 2.18). We observed in these animals the presence of antibodies against G. antigen (Table 2.VI). The prophylactic effect of BCG on an isogenically grafted K 36 (originally from spontaneous AkR) leukaemia, which will be described later (DORÉ *et al.*, 1970), as well as the favourable results of immunotherapy with BCG and tumour cells (MATHÉ *et al.*, 1973a) and those of RENOUX and RENOUX (1975) with levamisole on AkR mice with spontaneous leukaemia are complementary arguments against the existence of an "operational tolerance" in this model.

The favourable results of LACOUR *et al.* (1972, 1974), who treated mice carrying spontaneous mammary carcinoma with poly A-poly U, can also be registered as evidence against this concept in this best animal model of solid tumours.

3.1.3.2. Immunotherapy and Chemotherapy Interspersed

As shown in Figure 3.2, the tumour volume curve after active immunotherapy may go up, down, or present a plateau. If such a plateau does exist at an imperceptible stage of the disease, one is tempted to apply immunotherapy interspersed with chemotherapy, so that the latter may reduce the persisting cells to a number which can be eradicated by immunotherapy.

Hence, the author and his co-workers conducted experiments in which they compared the effect, on L1210 leukaemia of BCG following treatment by cyclophosphamide (CPM), to the effect of CPM following administration of BCG. While Figures 3.6 and 3.7 showed that BCG applied 6 days after CPM (80 mg/kg or 403 mg/kg once) considerably improved the number of animals cured (MATHÉ *et al.*, 1974a), one can see in Figure 3.13, that the result of the sequence BCG-CPM is poorer than that of CPM given alone (MATHÉ *et al.*, 1974a).

An experiment on the effect of BCG followed by CPM on allogeneic skin graft rejection suggests the mechanism of this phenomenon: BCG given before CPM considerably potentiates the immunosuppressive effect of this compound (Fig. 3.14) (MATHÉ *et al.*, 1974b): this potentiation may be explained by the entry of many quiescent lymphocytes into the cell cycle under the effect of BCG, as suggested by the observations the author made with KHALIL *et al.* (1975). As cyclophosphamide is a cyclodependent

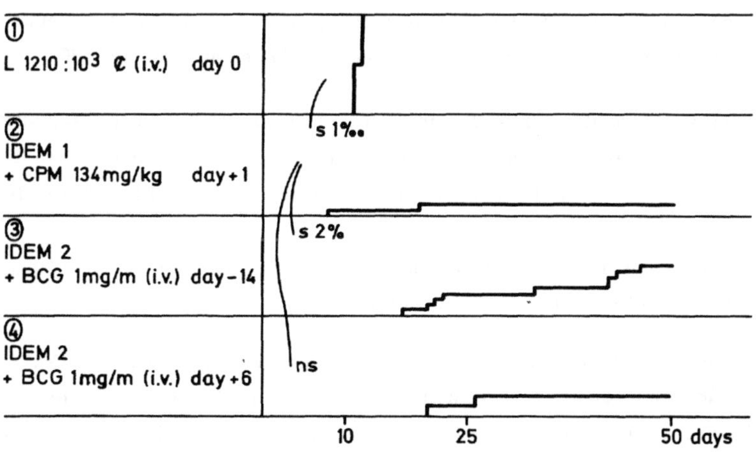

Fig. 3.13. Survival of mice carrying L1210 leukaemia. Deterioration by BCG given before 134 mg/kg of cyclophosphamide of its antileukaemic effect (MATHÉ et al., 1974a)

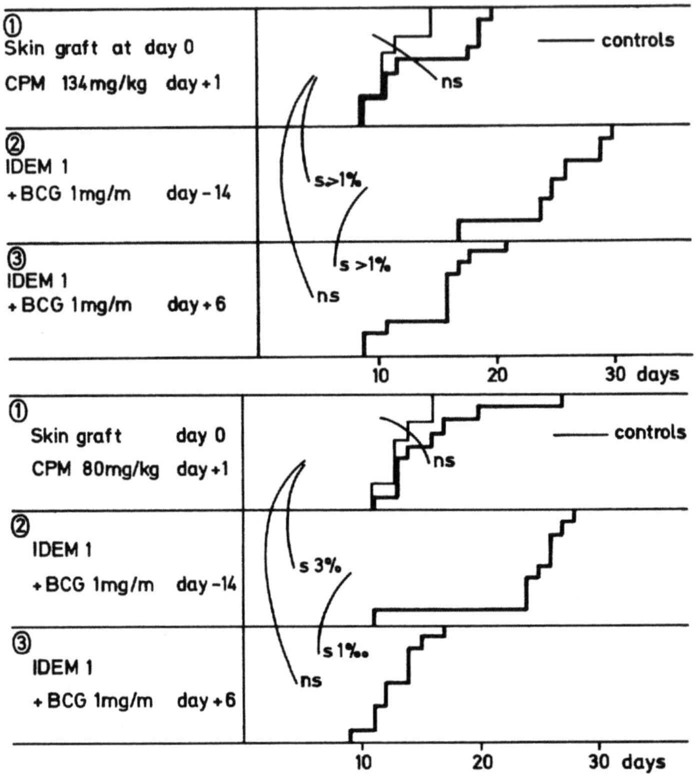

Fig. 3.14. BCG given before cyclophosphamide (CPM) increases significantly allogeneic skin graft survival compared to CPM alone (MATHÉ et al., 1974 b)

cytostatic (BRUCE *et al.*, 1966; van PUTTEN, 1972), it is reasonable to postulate that more lymphocytes are sensitive to it, and are,therefore,killed by the same dose.

As far as the possible deterioration by immunosuppression of the CPM anti-tumour effect is concerned, this is proved by its diminished antileukaemic action when an antithymocyte serum is administered at the same time (MATHÉ *et al.*, 1975a).

In an experiment on L1210 leukaemia, the author interspersed CPM and BCG: when they were given twice or three times, their effect was poorer than that obtained by one injection of CPM followed by one injection of BCG (Fig. 3.15) MATHÉ *et al.*, 1975a).

This unfavourable effect of the sequential combination of BCG and CPM has been confirmed by VASAREVIC *et al.* (1976).

The author conducted a similar experiment on Lewis tumour (MATHÉ *et al.*, 1975a). As already mentioned, BCG administered alone at day 6 after the inoculation of the neoplastic cells does not prolong survival. Nor is this improved when BCG is administered after CPM. When BCG is applied twice and interspersed with two injections of the cytostatic, it diminishes CPM effect (Fig. 3.16).

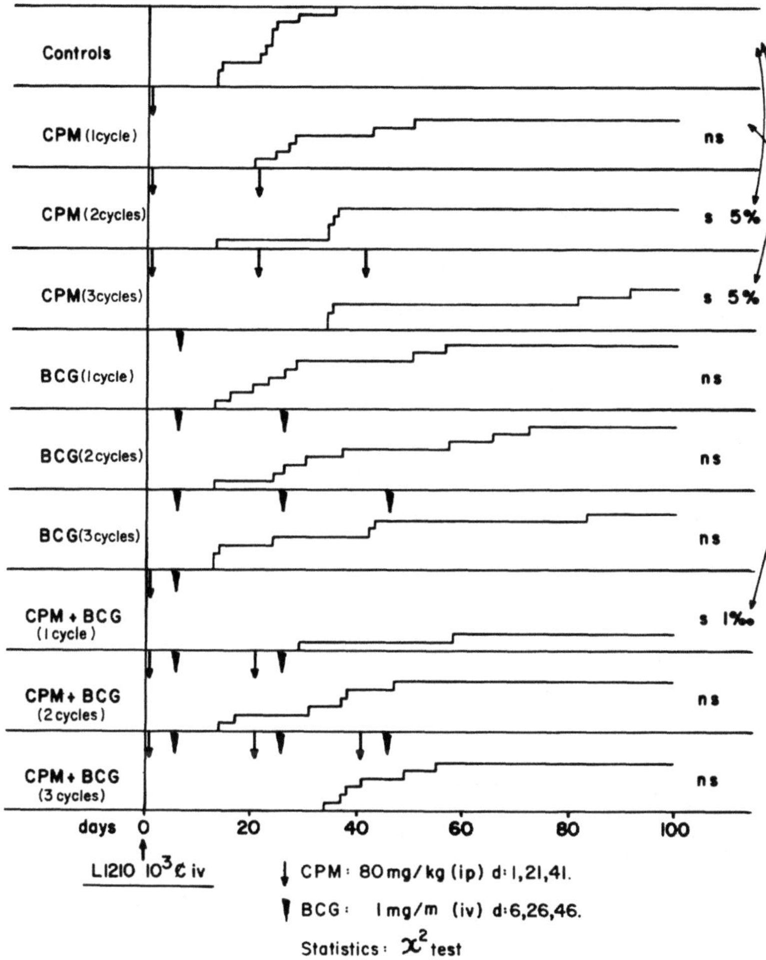

Fig. 3.15. Interspersion of cyclophosphamide (CPM) and BCG for treatment of L1210 leukaemia: when their sequential combination is given 2–3 times, the effect is poorer than that obtained by one injection of CPM foloowed by one administration of BCG (MATHÉ et al., 1975a)

On the other hand, in searching for arguments in favour of the interspersion of immunotherapy and chemotherapy, one wonders if chemotherapy does not give better results in subjects hyperimmunized against tumours. This hypothesis is sustained by the results obtained by GIULIANI *et al.* (1974) with adriamycin on two kinds of tumours derived from a Moloney virus-induced sarcoma: the effect is favourable against the neoplasias which are highly antigenic, while it is nil on the tumours which are not.

Active immunotherapy with C. parvum has also been shown to be able to potentiate the effect of a further local radiotherapy of grafted rat (MOROSON and SCHECHTER, 1976) and mouse (SUIT *et al.*, 1975) fibrosarcomas.

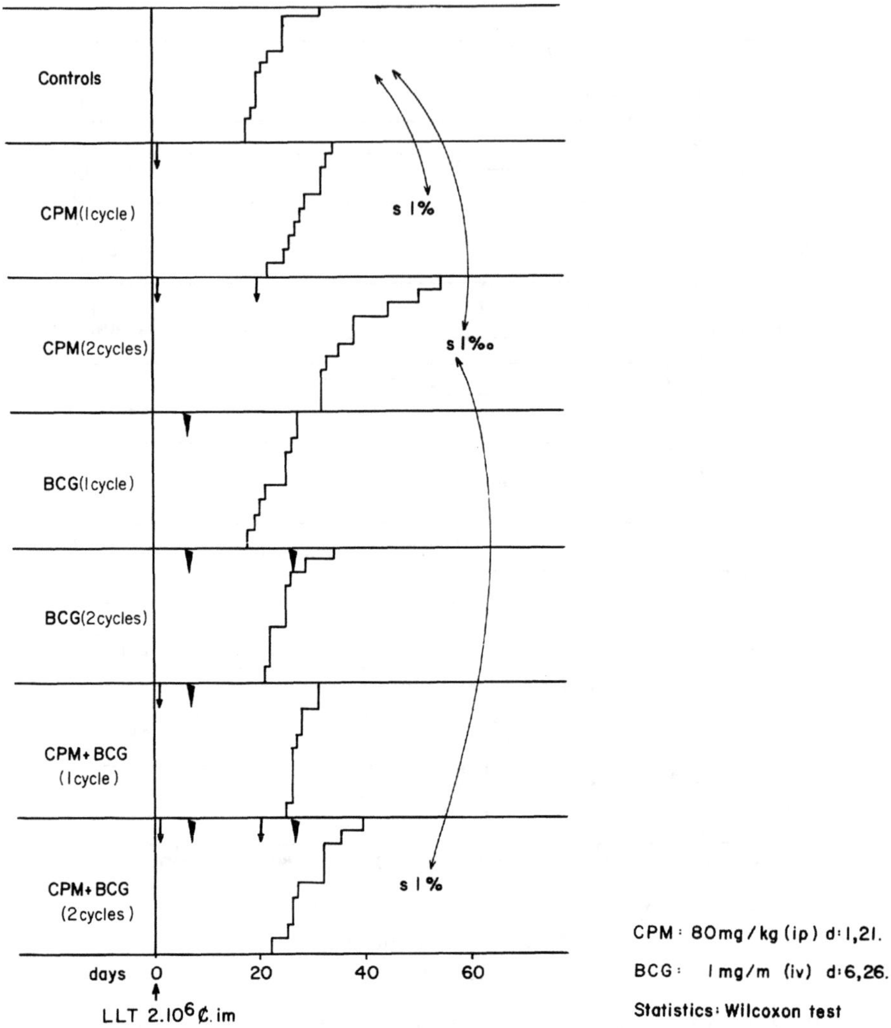

Fig. 3.16. Interspersion of cyclophosphamide (CPM) and BCG for treatment of Lewis tumour: BCG given after CPM does not improve its effect. When applied twice and interspersed with 2 injections of CPM, it deteriorates their effect (MATHÉ et al., 1975a)

Another theoretical argument in favour of the interspersion is that most chemotherapy agents work on cells in S phase or M phase (BRUCE *et al.*, 1966; van PUTTEN, 1972); immunotherapy can be supposed to act minimally during S phase because tumour-associated antigens are very poorly expressed, while it should work maximally during G1 phase, during which time they are intensively expressed (CIKES, 1971; OLSSON and MATHÉ, not published).

Finally, some adjuvants act on stem cells, and some, especially BCG, may favour haematological restoration as demonstrated by POUILLART *et al.* (1975) with the in vivo and in vitro colony-forming unit methods, including lymphoid restoration as the author also showed by studying lymphocytosis after an injection of cyclophosphamide whether or not it is followed by an administration of BCG (MATHÉ, 1974a).

3.1.3.3. Negative Results

Not all experiments on active systemic immunotherapy in which the results failed will be mentioned. However, the main "operational" factors of failure will be analyzed. The possible mechanisms will be discussed later (Chapter 9).

Large Tumours. With the exception of the few favourable results mentioned above on large tumours, the effect of active immunotherapy in such conditions has always been, or almost always is, nil.

We have observed no significant difference between the growth of control untreated methylcholanthrene-induced sarcomas and that of such tumours treated, when established,with BCG or C. parvum or poly I-poly C (not published).

Applying systemic BCG to 3-methylcholanthrene induced tumours in mice at the time they had become perceptible, TOKUNAGA *et al.* (1974) obtained no regression.

Type of Immunotherapy Applied. The fact that the effect of BCG administered alone for L1210 leukaemia immunotherapy is poor (MATHÉ, 1968; MATHÉ *et al.*, 1969b) may explain the negative results we have obtained with several other adjuvants given alone one day after the graft of 10^3 L1210 leukaemia cells. As a matter of fact, Mycobacteria cheiloni, Corynebacterium parvum, and Bordetella pertussis were ineffective, while irradiated cells were efficient (MATHÉ *et al.*, 1969b). More recently, PARR (1972) wrote of the effectiveness of a combination of B. pertussis and irradiated cells on L 5178 lymphosarcoma.

3.1.3.4. The Possibility of Neoplasia Growth Enhancement

The application of immunotherapy may, rather than delaying or suppressing tumour growth, accelerate or increase it. We saw in chapter 2 that the latter phenomenon was first described in allogeneic tumour grafts by KALISS (1952, 1957, 1958), who induced it by preimmunizing animals with special preparations of tumour cells (for example by lyophilized materials), and was able to transmit the phenomenon passively by serum (KALISS, 1956, 1957). The evolution of ideas about its mechanism was indicated (see p. 50).

Postimmunotherapy enhancement has been described in the case of established solid tumours. The literature mentions this type of effect observed after treatment with BCG (PIESSENS *et al.*, 1970) and C. parvum (MATHÉ *et al.*, 1969b).

The special growth acceleration of Lewis tumour by BCG interspersed with cyclophosphamide (CPM) compared to the evolution after CPM alone was also mentioned (p. 93) (MATHÉ *et al.*, 1975a).

It is essential to determine the factors which induce such an unfavourable effect.

The *variety of neoplasias* is classically the most important: growth enhancement has never or rarely (DORÉ *et al.*, 1974) been mentioned after immunotherapy of leukaemia, while the experiments mentioned above concern solid tumours.

The *volume of the tumour at the time of treatment* may be another essential factor: the mechanisms will be considered later (chapter 9).

3.2. Clinical

The author's first trials were conducted on leukaemias, since these neoplasias are, in the laboratory, less sensitive to growth enhancement than solid tumours which were described in 3.1.

3.2.1. Leukaemias and Hematosarcomas

The experimental observations described above had indicated that active immunotherapy in treatment of L1210 leukaemia is best suited for situations where the total number of neoplastic cells is the smallest possible, in other words, for the residual disease left by a first treatment.

It seemed to the authors that the most reasonable approach was, therefore, to try active immunotherapy as a supplementary treatment of leukaemia, whose cell population has been maximally reduced by chemotherapy (or chemoradiotherapy).

3.2.1.1. Residual Imperceptible Disease

3.2.1.1.1. Acute Lymphoid Leukaemia

The only neoplasia which was sensitive enough, in 1963, to chemotherapy in order to considerably reduce tumour cell population was acute lymphoid leukaemia (ALL) (MATHÉ and KENIS, 1975).

At that time, the results (expressed in durations of remission and of survival) obtained by chemotherapy alone, were poor (HOLLAND, 1968): thus, there was no ethical problem in trying a new method.

Scientifically, however, this disease was suspected of being an unwise choice for trial of active immunotherapy, because of possible immune tolerance. We were aware of its supposed existence in spontaeous leukaemia of AkR mice as the experimental model of human ALL . In the preceding chapter, the arguments mentioned were mainly those which were collected with DORÉ and AJURIA (1970), which made the authors doubt this alleged tolerance towards leukaemia-associated antigens, the existence of which was discussed on p. 14.

The experiments quoted above, suggest that the optimal condition for immunotherapy applied to man is that the patients should carry the smallest possible number of leukaemic cells. To achieve this, the cell number was reduced by chemotherapy, first to induce remission, then an attempt was made to obtain a further reduction by sequentially administering all the different forms of chemotherapeutic drugs available and CNS radiotherapy (MATHÉ *et al.*, 1968, 1969a, 1976a).

In the first trial (MATHÉ *et al.*, 1968, 1969a), 30 patients, whose ages varied from 3 to 80 years, were treated in this way (Fig. 3.17). They received, after induction of a remission, a sequential complementary cell-reducing chemotherapy (one drug at a time) combined with intrathecal administration of methotrexate and meningeal radiotherapy. They were then divided into 4 groups at random. The first group of 10 control patients did not receive any further treatment after complementary chemotherapy was stopped. In the second, 8 patients were treated with fresh living Pasteur Institute BCG applied, at the dose of 75 mg, on scarified areas on every 4th day for one month, and then on every 7th day (the method will be described in detail on p. 270). In the third group, 5 patients received each week, both intradermally and subcutaneously, 4×10^7 leukaemic cells, which had been obtained from a pool of allogeneic donors suffering from ALL. These cells had been preserved at -70° C in DMSO. For the first 6 injections, the cells were treated with a 4% formol solution to inactivate any possible virus and, for the remaining injections, they were irradiated with 4,000 rads in vitro. In the final group, 7 patients were given both forms of immunotherapy (BCG plus cells). The total length of immunotherapy was 5 years.

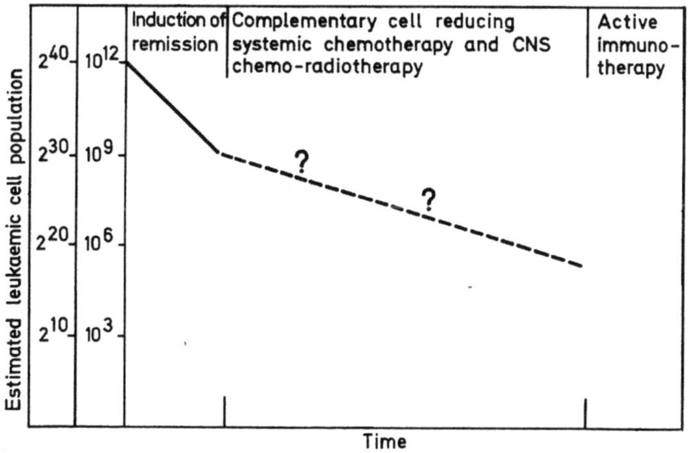

Fig. 3.17. Principle of our treatment protocols of acute lymphoid leukaemia. CNS: central nervous system

Each of the 10 patients who were left without treatment after chemotherapy was stopped, relapsed (Fig. 3.18). The average duration of the remissions was 66 days, the median lying between 70 and 77 days. The limits were 30 and 130 days. On the 130th day, only 9 of the 20 patients given immunotherapy had relapsed, and the difference between these two groups is highly significant ($X^2 = 6.18$, p 0.02). Later, 4 other patients had relapsed.

An examination of the relapses gives rise to the following ideas: (a) the majority of them appeared to be early: 9 before the 100th day and 5 before the 30th day. This is comparable to the experimental observations in mice, and suggests that the number of tumour cells left after the complementary cell-reducing treatment was greater than the maximum number which could be controlled by active immunotherapy; (b) four relapses were late in onset. One occurred on the 210th day, another on the 315th day, and another on the 950th day (which is comparable with those exceptional late relapses noted in experiments with mice) (Fig. 3.2). The fourth, which occurred on the 324th day, was in an infant in whom the BCG treatment had been stopped 44 days previously (due to a phlyctenular keratoconjunctivitis which had required the treatment to be stopped several times). This last patient, when considering the possibility of late relapse(s), answers the question as to whether immunotherapy should be continued for a very long time. Seven patients are still in remission, six more than eight years after chemotherapy has been stopped, and one for more than 12 years (MATHÉ *et al.*, 1975b, 1976a).

Figure 3.18 shows the actuarial curves, demonstrating differences between the groups submitted to immunotherapy and the control group. It shows that the shape of the

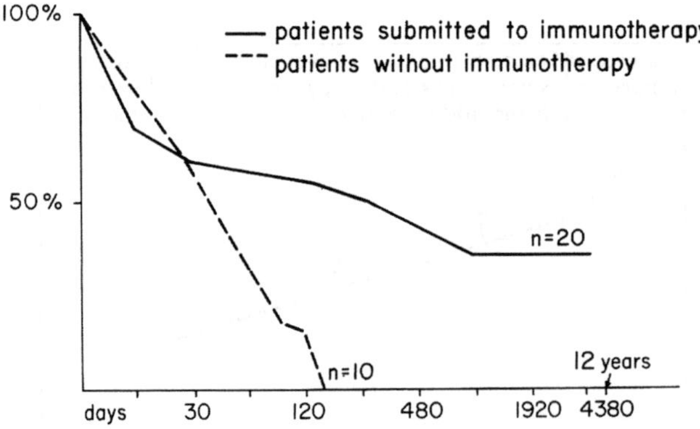

Fig. 3.18. First controlled active immunotherapy trial of acute lymphoid leukaemia (ALL) in patients of all ages. Actuarial duration of first remission after immunotherapy is started. Note that time scale is geometrical (MATHÉ et al., 1968, 1976a)

actuarial curve of the cumulative remission duration of the patients given immunotherapy tends to straighten out and continue as a plateau.

There was no significant difference between the group given BCG (5 relapses out of 8 patients; treatment had been stopped in one), the one given leukaemic cells alone (3 relapses in 5 patients), and the group given both these forms of immunotherapy (5 relapses in 7 patients).

The follow-up has now continued for 8 to 12 years, and the plateau between day 960 and 12 years is interpreted as the statistical expression of cure expectancy for 7 patients.

Because of progress in chemotherapy, we had to modify the procedures. The new drugs efficient in ALL and made available between 1963 and now, have been introduced into the systemic chemotherapy, the duration of which was variable between the successive protocols (MATHÉ and KENIS, 1975) (Table 3.II.). Total CNS radiotherapy (1500 rads) (prot. 1-8) POUILLART *et al.*, 1972; MATHÉ *et al.*, 1975c) was replaced, in 1971, by cranial irradiation (2400 rads) (prot. 9-12).

As far as immunotherapy is concerned, BCG given alone was compared with BCG plus C. parvum after 1968 (prot. 5-8) or C. granulosum after 1972 (prot. 9-10) MATHÉ *et al.*, 1975d), and BCG plus C. parvum was compared to BCG plus C. parvum plus poly I-poly C in 1970 (MATHÉ *et al.*, 1970a) (the latter was only administered for one month) (prot. 7 and 8). A summary of the main characteristics of successive immunotherapy protocols is given in Table 3.I.

The value of the combination of vincristine (as oncostatics, one injection each month) or of the administration of adamantadine (an antiviral agent according to GALBRAITH *et al.*, 1969) was also estimated (MATHÉ *et al.*, 1971b) (prot. 5-7).

The overall results, recently evaluated for a series of 100 patients whose cytological type could be determined (MATHÉ *et al.*, 1971, 1973b) and where the follow-up was long enough, are represented on Fig. 3.19 (MATHÉ *et al.*, 1976a). For patients of all ages, the median duration of survival is more than 10 years. Most interesting is the observation that the remission and survival curves, after a descent, flatten out into a "plateau" at about the 48th month.

These results are far superior to those recently published by ROSEN and NOWAK (1974) concerning a large number of patients "treated conservatively by present standards", that is to say by maintenance chemotherapy, in a group which comprises only children (while the author's group comprised patients of all ages): among 269 children with ALL, diagnosed between 1957 and 1967, 21 (7.8%) have lived for more than 5 years.

Table 3.II. Successive protocols[a] applied at Villejuif between 1963 and 1976 to acute lymphoid leukaemia patients

No.	Duration of chemotherapy in months before immunotherapy	Intensity of the chemotherapy before immunotherapy [b]	Treatment of central nervous system			Immunotherapy		
			Chemotherapy I.T.	Radiotherapy Brain to S_2	Radiotherapy Brain only	Specific	Non specific	With chemotherapy
1 (1963)	24	1	+	+	–	Leukaemic cells	BCG	–
4 (1967)	3	3	+	+	–	"	BCG	–
5 (1968)	7	3	+	+	–	"	BCG +C.Parvum	Amantadine or Vincristine
6 (1969)	5	2	+	+	–	"	BCG +C.Parvum	Amantadine or Vincristine
7 (1970)	5	2	+	+	–	"	BCG +C. Parvum ±Poly IC	
8 (1970)	5	2	+	+	–	"	BCG +C. Parvum ±Poly IC	
9 (1971)	9	2	+	–	+	"	BCG ±C. Granulosum	
10 (1973)	9	2	+	–	+	"	BCG ±C. Granulosum	
11 (1973)	20	2 to 3	+		+	Cultivated L.cells vs. preserved L.C.	BCG	
12 (1974)	18–24	3 to 2	+		+	"	"	

a All were phase III trials (for definitions, see MATHÉ 1973, 1974b)
b Maximum number of drugs given in each period.

Fig. 3.19. Acute lymphoid leukaemia. Cumulative duration of survival of 100 patients (mentioned in the text) after active immunotherapy is started. Note that the time scale is geometric (MATHÉ et al., 1976a)

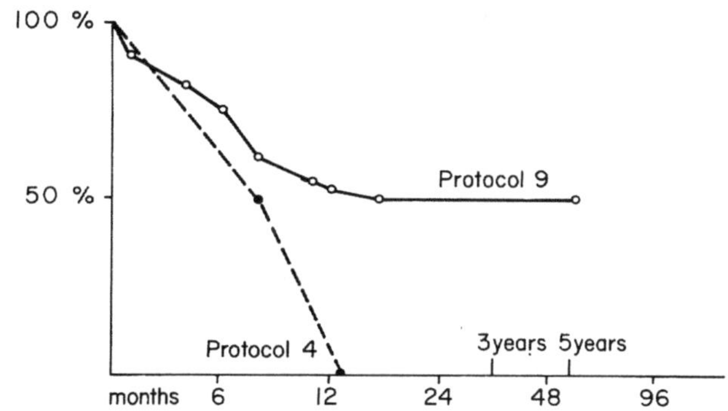

Fig. 3.20. Cumulative duration of first remission for two protocols: protocol 4 with short and intensive preimmunotherapy chemotherapy, and protocol 9 with moderate but long preimmunotherapy chemotherapy

Comparing different systemic preimmunochemotherapy or chemoradiotherapy protocols several remarks can be made: a) protocol 4 (see Table 3.II), in which the treatment was intensive, combining 3 drugs (prednisone, vincristine, daunorubicine), but short (3 months), was the worst (Fig. 3.20); protocol 9, in which chemotherapy was moderate, combining only 2 drugs at a time, but longer (9 months), gives a much better result (Fig. 3.20). b) The necessity of intrathecal chemotherapy and spinal or cranial radiotherapy is indicated on Table 3.III, which also shows that methotrexate alone is as efficient as its combination with cytosine arabinoside (POUILLART *et al.*, 1972; MATHÉ *et al.*, 1975c).

Comparing the patients receiving immunotherapy alone and those receiving it plus vincristine (1 mg/m2/month), or adamantadine (100 mg/m2/day), it appears that these two drugs diminished the results (Fig. 3.21) (MATHÉ *et al.*, 1971b): vincristine, probably because, as an oncostatic, it induces immunodepression, and it was seen that chemotherapy immunosuppression can be enhanced by combined immunostimulation (MATHÉ *et al.*, 1974b, 1974c); adamantadine for the same reason, as BREDT and MARDINEY (1969) showed that it is an immunosuppressive agent.

As far as the efficacy of C. parvum, which was replaced secondarily by C. granulosum (1 mg/week/i.m.) administered in complement to BCG, is concerned, there is no significant difference in favour of their additon (MATHE *et al.*, 1975d).

The same can be said for poly I-poly C (1 mg/m2/day x 30 days, i.v.), but three patients who had relapsed under immunotherapy with BCG and cells between one and one

Table 3.III. Incidence of meningeal localisations according to the CNS prophylactic treatment (POUILLART *et al.*, 1972; MATHÉ *et al.*, 1975c)

Treatment	No meningeal perceptible localization	Meningeal localisation
Controls	50	26
Intrathecal (IT) methotrexate alone	28	9
IT methotrexate + irradiation	23	3
IT methotrexate + IT cytosine arabinoside + irradiation	30	2

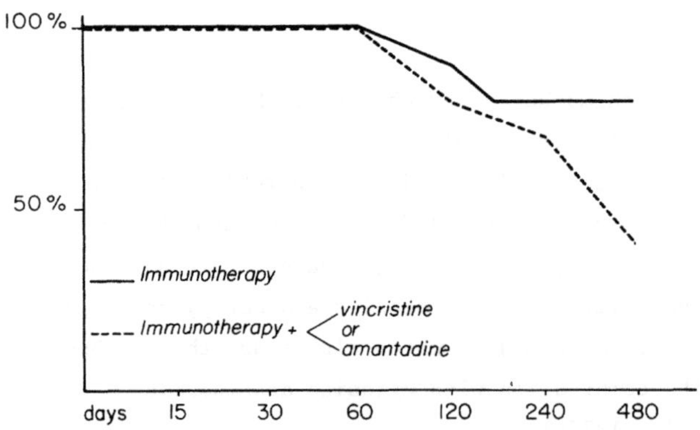

Fig. 3.21. Actuarial curves of duration of first complete remission in patients receiving immunotherapy alone or immunotherapy combined with vincristine or adamantadine (MATHÉ et al., 1971b)

and a half years, and who were submitted again to immunotherapy comprising cells, BCG, and poly I-poly C, have not yet relapsed after 3 years.

The search for *factors of prognosis in ALL patients submitted to active immunotherapy* (MATHÉ *et al.*, 1976a) allowed the author in his last statistics, to eliminate sex (Fig. 3.22) (the difference in favour of females is not significant as far as survival is concerned) and age (Fig. 3.23), and to retain: a) tumour volume (estimated by the number of circulating leukaemic cells, and/or the presence or absence of adenomegaly and/or splenomegaly and/or hepatomegaly) (Fig. 3.24); b) the meningeal localization (Fig. 3.25); and c) the cytological type.

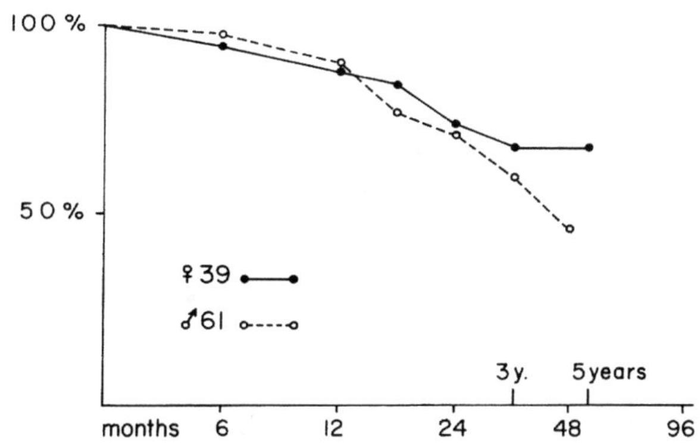

Fig. 3.22. Survival of acute lymphoid leukaemia submitted to active immunotherapy according to sex (MATHE et al., 1976a)

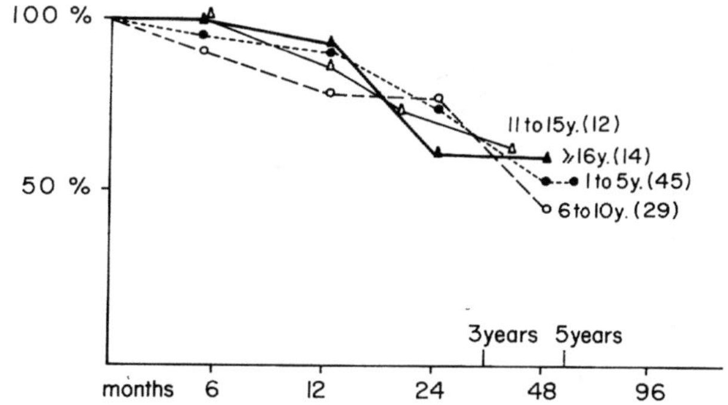

Fig. 3.23. Acute lymphoid leukaemia. Cumulative duration of survival of 100 patients (submitted to active immunotherapy) according to age (MATHÉ et al., 1976a)

At the time of diagnosis, the acute lymphoid leukaemia is classified as one of 5 categories (MATHÉ *et al.*, 1971c; MATHÉ *et al.*, 1973b). Fig. 3.26 illustrates their characteristics. The cells conventionally considered as typical "lymphoblasts" are called "microlymphoblasts" when their diameter is less than 11 μ, and "macrolymphoblasts" when it is greater. The cells more differentiated than the typical lymphoblasts are called "prolymphocytes" and those less differentiated have acquired, by convention, the name "prolymphoblasts", often called, in english literature, "undifferentiated". The leukaemias are given the name of the cell which predominates. A fifth type has recently been described under the name of acute lymphoid immunoblastic leukaemia (MATHÉ *et al.*, 1974c), but its sensitivity to chemotherapy is so poor that the 5 patients who were observed do not come into the statistics concerning the prognostic factors in ALL submitted to active immunotherapy.

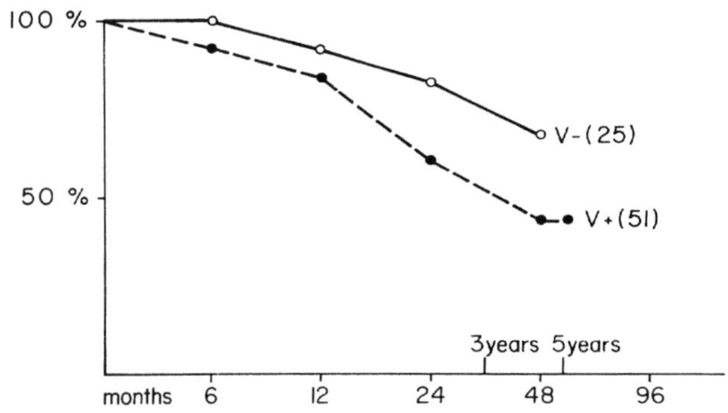

Fig. 3.24. Acute lymphoid leukaemia. Cumulative duration of survival of 76 patients (submitted to active immunotherapy) according to volume of the neoplasia (V) estimated by the number of circulating leukaemic cells and the presence (V+) or absence (V-) of adenosplenohepatomegaly (MATHÉ et al., 1976a)

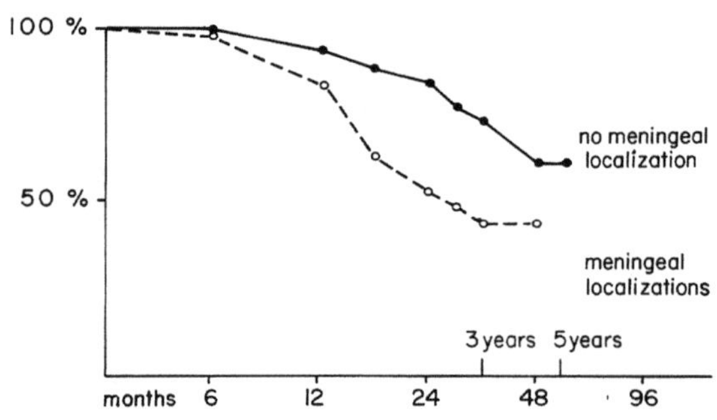

Fig. 3.25. Acute lymphoid leukaemia. Cumulative duration of survival of patients submitted to active immunotherapy according to the presence or absence of detectable meningeal localization at first presentation of disease or at first relapse (MATHÉ et al., 1976a)

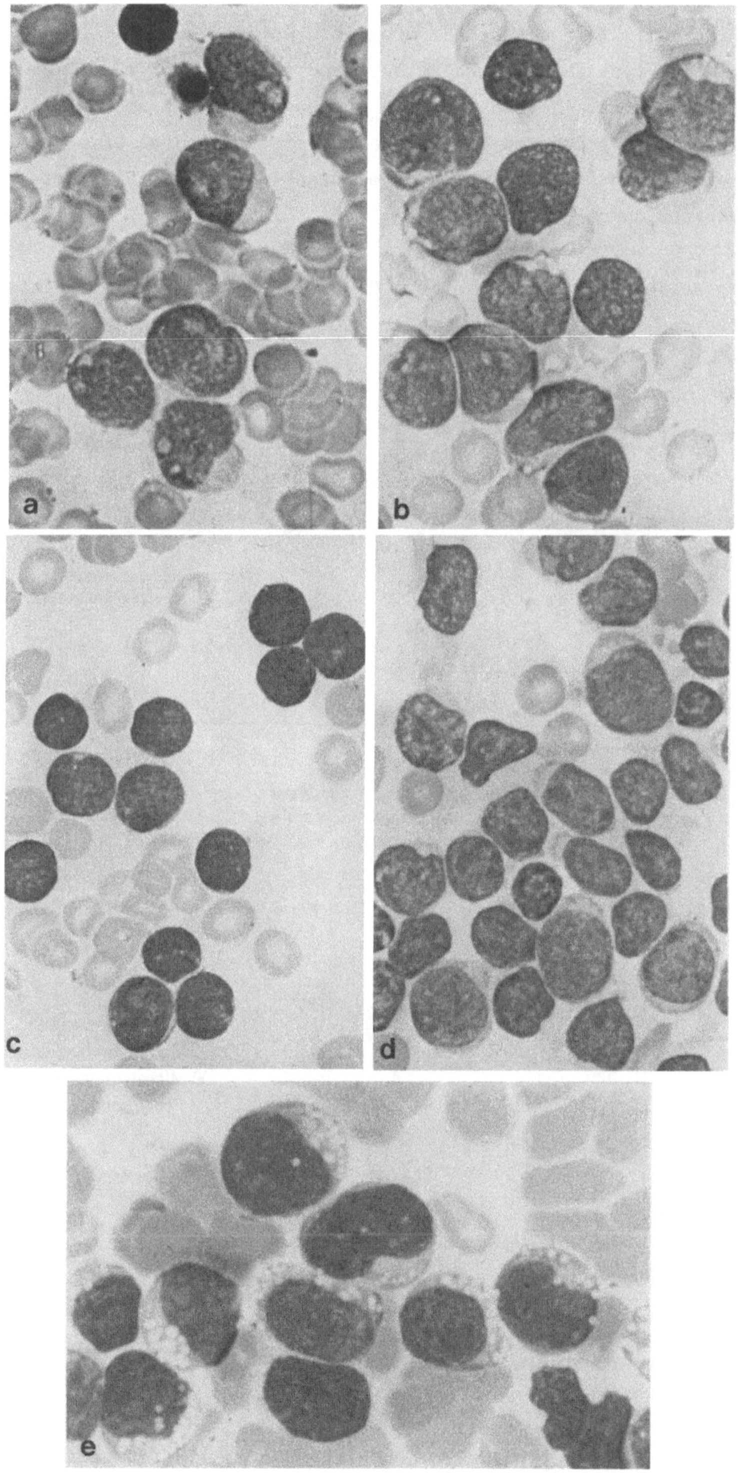

Fig. 3.26. Morphological aspects of the 5 types of acute lymphoid leukaemia: a) prolymphoblastic; b) macrolymphoblastic; c) microlymphoblastic; d) prolymphocytic; e) immunoblastic (MATHÉ et al., 1971c; MATHÉ et al., 1973b). These types are those used in WHO nomenclature (MATHÉ and RAPPAPORT, 1976)

Figure 3.27 shows that the prognosis is much more favourable for the microlymphoblastic type than for the others and that the so-called prolymphoblastic type is the worst. About 85% of the patients of the former category belonged to the population represented by the plateau of their curve of cumulative survival duration covering between 3-5 years. The author considers this plateau as the statistical expression of cure expectancy.

In another statistical study, HAUSS (1975) searched for links between the different factors mentioned above, which seem to play a role in prognosis, and he found that

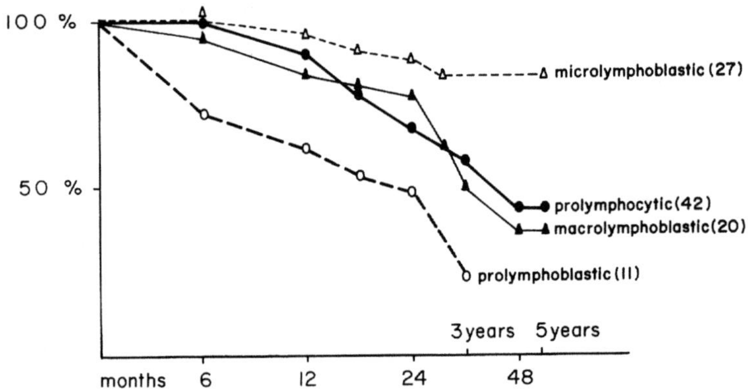

Fig. 3.27. Acute lymphoid leukaemia. Cumulative duration of 100 patients (submitted to active immunotherapy) according to cytological types. Note that time scale is geometrical (MATHÉ et al., 1976a)

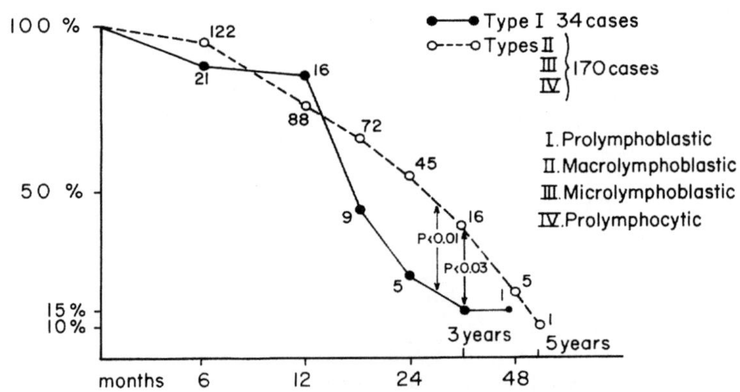

Fig. 3.28. Acute lymphoid leukaemia. Cumulative duration of survival of Jean Bernard's patients submitted to maintenance chemotherapy according to cytological types (BERNARD et al., 1975; JACQUILLAT et al., 1975)

only the cytological type is related to all the other prognostic factors. Of all these factors taken as constant, the cytological type keeps its prognostic value;

on the other hand, in the case of a constant cytological type, the other factors lose their prognostic value.

It is interesting to note that the prognostic value of this cytological typing is also found by Jean Bernard's group (BERNARD *et al.*, 1975; JACQUILLAT *et al.*, 1975) for the prolymphoblastic or "undifferentiated" variety, but not for the other three forms in patients submitted to maintenance chemotherapy (Fig. 3.28). This may be due to differences between immunotherapy and chemotherapy sensitivities. Hence the high cure rate (85%) of the microlymphoblastic type appears to be due to active immunotherapy. Three important remarks have to be made when comparing immunotherapy and maintenance chemotherapy results:

a) Late relapses (after 4 years) have not been observed in the patients submitted to immunotherapy for 5 years, while their incidence has been noticeable after 4 years in those submitted to maintenance chemotherapy (BERNARD, 1975).

b) The incidence of remissions induced by chemotherapy is as high in patients who relapsed under immunotherapy (92%) as it is in patients treated for the first presentation of the disease, and the chance for good prognosis is still remarkable: the curve of cumulative survival duration of the microlymphoblastic patients who relapsed once form a plateau (cure expectancy) for more than 50% (MATHÉ *et al.*, 1976a). This is not the case for patients who relapse under so-called maintenance chemotherapy: the explanation lies in the fact that the patients submitted to immunotherapy are not resistant to the drugs efficient against ALL, while the patients submitted to maintenance chemotherapy are.

c) Active immunotherapy is remarkable by its total innocuity compared to the risk of maintenance chemotherapy. It has produced, in the experience of the author, which today concerns 277 patients (MATHÉ *et al.*,1976a; SCHWARZENBERG *et al.*, 1976), no deaths (chapter 8 on Toxicology). Other groups who submitted their patients to more vigorous and lengthy chemotherapies have published deaths in remission, due mainly to Pneumocystis carinii, herpes viruses, cytomegalovirus, and fungal infections, the incidence varying from 5% to 15% (Table 3.IV).

The most expensive protocol, as far as human life and comfort is concerned, was that of the Memphis group using methotrexate (40-80 mg/m2/w) for "maintenance" of remission, which induced "crippling or fatal leucoencephalopathy in 50% of the patients and failed to produce long-lasting remission in 70% of all patients" (AUR *et al.*, 1975).

Moreover, sterility (MATHÉ and KENIS, 1975; BUCHANAN *et al.*, 1975) and second neoplasias (CANELLOS *et al.*, 1974; BELPOMME *et al.*,1974a), now reported in patients submitted to long-term chemotherapy, may be feared after such treatment, especially when applied to children.

As seen above, no ALL "control" patients without immunotherapy were kept in our further trials after a significant difference in the 1963 trial had been observed. This is for obvious ethical reasons (MATHÉ, 1973, 1974b).

Other groups have conducted other comparative trials on ALL.

A MRC cooperative group (1971) was not able to find any difference between postchemotherapy controls and immunotherapy patients. This may have been due to: a) the weakness of the immunotherapy form employed (BCG only) which, moreover, was Glaxo BCG, an inefficient preparation in the author's experimental screening (MATHÉ *et al.*, 1973c; HALLE-PANNENKO *et al.*, 1976), b) the weakness of preimmunotherapy oncostatic cell reduction (in particular, the patients received neither intrathecal chemotherapy nor CNS radiotherapy), c) the heterogeneity of the patients included in the protocol (the smears were not reviewed in a central laboratory either before inclusion of the patients in the protocol, or after "remission" was obtained) and of the quality of the treatments applied in the different services.

Table 3.IV. Comparative toxic cost of active immunotherapy and of maintenance chemotherapy in acute lymphoid leukemia treatment

Active immunotherapy	Deaths	Severe not lethal toxicity
G. MATHÉ *et al.* (1976a)	0/277 patients	
"Maintenance chemotherapy"		
J.V. SIMONE (1975) (Protocol V)	3/31 patients	
" " " " (Protocol VI)	5/94 "	
" " " " (Protocol VIa)	2/10 "	
" " " " (Protocol VII)	4/94 "	
" " " " (Protocol VIII)	2/92 " 2/17 "	
R. AUR and J.V. SIMONE (1975)	3/20	6 Crippling leucoencephalopathy/20
Cl. JACQUILLAT and J. BERNARD (1975) (Protocol 06 LA/66)	6/81 children 6/21 adults	
F. LAMPERT *et al.* (1975) (Early irradiation) (Late irradiation)	4/79 patients 1/19 "	
F. MANDELLI *et al.* (1975)	2/30 "	
T.S. GEE *et al.* (1974)	3/74 children	
A.C. SMYTH *et al.* (1975)	2/17 adults	

More interesting are the recent results of the EORTC Haemopathies Working Party (1975), which found that active immunotherapy (with living fresh Brussels Pasteur Institute BCG) is more, or at least as efficient in maintaining the remission of patients after 2 years (62%) than 6-mercaptopurine and methotrexate chemotherapy, the most efficient chemotherapy modality for remission maintenance (55%) (this difference is not significant), and that its lethal cost is nil (0 death out of 21 patients) compared to that of chemotherapy (3 deaths/17 patients).

Also interesting are the preliminary results recently published by EKERT and JOSE (1975) who compared the duration of remission in ALL children receiving maintenance chemotherapy (with 6-mercaptopurine and cyclophosphamide) with those of children submitted to BCG interspersed with the same chemotherapy: the proportion of subjects in remission was at time of publication higher in the BCG group (6/7) than in the maintenance chemotherapy group (6/12).

ALL Immunotherapy and Immunological Parameters. Over the years and in successive protocols, modifications induced by immunotherapy have been sought. Before mentioning the results, it is necessary to underline the difficulty in finding valuable controls for comparison with patients under immunotherapy because, before being submitted to this treatment, they receive cell-reduction chemotherapy.

In a first study of the kinetics of different test result modifications, SCHNEIDER *et al.* (1974) found an increase in the number of immunoblasts in the peripheral blood and in the percentage of recall antigen skin responders, starting at the beginning of active immunotherapy and maintained during the latter, while the numbers remained low during chemotherapy (Fig. 3.29).

In a second study, OLDHAM *et al.* (1976) found more reactivity to PHA, PWM and PPD in vitro in patients submitted to immunotherapy than in those still receiving chemotherapy (Table 3.V).

The last study conducted by JOSEPH *et al.* (1976) in the author's laboratory concerned the counting of sheep red blood cell rosette-making (T) lymphocytes, membrane immunoglobulin secreting (B) lymphocytes and peroxydase positive mononuclear cells (monocytes) in patients submitted to immunotherapy compared to patients receiving cell reduction chemotherapy and to normal subjects. As seen in Table 3.VI. there is a significant increase of null cells, a population of mononuclears in which K-cells belong, according to CEROTTINI (pers. comm.).

This finding could be compared with that published by MacLENNAN (1975) which showed, "in a small group" of patients given immunotherapy with BCG and irradiated allogeneic lymphoid leukaemia cells, the significantly greater recovery of peripheral blood K-cells, previously decreased by chemotherapy, than in controls.

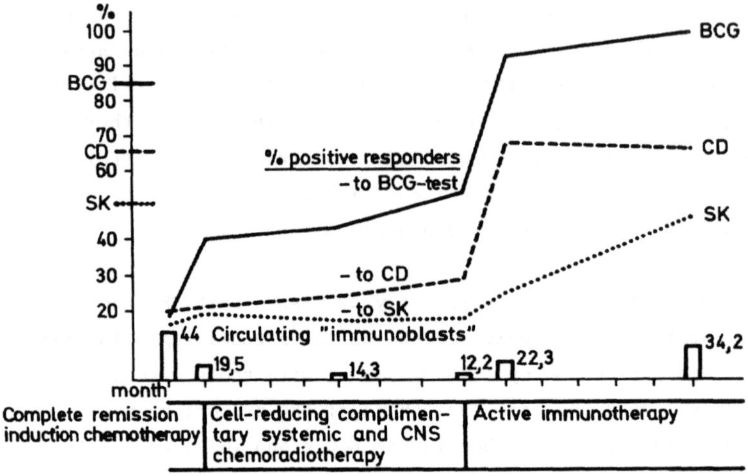

Fig. 3.29. Kinetics of some nonspecific immune responses studied by first immunological investigation battery in patients during preimmunotherapy chemotherapy, between chemotherapy and immunotherapy, and during immunotherapy. CD = candidine, SK = streptokinase (SCHNEIDER et al., 1974)

Table 3.V. Changes in PHA lymphocyte reactivity in patients submitted to active immunotherapy after (intermittent) chemotherapy (OLDHAM *et al.*, 1976)

		Lymphocyte stimulation		
		Mean net CPM[a]		Mean stim. index
		PHA (c)	PWM	PPD (d)
Study A	Intermittent chemotherapy	18,000	14,000	4.5
	Immunotherapy 60 days	56,000	26,000	42
Study B	Intermittent chemotherapy	16,000	11,000	3.6
	Immunotherapy 120 days	49,000	44,000	108

[a] CPM = counts per minute

Also interesting is the work of EKERT *et al.* (1975), on children with ALL who had been in complete remission on continuous chemotherapy for at least 12 months and were treated with intermittent courses of chemotherapy alternating with BCG inocu-

Table 3.VI. Increase of null cell in the blood of acute lymphoid leukaemia patients submitted to active immunotherapy summary of results. (JOSEPH *et al.*, 1976)

	Mononuclear cells	E_{ABRFC} [a]	mIg positive [b]	Peroxidase positive	"Null" [c]
Chemotherapy	1470	696	455	344	104 (7%)
Immunotherapy	2714	1173	467	198	883 (33%)
Normal	2356	1299	381	396	304 (12%)
Chemotherapy/ immunotherapy	p = 0.001	p = 0.01	N.S.	N.S.	p = 0.001
Normal/ chemotherapy	p = 0.02	P = 0.01	N.S.	N.S.	N.S.
Normal/ immunotherapy	N.S.	N.S.	N.S.	N.S.	p = 0.01

[a] spontaneous E rosettes with sheep red blood cells in the presence of decomplemented human AB serum

[b] membrane immunoglobulin positive cells

[c] cells without detectable T or B membrane markers

N.S. not significant.

lation during the drug-free intervals. Antibody titres and blastogenic responses to soluble leukaemic blast-cell membrane-derived antigen(s) increased in 8 out of 12 and 6 out of 7 children, respectively, during the BCG phase.

Conclusions for a Rational Protocol. From these observations, the following conclusions can be drawn: (a) the cell-reducing period should be "not too long" and "not too intensive" in order to avoid, as much as possible, the above-mentioned complications of oncostatics (Table 3.IV); (b) it should not be too short, since the author's protocol 4, which comprised only a 3 month preimmunotherapy chemotherapy period, gave the worst results; (c) it should not be the same for all types of ALL, as their immunosensitivity, their chemosensitivity (MATHÉ *et al.*, 1975e, 1976a) and their prognosis are different from one another. Therefore, it is as unethical to administer a too intensive and potentially dangerous chemotherapy to the immunotherapy-sensitive and good prognostic types (first category), as it is not to apply enough chemotherapy to the immunotherapy poorly sensitive and less favourable prognostic types (second category).

The author's protocol No. 12 (Fig. 3.30) is based on these notions. Remission chemotherapy starts identically for all patients, as the combination of prednisone, vincristine, and asparaginase used has given 100% so-called complete remission(s) (MATHÉ *et al.*, 1976b, 1976c), except in the recently described and exceptional immunoblastic type (MATHÉ *et al.*, 1975f), but a complementary so-called remission induction "reinforcement" cycle combining adriamycine, vincristine, and cytosine arabinoside, is applied in the case of the second category.

The duration of the preimmunotherapy period is 17 months for the first category, while it is 23 in the second.

This period comprises, for all patients: a) a meningeal methotrexate chemotherapy and a cranial radiotherapy, b) a long phase of complementary cell-reducing chemotherapy during which 6-mercaptopurine and methotrexate are administered, and c) finally an intermittent chemotherapy phase. The latter is applied at that time because it was shown that this form of application of oncostatics is much less immunosuppressive than the continuous one (MATHÉ *et al.*, 1970b): it comprises 3 cycles combining vincristine and methotrexate.

A gap of 10 days is left between chemotherapy and immunotherapy: its necessity was discussed above.

It can be seen that all chemotherapy compounds efficient in ALL (MATHÉ and KENIS, 1975) are administered, except daunorubicine: adriamycin, to which the sensitivity

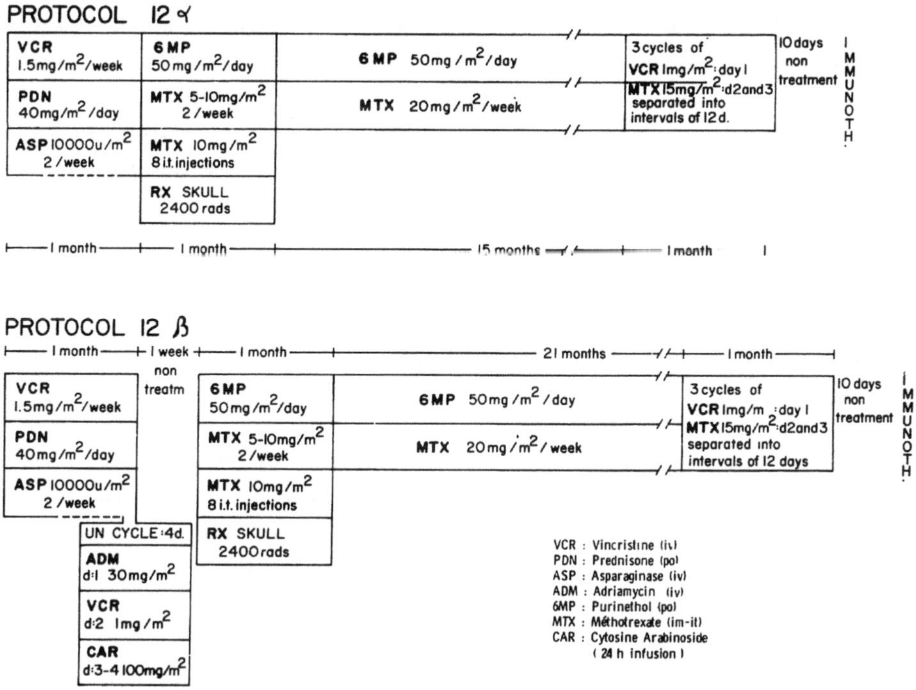

Fig. 3.30 Protocol No. 12 for acute lymphoid leukaemia treatment. α) patients with good predicted prognosis; β) patients with more severe predicted prognosis

of ALL cells is probably identical (MATHÉ *et al.*, 1972), has been preferred since the author has found it less toxic.

As far as the duration of active immunotherapy is concerned, it is maintained for 5 years: immuno-intervention should not only be considered as an immunotherapeutic weapon to kill the last leukaemic cells (MATHÉ, 1967), but as a relapse potential immunoprophylactic means: SPIEGELMAN (1974) and MAK *et al.* (1974) found, in the lymphocytes of patients in long-term remission only maintained by chemotherapy, the same biochemical stigmata of an oncorna virus and in the same quantity as in leukaemic cells during the perceptible phase of the disease.

It seems necessary, at this point, to underline the interest of classifying each patient according to the parameters which have been shown to determine prognostic and therapeutic sensitivity, especially the cytological types, in order not only to give enough treatment to the patients with a severe prognosis, and not too much to those with a spontaneously good one, but also to compare, for the benefit of the patients, the results of different centres and different protocols. The author has proposed, as head of WHO International Reference Center for neoplastic diseases of haematopoietic and lymphoid tissues (MATHÉ and RAPPAPORT, 1976), a classification based on Giemsa staining, which can be used in all laboratories throughout the world, and which is the best, if not the only, prognostic and therapeutic sensitivity factor. Cytochemistry (HAYHOE and FLEMANS, 1970) is of no complementary help; scanning electron microscopy (DANTCHEV, 1976) and immunological markers (BELPOMME *et al.*, 1974b; BELPOMME *et al.*, 1976; BROUET *et al.*, 1975) may be useful in the future.

3.2.1.1.2. Acute Myeloid Leukaemia

The results of several trials on active immunotherapy in acute myeloid leukaemia (AML) patients are now available.

The first one was conducted by POWLES *et al.*(1973, 1974). The patients were divided into two groups before their first treatment: one which would receive chemotherapy alone, in case of complete remission (CR), and the other the same chemotherapy interspersed with immunotherapy. They were then given remission induction treatment, and 45 out of 107 attained CR.

These 45 subjects then received complementary chemotherapy consisting of a five day treatment every 28 days. Patients receiving supplementary immunotherapy were also given multiple weekly intradermal injections of irradiated, stored AML cells, and Glaxo BCG using a heaf gun.

Nineteen patients received only chemotherapy during remission: 7 of them were still alive at the time of publication (the median survival was, after attaining remission, 43 weeks), only 5 were still in their first remission (the median remission length was 27 weeks). Out of 23 patients allocated to receive immunotherapy interspersed with chemotherapy, 16 remained alive (the median was 78 weeks); 8 were still in their first remission (the median was 45 weeks). The difference in survival of the two groups was significant with a p value of 0.003 (Figs. 3.31 and 3.32).

In a separate study (POWLES, 1974), 9 patients received the same immunotherapy after chemotherapy. The median remission length was 46 weeks, and the median survival 83 weeks (Figs. 3.33 and 3.34).

A trial on AML has been published by the South-Eastern Group of the U.S. (VOGLER and CHAN, 1974). After remission induction and "consolidation" complementary chemotherapies, 41 patients with AML were put at random into two groups: the first received Tice-strain BCG (see p. 245) given by the tine technique twice weekly for 4 weeks, followed, 2 weeks later, by methotrexate (30 mg per sq.m. twice weekly);

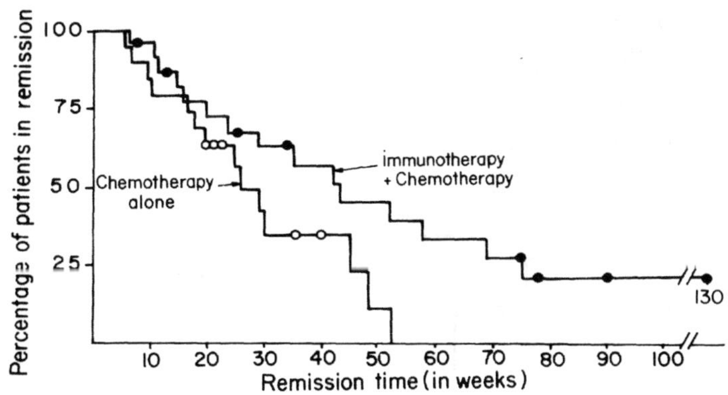

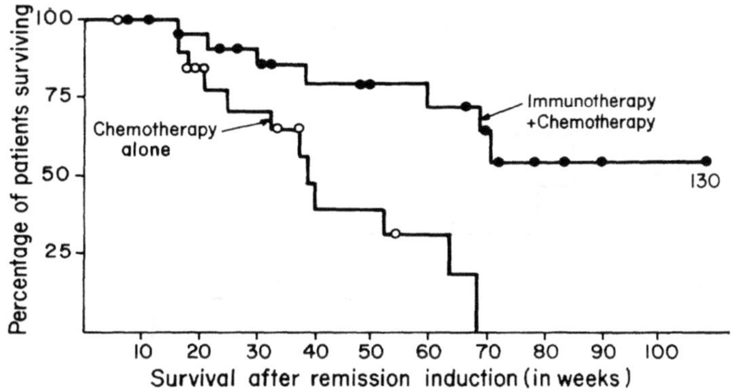

Figs. 3.31 and 3.32. Acute myeloid leukaemia. Powles' first trial comparing remission and survival durations of patients receiving, after remission induction, chemotherapy alone or chemotherapy interspersed with immunotherapy. Follow-up to May 31st, 1973 (POWLES et al., 1973)

the second was treated by methotrexate alone. The median duration of remission from its onset was 39.4 weeks in the BCG group, and 26 weeks in the methotrexate group. The median duration of remission after consolidation chemotherapy was 20.4 weeks for the BCG group and 9.7 weeks for the methotrexate one (Figs. 3.35 and 3.36).

A third trial on acute myeloid leukaemia has been conducted by GUTTERMAN *et al.* (1974a). Fourteen consecutive patients with AML received intermittent chemo-immunotherapy with cytosine arabinoside, vincristine, and prednisone, the same combination they had used for induction, except that the administration was limited to five day courses repeated every 24-28 days; on days 7, 17 and 21 of each monthly cycle, fresh liquid Pasteur strain BCG was given on scarifications in a dose of 6 x 10^8 viable units (150 mg). The durations of remission were compared with those of a similar group of 21 consecutive patients maintained on the same chemotherapy as above but without BCG. Ten of the 14 chemo-immunotherapy patients remain in remission at the

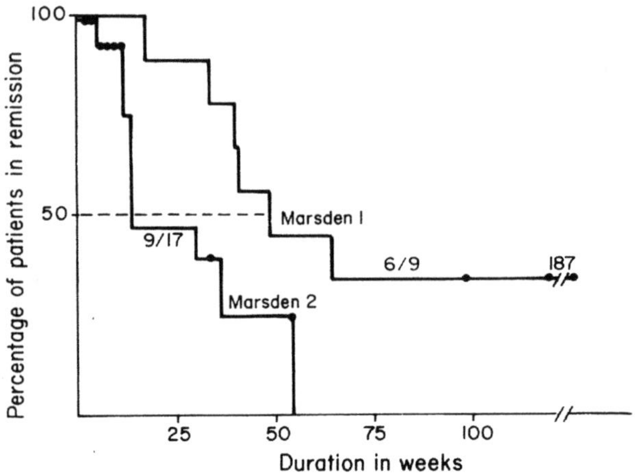

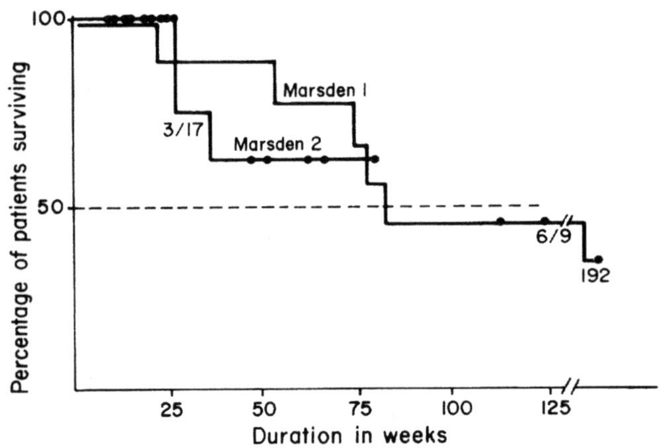

Figs. 3.33 and 3.34. Acute myeloid leukaemia (Powles' second trial). Remission and survival duration of patients receiving immunotherapy alone after chemotherapy. (POWLES, 1974)

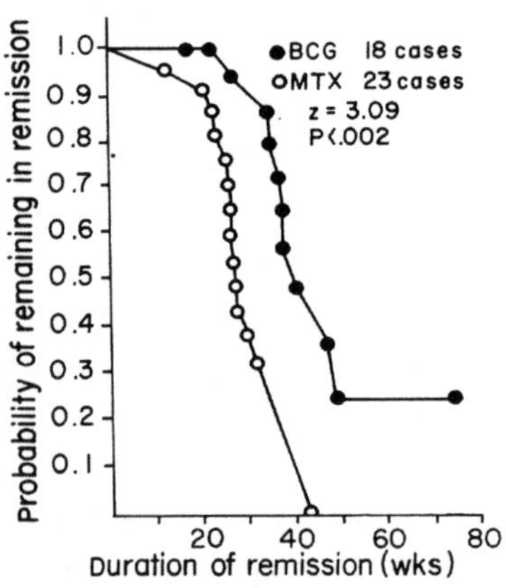

Fig. 3.35. South-Eastern Group trial on immunotherapy of acute myeloid leukaemia. Duration of remission from its onset to time of publication (VOGLER and CHAN, 1974)

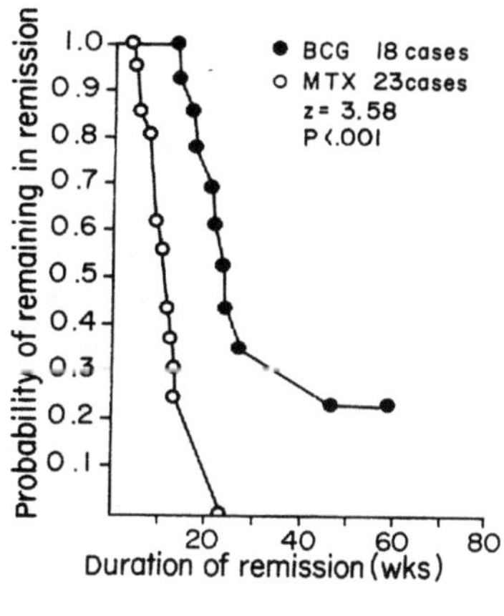

Fig. 3.36. South-Eastern Group trial on immunotherapy of acute myeloid leukaemia. Duration of remission from onset of maintenance phase. (VOGLER and CHAN, 1974)

time of publication with an estimated median duration of 72 weeks. Only 9 of the 21 maintained on chemotherapy alone remained in remission with a median duration of 60 weeks (Fig. 3.37). The immunotherapy effect was significant (p = 0.04).

Four other immunotherapy noncomparative trials on AML will be described.

One has been conducted in Manchester (FREEMAN *et al.*, 1973). Thirty-two patients suffering from this neoplasia were initially treated with daunorubicin (DRB) and cytosine arabinosine (CAR); 8 achieved complete remission. They were given a cell reduction course of cyclophosphamide (CPM) and thioguanine. Seven were then actively immunized with 10^9 irradiated allogeneic AML cells and BCG (Glaxo, heaf gun) at weekly intervals. The duration of the first remission of the patients under immunotherapy was 14-56 weeks. Two of the patients were still in their first remission at time of publication (38 and 56 weeks respectively). The other 5 patients had all relapsed at periods ranging from 14-28 weeks (mean 20 weeks), and four of them

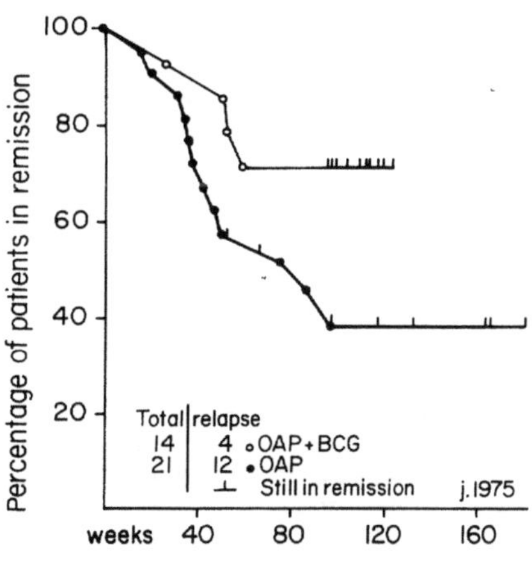

Fig. 3.37. Gutterman's trial on immunotherapy of acute myeloid leukaemia. Duration of remission at time of publication of two groups of patients: one group receiving only chemotherapy, the other chemotherapy and BCG (GUTTERMAN et al., 1974a)

achieved a second remission with DRB and CAR. At the time of publication, one patient who was under immunotherapy only had died 43 1/2 weeks after the beginning of his disease. The remaining six were alive for 43-91 weeks (with a mean of 59 weeks).

The other trial has been conducted by the author's group (MATHÉ *et al.*, 1975g). In a phase II trial, 18 patients were submitted to the following sequential treatment: chemotherapy to induce a remission, complementary cell-reducing chemotherapy and active immunotherapy comprising fresh living Pasteur BCG on scarification and, separately, AML pooled irradiated and preserved cells (as in ALL protocol). The median duration of first remission is 64 weeks as seen in Fig. 3.38 and that of survival is 130 weeks (Fig. 3.39). The medians given for comparison on Table 3.VII are those belonging to other trials based on the same chemotherapy principle, but without immunotherapy, and specially those of two EORTC trials (EORTC Leukaemia and Haematosarcoma Group, 1973, 1974, 1975): our immunotherapy medians are significantly longer than the chemotherapy ones, especially those called "Villejuif" and "EORTC", to which they are comparable.

CUTTNER *et al.* (1975) treated 6 AML patients in remission with monthly 5 day courses of cytosine arabinoside, thioguanine, cyclophosphamide, CCNU, or daunorubicine, and five patients with the same chemotherapy regimen combined with monthly intradermal

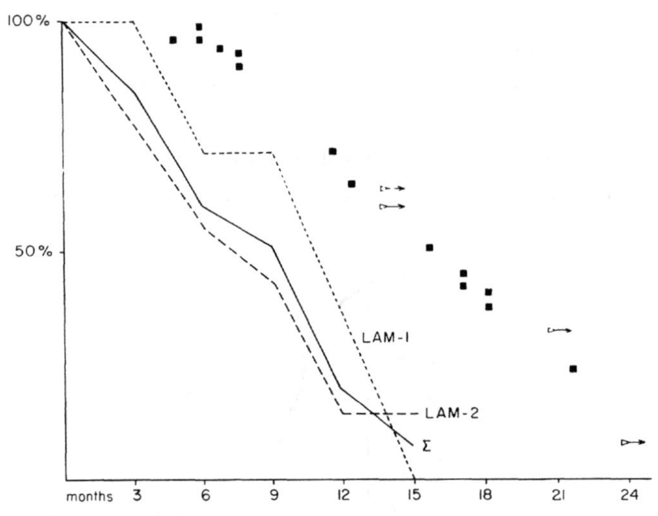

Fig. 3.38. Acute myeloid leukaemia phase II active immunotherapy trial. Cumulative duration of first complete remission: ▷→ patients in remission; ▶→ ■ : patients who relapsed (MATHE et al., 1975g). LAM-1 and LAM- 2: curves of cumulative duration of first remission of Villejuif patients introduced into the maintenance chemotherapy EORTC trials (EORTC Leukaemia-Haematosarcoma Group, 1973, 1974, 1975)

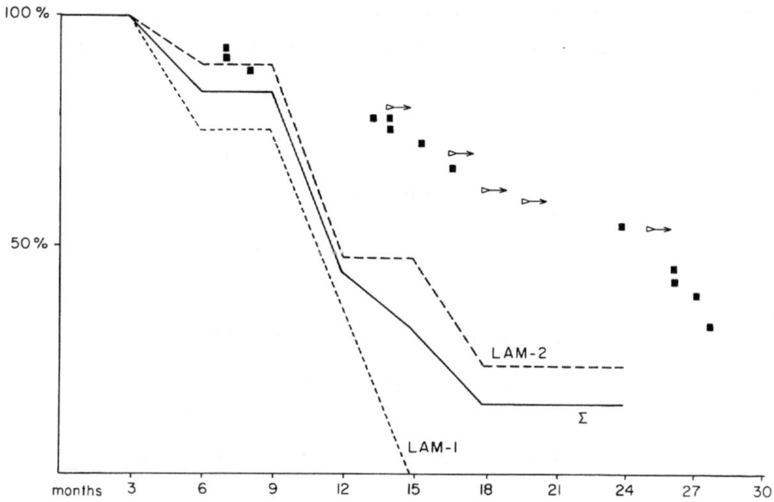

Fig. 3.39. Acute myeloid leukaemia phase II active immunotherapy trial. Cumulative duration of survival after first complete remission. ⊳→ *patients in remission;* ■ *patients who died (MATHÉ et al., 1975g). LAM-1 and LAM-2: curves of cumulative duration of survival of Villejuif patients introduced into the maintenance chemotherapy trials (EORTC Leukaemia-Haematosarcoma Group, 1973, 1974, 1975)*

injections of 1 mg MER divided between 5-10 sites. While the subjects of the first group relapsed after 63 to 162 days, the remissions of those of the second persist from 92 + to 193 + days.

The same group (BEKESI and HOLLAND, 1975) has submitted AML patients in postchemotherapy remission to active immunotherapy with neuraminidase-treated allogeneic myeloblasts. While, in previous patients, the median remission duration of chemotherapy alone was 20 weeks for 7 patients, 5 out of 7 subjects receiving this form of immunotherapy had been, at the time of publication, in remission for from 56 to 97 weeks.

Conclusion for a Rational Protocol. The conclusion of the preceeding results is that they demonstrate or illustrate the efficiency of active immunotherapy in AML. But cure expectancy is very small or nil, according to the different trials.

Two directions can be followed for further studies: (a) Intensification of immunotherapy by using the best adjuvant (which may not have been the case in several trials mentioned) and tumour cells (in the optimal manner) which were not administered in most trials; (b) intensification of preimmunotherapy chemotherapy, which was never long or maximal (this could necessitate the patients staying in aseptic rooms, as the EORTC leukaemia and Haematosarcoma Group (1975) showed that the

Table 3.VII. Acute myeloid leukemia active immunotherapy. Comparison of the results of our trial of immunotherapy, following cell reduction chemotherapy, to those of other trials

	Medians		
	Of first complete remissions (FCR)		Of survival after FCR
Maintenance chemotherapy	EORTC first trial (1973)	32 weeks	55 weeks
	EORTC second trial (1974)	34 weeks	N.D.[a]
	EORTC Villejuif patients in both trials	26 weeks	52 weeks
	BEKESI and HOLLAND (1975)	20 weeks	
	VOGLER *et al.* (1974)	26 weeks	
	POWLES (1974)	26 weeks	46 weeks
	GUTTERMAN *et al.* (1974a)	60 weeks	
Chemotherapy interspersed with immunotherapy	GUTTERMAN *et al.* (1974)	72 weeks	
	POWLES (1974)	58 weeks	77 weeks
Chemotherapy followed by immunotherapy	POWLES (1974)	48 weeks	83 weeks
	VOGLER *et al.* (1974)	39.4 weeks	
	MATHÉ *et al.* (1975g)	60 weeks	104 weeks

[a] N.D. = not yet determined

results of chemotherapy in AML were much better in services with haemato-oncologic intensive care units than in conventional ones).

If, when these two precautions are applied, the plateau of the curve of cumulative duration of remission is still very low, it can be reasonably supposed that the number of cells left by chemotherapy is too high to be eradicated by immunotherapy. Hence, one could be tempted to intersperse immunotherapy and chemotherapy, the object being to decrease the number of leukaemic elements persisting to a level possibly accessible to further immunotherapy. However, such an interspersion needs an immunological monitoring of the patients, at least during the first interspersed cycles, to check that chemotherapy is not diminishing the immune response too intensively, and to help in the choice of the best intervals.

3.2.1.1.3. Blastic Crisis of Chronic Myeloid Leukaemia

SOKAL and AUNGST (1974) have treated 15 patients with chronic myeloid leukaemia in blastic crisis with BCG and cells from a permanent line resulting from a culture of AML blood (these cells reacted with a primate antiserum "specific" for myeloid leukaemia, and not with antisera "specific" for lymphoid leukaemia). There was no evidence of benefit amongst those in whom chemotherapeutic remission was not achieved, but the survival of six subjects whose blastic crisis was reversed was 6, 10, 10, 18, 38, and 40 months after the initial immunotherapy, which is longer than one could expect from chemotherapy alone (SCHWARZENBERG *et al.*, 1973).

This result is in favour of the concept according to which immunotherapy can only work when the population of tumour cells is minimal.

3.2.1.1.4. Lymphosarcomas

Leukaemic Terminal Lymphosarcoma. Lymphosarcoma (LS) is often converted at the end of its evolution into a lymphoid leukaemia (MATHÉ *et al.*, 1975h).

Patients at this terminal stage of LS were submitted to a protocol similar to the one applied to primary acute lymphoid leukaemia. Radiotherapy was administered not only to the central nervous system, but to the sites where there were important masses between remission induction and complementary chemotherapy. An example of such treatment is given in Fig. 3.40 which also shows that, under active immunotherapy, a patient who was at an apparently terminal phase, has been in complete remission for more than five years.

Figure 3.41 shows the cumulative duration of complete remission of 9 patients who were submitted to such a treatment: four are in apparently complete remission for more than three years (MATHÉ *et al.*,1975h). This result is far superior to that of patients who entered into remission and were submitted only to maintenance chemotherapy.

Nonleukaemic Lymphosarcomas. THOMAS *et al.* (1974) have reported the observation of 7 patients with non Hodgkin's "lymphoma" (lymphosarcomas or reticulosarcoma). They were in complete remission after "initial standard therapy", and were submitted to repeated BCG applications as complementary therapy. Six of the 7 are alive, and five have had no recurrence at the time of publication, after 44, 32, 29, and 25 months. BCG was given on scarifications and orally.

Burkitt's (African) Lymphosarcoma. MAGRATH *et al.* (1974) compared the duration of remission in 9 patients treated with Pasteur Institute BCG applied on scarifications after chemotherapy by cyclophosphamide. At the time of publication,

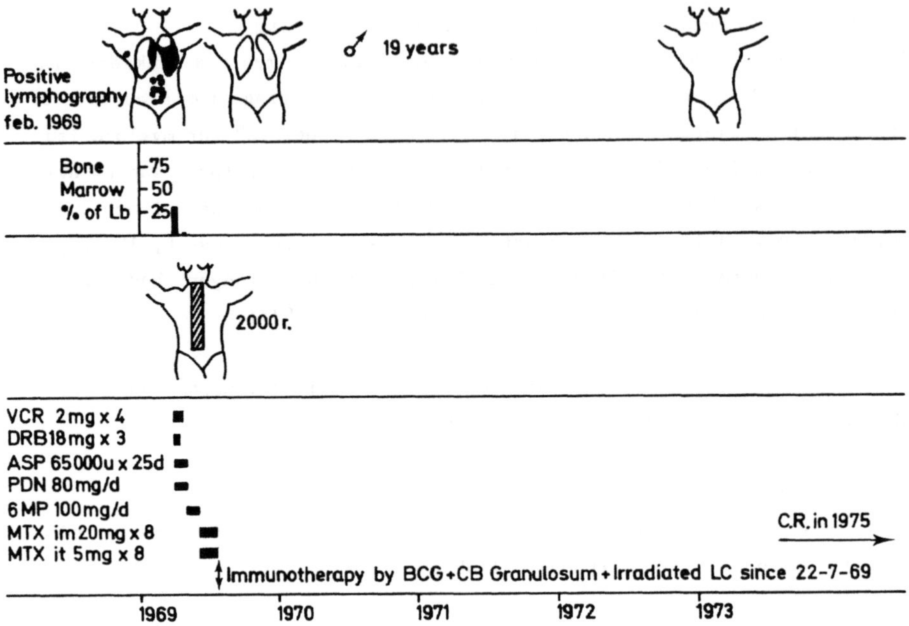

Fig. 3.40. Terminal leukaemic lymphosarcoma submitted to chemoradiotherapy, then active immunotherapy. In complete remission for five years (MATHÉ et al., 1975h)

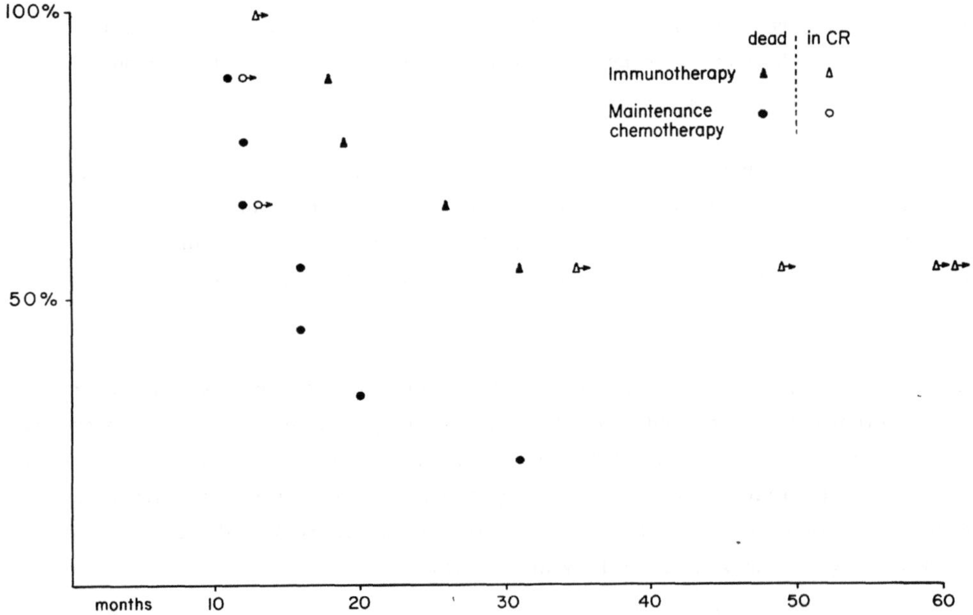

Fig. 3.41. Cumulative duration of apparently complete remission (upper curve) of patients submitted to active immunotherapy after cell-reduction chemoradiotherapy for terminal lymphosarcoma converted into leukaemia. The second curve concerns patients with similar disease treated with chemoradiotherapy without immunotherapy (MATHÉ et al., 1975h)

five out of 9 were still in remission on day 50 in the immunotherapy group, versus 0 in the control group. The follow-up was short at the time of publication and such a result cannot be considered as significant, but only as encouraging.

In a later paper, MAGRATH and ZIEGLER (1976) published the result of a randomised trial in which there was no difference between the BCG treated patients and the control patients. This absence of BCG effect may be due to the use of a lyophilised preparation which, in our experimental screening, does not exert any antitumour effect (MATHÉ *et al.*, 1973c). It may also be due to the premature application of BCG immediately after remission induction without any complementary cell reducing chemotherapy as used in our acute lymphoid leukaemia protocols. .

3.2.1.1.5. Hodgkin's Disease

There is only one trial in the literature concerning immunotherapy in Hodgkin's disease (HD) patients.

HOERNI *et al.* (1976) submitted to BCG, 19 patients to be compared with 13 "randomized" controls treated in stage III or IV successively with chemotherapy (vinblastine, cyclophosphamide, and procarbazine), then radiotherapy, then chemotherapy, and finally BCG on scarifications. Between 4 and 5 years after starting immunotherapy, 85% of the patients are still in perfect remission as against 62% of the controls.

3.2.1.2. Treatment of Perceptible Disease

3.2.1.2.1. Acute Lymphoid Leukaemia

After observing that poly I-poly C is an efficient adjuvant in mice against L1210 leukaemia (MATHÉ *et al.*, 1971d), the author submitted this polynucleotide to a phase I and then to a phase II trial (see for definition, MATHE, 1973, 1974b), and we chose, for the latter, acute lymphoid leukaemia (ALL) patients in relapse. The drug was infused i.v. daily at the dose of 1 mg/m2/day. Six apparently complete remissions out of 17 were obtained (Table 3.VIII) (Fig. 3.42). It should be emphasized that the remissions were only obtained in patients in whom the leukaemic cell percentage in the bone marrow was $\leqslant$30% (10^{10}) (MATHÉ *et al.*, 1970a).

In a few patients with ALL starting a relapse under BCG, an increase in the dose of the BCG was able to reinduce an apparently complete remission (Fig. 3.43) (not published).

SKURKOVICH *et al.* (1969), working in the USSR, published an article in "Nature" claiming that twelve children with acute leukaemia in relapse, obtained clinical benefit from active immunization with allogeneic live leukaemic cells combined with passive immunization with plasma and adoptive immunotherapy with leukocytes obtained during remission. Five patients entered a partial remission, but it is difficult to attribute this to one or several administered immunotherapy types, as these five patients not only had them simultaneously, but received chemotherapy at the same time.

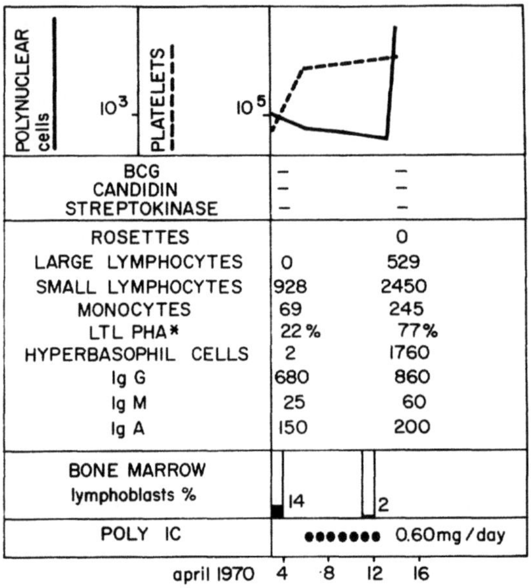

* *Lymphocytes transformation level "in vitro" in presence of phytohemagglutinin.*

Fig. 3.42. Remission obtained with poly I-poly C in patient with acute lymphoid leukaemia and treated at the very beginning of a relapse (MATHÉ et al., 1970a)

Table 3.VIII. ALL patients in relapse submitted to poly I-poly C. Phase II trial MATHÉ *et al.*, 1970a)

Bone marrow status	No. patients	Complete remission	Failure
Leukaemic cells ⩽ 30% (15-30)	11	6	5
Leukaemic cells > 30% (40-97)	14	0	14

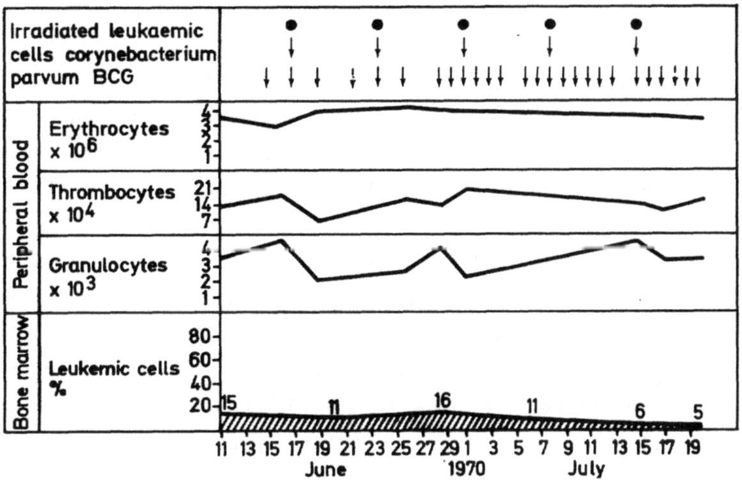

Fig. 3.43. Reinduction of remission by applying BCG more frequently in a patient who was relapsing under immunotherapy with BCG, C. parvum, and leukaemic cells (Pers. Obs. unpubl.)

3.2.2. Solid Tumours

There has been less clinical work on active systemic immunotherapy for solid tumours than for leukaemia and haematosarcoma, at least applied to residual imperceptible disease.

3.2.2.1. Treatment of Residual Imperceptible Disease

In the case of these neoplasias, the perceptible tumour is usually removed by surgery.

3.2.2.1.1. Special Case of Placental Choriocarcinoma

Placental choriocarcinoma is a semi-allogeneic grafted tumour, and it was possible to demonstrate the presence of antibodies in the serum of the patients against the nucleated cells of their husbands (MATHÉ *et al.*, 1964). Hence, BAGSHAWE and GOLDIN (1971) expected much from the immunization of such patients with skin transplants from their husbands combined with BCG. They were disappointed, as there was no unequivocal evidence of benefit (pers. comm. 1975).

3.2.2.1.2. Melanoma

The autologous solid tumour which has been submitted most often to immunotherapy, in order to eradicate the residual cells left by surgery, is melanoma.

The first study was conducted by BLUMING *et al.* (1972, 1974). They applied either Pasteur BCG on scarifications or Glaxo BCG by i.d. inoculation, on African patients with primary foot melanoma, previously removed by surgery. While the subjects who received Glaxo BCG ID relapsed in 30 weeks, those who had been treated with the Pasteur Institute preparation, applied on scarifications, benefited from an improved disease-free interval for 60 weeks at the time of publication (Fig. 3.44).

GUTTERMAN *et al.* (1973, 1975) submitted melanoma patients, who had metastases which had spread to regional nodes (stage III) or to distant sites (stage IV), to BCG applied on scarifications after surgical removal of all perceptible disease. At the time of the last presentation of the authors, the medians of remission and survival of 52 stage III patients were respectively 15 months and 18 months compared to 8 months ($p = 0{,}03$) and 15 months ($p = 0.003$) for historical controls (Table 3.IX). The medians were, in the 15 stage IV A group, 10 months for remission and 22 months for survival, compared to 4 months ($p = 0.007$) and 10 months ($p = 0.05$) for the historical control patients.

IKONOPISOV (1972) treated patients with melanoma, who had regional but not visceral disease, with surgery alone or surgery with BCG. A significantly lower recurrence rate was noted in the immunotherapy group; the result was particularly remarkable in two patients who had received tumour cells with BCG.

MORTON *et al.* (1976) and his group (GROOMS and MORTON, 1975; HOLMES *et al.*, 1975) published papers showing the decrease of recurrence rate and the prolongation of survival by

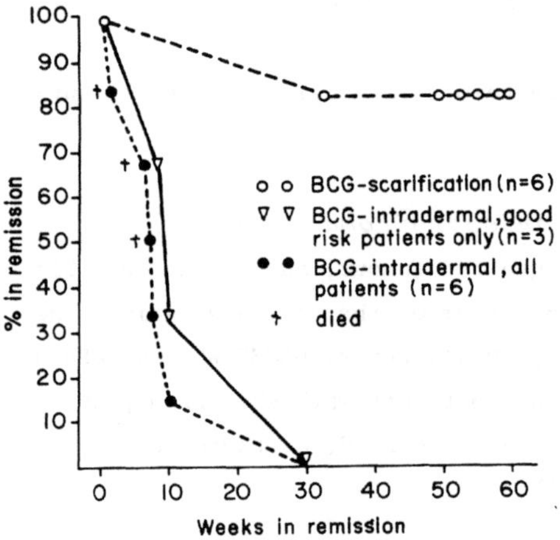

Fig. 3.44. Primary melanoma African patients submitted to active immunotherapy by Pasteur BCG on scarifications or Glaxo ID BCG (BLUMING et al., 1972, 1974)

Table 3.IX. BCG on scarifications in melanoma patients stages III and IV after removal of all known disease (GUTTERMAN *et al.*, 1973, 1975)

Medians	BCG	Controls	Statistics
Of remission	15 months	8 months	p = 0.03
Of survival	18 months	15 months	p = 0.003

BCG immunotherapy following surgery for stages II and III. Figures 3.45 and 3.46 illustrate the results they obtained in the University of California, Los Angeles (UCLA), on more than 151 stage II patients submitted to local excision of the primary tumour and resection of regional lymph nodes, compared to UCLA and M.D. Anderson historical controls (HOLMES *et al.*, 1975).

The same group also observed that while non-neurological metastases were rare after BCG, the brain was the most frequent and sometimes the only site of recurrence in

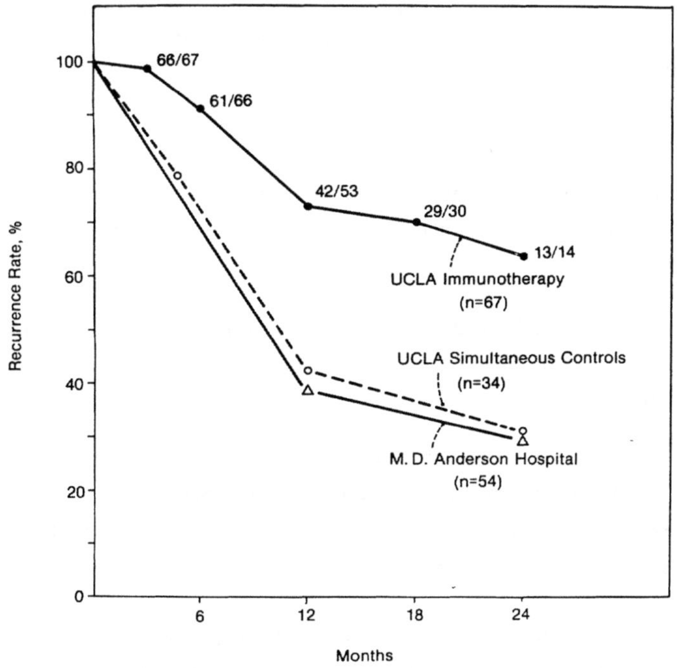

Fig. 3.45. Recurrence rate, stage II melanoma. Percentage of those patients free of recurrence after regional lymphadenectomy with postoperative immunotherapy. Numerators are numbers of patients who remained free of disease; denominators are numbers of patients at risk (HOLMES et al., 1975)

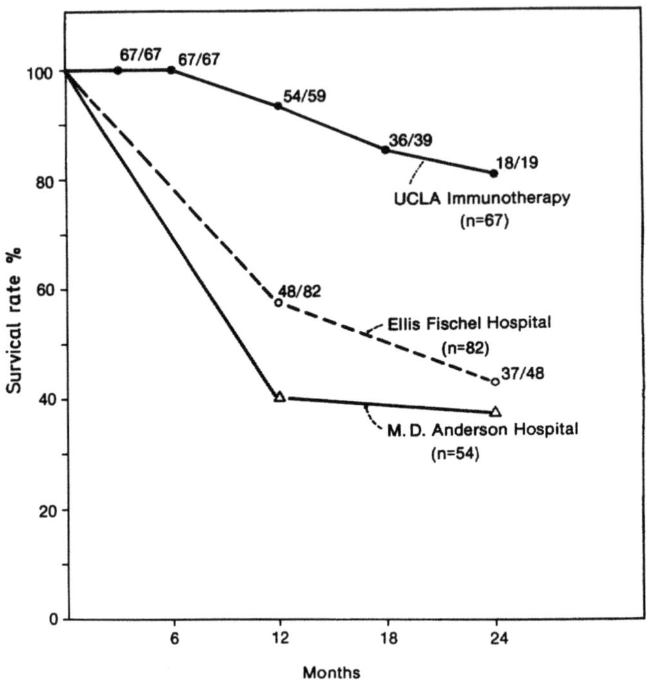

Fig. 3.46. Survival rate, stage II melanoma. Percentage of patients surviving after regional lymphadenectomy alone or lymphadenectomy with postoperative immunotherapy. Numerators are numbers of patients who remained free of disease; denominators are numbers of patients at risk (HOLMES et al., 1975)

stage II disease. They suggested that this organ may be an immunologically inaccessible site, as indicated by noncompatible tumour grafts in animals (MATHÉ and AMIEL, 1962).

McCULLOCH *et al.* (1975) published preliminary encouraging data concerning BCG immunotherapy after surgical resection of local or metastatic recurrences in melanoma patients.

The author has also submitted patients to BCG after surgery for melanoma at the first presentation of the disease: his preliminary results are in favour of a more promising effect from active immunotherapy than from that of systematic complementary chemotherapy.

In conclusion, this form of immunotherapy is, from now on, indicated as a complementary treatment to surgery in cases of melanoma.

Moreover, using the microcytotoxicity test, BERKELHAMMER *et al.* (1975) found that melanoma patient lymphocytes, taken before treatment, killed less tumour cells than normal lymphocytes, but that the reactivity of these cells was increased in all their patients after immunotherapy.

3.2.2.1.3. Bronchus Cancer

TAKITA and BRUGAROLAS (1973) treated five patients with locally advanced bronchogenic carcinoma with autologous tumour "vaccine" after incomplete surgical resection. This "vaccine" was the result of a chemical coupling of the fresh autologous tumour cells to a highly antigenic foreign protein with bis-diazobenzidine for two patients, and autologous cells coated with concanavallin for three others. In both cases it was injected with Freund's adjuvant. An improvement in the survival of the treated group was observed, as indicated by the comparison with a control group.

PIKES (1974) submitted several forms of bronchus cancer to Glaxo BCG injected with the multipuncture gun: his conclusion is that this adjuvant has some effect in certain patients with carcinoma in whom radical radiotherapy could be given. With POUILLART *et al.* (not published) we are conducting a randomized trial on bronchus cancer patients submitted to a radical operation. The present preliminary data shows a longer survival of the patients treated by BCG than in the controls. AMERY (1975) conducted a trial with levamisole in resectable bronchial cancer. He claims a much longer median of disease-free interval in the tetramisole epidermoid carcinoma group than in other types. Though the "randomization" of the subjects was not very strict, this observation can be considered as encouraging.

3.2.2.1.4. Mammary carcinoma

It is regrettable that, when ROJAS *et al.* (1976) obtained remarkable results with levamisole, in women who were operated on for breast cancer, a strict randomization was not used in the trial.

3.2.2.1.5. Osteosarcoma

MARCOVE *et al.* (1974) have conducted a trial in which they collected tumour tissues at surgery. The patients who were free of disease after one month received repeated injections of an aqueous tumour extract, irradiated with ultraviolet light.

The disease-free interval and survival of the immunized patients are significantly longer than those of historical controls.

TOWSEND and EILBER (1975) have treated patients with BCG and allogeneic tumour cells for primary or metastatic tumours which were completely resected. Sixteen out of 27 (59%) were alive at the time of publication, without a perceptible tumour, while only 6 out of 32 (21%) subjects who had not been submitted to immunotherapy were free of disease at the time of publication.

MORTON (pers. comm.) also mentioned an improvement in the prognosis of osteosarcoma patients submitted to BCG after the primary treatment.

3.2.2.2. *Treatment of Perceptible Disease*

If the trials of active systemic immunotherapy have been rare in the case of solid tumours for residual imperceptible neoplasias (the best condition for this new weapon against cancer to be effective), they have been numerous in the case of inoperable, and/or radioresistant solid tumours, the conditions for which the experimental data are not very encouraging.

3.2.2.2.1. *Melanoma*

GUTTERMAN *et al.* (1974b) have treated 89 melanoma patients with perceptible tumours. They administered fresh living Pasteur Institute BCG on scarifications; they applied 6×10^8 organisms on days 7, 12, 17 after each chemotherapy course with imidazole carboxamide (DTC). They obtained 55% regressions (Rg) superior to 50%, versus 18% in patients with lymph node involvement only submitted to chemotherapy. In the case of visceral metastasis, they obtained 17% Rg versus 13% with DTC alone. The duration of remission and survival was significantly longer for patients (both with and without visceral metastasis) treated with chemoimmunotherapy than for those treated with chemotherapy alone (Fig. 3.47). A better prognosis was associated with immunocompetence before treatment or an increase of immunocompetence during treatment.

THOMAS *et al.* (1974) treated 22 melanoma patients with BCG. They obtained 8 satisfactory responses (6-30 months). Two subjects with metastases in the small bowel only, seem to have responded to oral BCG. The only patients who responded were those who showed immunological reactivity to the stimulating agent. However, the best clinical results were not associated with a high degree of immune response in vivo and in vitro.

It is unfortunate that there is no trial in the literature based on the experimental data and comprising sequentially, as in acute lymphoid leukaemia, induction treatment, complementary chemo (± radio) therapy, active immunotherapy. A trial applying this principle is at present being conducted by the author's group.

RUMKE *et al.* (1974) selected three patients with measurable lung melanoma metastases and treated them with viable Pasteur Institute BCG applied on scarifications. The growth curve of heterotopias, which was a strain line, decreased after application of BCG (from about 20%), but, after three weeks, started to grow again even faster than before, despite continuation of BCG.

GRANT *et al.* (1974), treating two melanoma patients with BCG on scarifications combined with autologous irradiated cells, observed a four week regression of an orbital tumour mass.

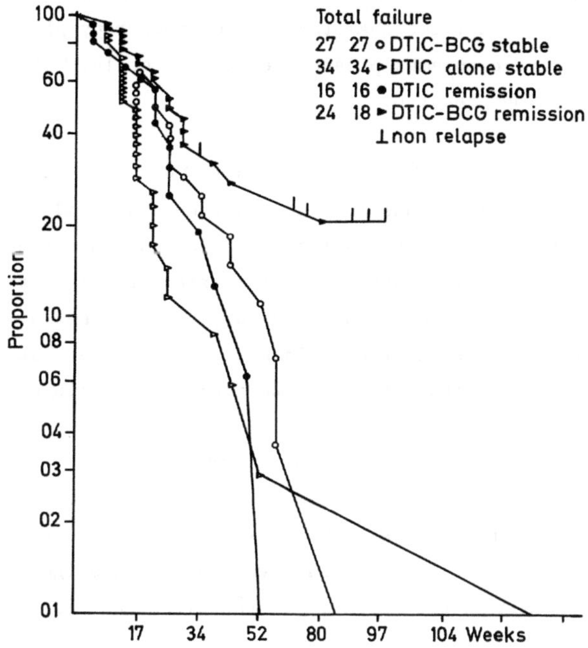

Fig. 3.47. Chemoimmunotherapy of disseminated malignant melanoma with dimethy-triazeno imidazole carboxamide and bacillus Calmette-Guérin: Duration of remission or stability with DTIC-BCG vs DTIC. For remission with DTIC-BCG vs DTIC, p = 0.05; for stability with DTIC-BCG vs, p = 0.008; and for remission plus stability with DTIC-BCG vs DTIC, p = 0.002 (GUTTERMAN et al., 1974b)

3.2.2.2.2. *Bronchus Cancer*

TAKITA and BRUGAROLAS (1973) have treated five patients with locally advanced bronchus carcinoma with autologous tumour cells modified according to CZAJKOWSKI *et al.* (1966) or coated with concavallin, and, on the same days, Freund's adjuvant: the patients were still alive after 18 months, compared to six deaths out of six controls at that time.

ISRAEL (1972, 1974) reported favourable results observed in the survival of bronchus cancer patients when Corynebacterium parvum was given at weekly intervals at a dose of 4 mg of heat-killed cultures to patients submitted to chemotherapy. The benefit was not an increase in response rate, but a prolongation of stabilisation periods: the mean survival from the start of therapy was 10 months in the treated group, versus 6 months in the control one. Such results should be considered with caution, as the author had published almost similar results obtained with chemotherapy ten years ago (ISRAEL *et al.*, 1965)("10 patients out of 11 had weathered the cape of 6 months").

ALTH *et al.* (1973), knowing that vitamin A is an adjuvant, compared bronchus cancer patients receiving chemotherapy, chemotherapy plus vitamin A, or vitamin A plus proteolytic enzymes plus pooled tumour cells. They claimed that they have obtained "promising" results.

3.2.2.2.3. Head and Neck Tumours

DONALDSON (1972) has submitted patients with advanced metastatic head and neck cancers to chemotherapy by methotrexate and immunotherapy with BCG. The initial results suggested an increased remission rate for patients treated with BCG compared to historical controls submitted only to chemotherapy. Further study does not confirm that BCG increases the incidence of remission, but that it may prolong it, compared to those results obtained with chemotherapy alone (GUTTERMAN *et al.*, 1974c).

3.2.2.2.4. Gynaecological Cancers

SONKIN *et al.* (1972) combined BCG and chemotherapy for a group of gynaecological cancers, the results of which were compared to those of patients receiving chemotherapy alone: the authors have observed a slightly better survival in the first group than in the second, but the difference is not significant.

3.2.2.2.5. Brain Tumours

TAKAKURA (quoted by Medical World News, 1975) who treated brain tumours by BCG, obtained a better survival rate than found in patients not receiving this treatment. But the BCG treatment was combined with surgery, radiotherapy, chemotherapy, and adoptive immunotherapy, and so no interpretation of this result is possible.

3.2.2.2.6 Various Tumours

VILLASOR (1965) submitted terminal cancer patients with breast, lung, colon, thyroid, and prostate cancers to chemotherapy and BCG. He wrote that the addition of BCG in the treatment of 43 patients increased the percentage of good responses, compared to the results obtained in 31 chemotherapy controls (74% versus 38%). It is difficult to accept such a conclusion: the heterogeneity of the diseases makes any comparison impossible.

MOERTEL *et al.* (1975) have treated metastatic gastro-intestinal carcinomas, which were progressing with previous chemotherapy, with MER alone: 3 out of 33 patients presented more than 50% regressions and 2 more than 25 % regressions.

3.2.3. Problems Common to Leukaemia and Solid Tumours

3.2.3.1. Immunotherapy and Chemotherapy Interspersed

The results claimed by several authors with immunotherapy and chemotherapy interspersed, or simultaneously combined, are convincingly favourable.

Is the effect of BCG or the other agents administered for immunotherapy, really immunotherapeutic?

The effect of some immunity adjuvants on stem cells which they have been shown to stimulate experimentally will be described (POUILLART *et al.*, 1975). This action is also demonstrabel in man. When we apply an intermittent chemotherapy with cycles separated by intervals for blood restoration, the latter is always accomplished after the same intervals if the cycles are identical. Applying BCG during certain intervals, and comparing them with the others, it has been noticed that the first are shorter than the latter; in other words, that haematological restoration is faster (MATHÉ, 1974a) (Fig. 3.48).

The question arises, therefore, as to whether or not the effect of adjuvants, when interspersed or simultaneously combined with chemotherapy, does not activate bone marrow recovery, hence allowing the administration of larger doses of chemotherapy.

The observations of GUTTERMAN *et al.* (974b) and of THOMAS *et al.* (1974) who obtained able results depending on the good immunological condition of the patients, are in favour of an immune effect of BCG, but they are not entirely convincing, as the results of chemotherapy are, for certain tumours, only favourable in patients with immune reactivity (POUILLART *et al.*, 1975).

3.2.3.2. Negative Results

It goes without saying that there must be many trials, of which negative results have not been published. About 50 patients with various advanced tumours have been submitted by us to BCG administered on scarifications or per os: no objective result has been observed.

3.2.3.3. Possibility of Neoplasia Growth Enhancement

LEVY *et al.* (1972) have observed a deterioration in the condition of melanoma patients after BCG therapy.

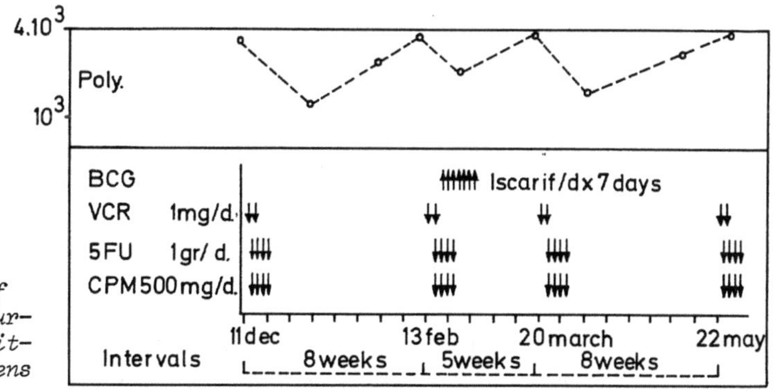

Fig. 3.48. Application of BCG on scarifications during intervals of intermittent chemotherapy shortens them (MATHÉ 1974a)

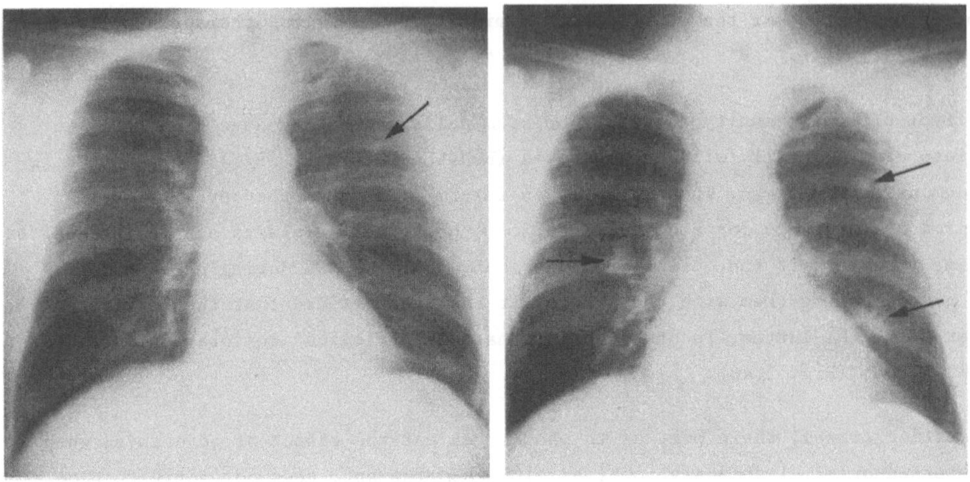

Fig. 3.49. Probable enhancement of tumour growth in melanoma patient treated by i.v. BCG (2 injections of 0.6 and 1.4 mg in 2 weeks): while hepatomegaly increased, cerebral manifestations appeared and chest films showed three nodules instead of one (pers. obs.)

After i.v. administration of BCG, and after the combination of BCG on scarifications and C. granulosum i.m., a growth of the tumour, which appeared to be faster than before treatment, was observed by us (Figs. 3.49 and 3.50).

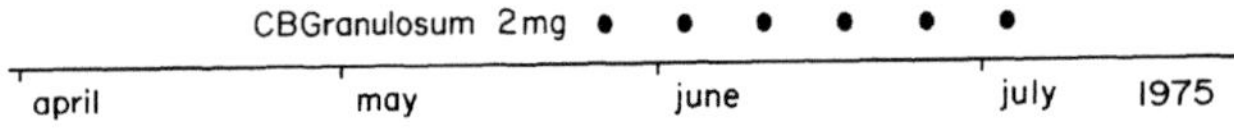

Fig. 3.50. Very rapid reappearance of metastases in melanoma patient submitted to combination of BCG and C. granulosum (immunological enhancement?) (pers. obs.)

In a sophisticated study, GUNVEN *et al.* (1973) treated a patient suffering from Burkitt's lymphosarcoma with BCG during a chemotherapy-induced remission: Epstein-Barr-Virus (EBV) associated membrane antibodies gradually declined during BCG treatment, 3-4 months prior to triple pertussis-diphtheria-tetanus (PDT) vaccination, after which they rapidly increased simultaneously with the appearance of multiple recurrent tumours. Hence, Bordetella pertussis may be, in man as in animals, a dangerous adjuvant, able to enhance tumour growth.

GUTTERMAN (pers. comm.) treated different types of nonhaematological sarcomas with a combination of chemotherapy and C. parvum: the percentage of remissions was decreased compared to those obtained with chemotherapy alone.

References

ALTH, G., DENCK, H., FISCHER, M., KARRER, K., KOKRON, O., KORIZEK, E., MICHSCHE, M., OGRIS, E., REIDER, C., TITSCHER, R., WRBA, H.: Cancer Chemoth. Rep. 4, 271 (1973).

AMERY, W.: Lancet 1, 574 (1975).

AMIEL, J.L., BERARDET, M.: Rev. Fr. Et. Clin. Biol. 14, 685 (1969).

AMIEL, J.L., BERARDET, M.: Europ. J. Cancer 6, 557 (1970).

AUR, R., VERZOSA, M., HUSTU, O., SIMONE, J., BARKER, L.: Proc. Amer. Assoc. Cancer Res. 16, 92 (abstract No. 365) (1975).

BAGSHAWE, K.D., GOLDING, P.R.: Immunity and tolerance in oncogenesis. Perugia: University of Perugia 1971.

BALDWIN, R.W., PIMM, M.V.: Int. J. Cancer 12, 420 (1973).

BALDWIN, R.W., PIMM, M.V.: Brit. J. Cancer 30, 473 (1974).

BANSAL, S.C., SJOGREN, H.O.: Int. J. Cancer 9, 490 (1972).

BARSKI, G., YOUN, J.K.: J. Nat. Cancer Inst. 43, 111 (1969).

BARSKI, G., YOUN, J.K., BELEHRADEK, J. jr., LE FRANCOIS, D.: Ann. Inst. Pasteur. 122, 633 (1972).

BEKESI, J.G., HOLLAND, J.F., YATES, J.W., HENDERSON, E., FLEMINGER, R.: Proc. Amer. Assoc. Cancer Res. 16, 121 (abstract No. 481) (1975).

BELPOMME, D., CARDE, P., OLDHAM, R.K., MATHÉ, G., JACQUILLAT, C., CHELLOUL, N., WEIL, M., AUCLERC, G., WEISBERGER, G., TANZER, T., BERNARD, J.: p. 115 in "Complications of Cancer Chemotherapy" (G. Mathé, R.K. Oldham, eds.) Vol 49 R.R.C.R., Heidelberg: Springer-Verlag 1974a.

BELPOMME, D., DANTCHEV, C., DU RUSQUEC, E., GRANDJON, D., HUCHET, R., POUILLART, P., SCHWARZENBERG, L., AMIEL, J.L., MATHÉ, G.: Biomedicine 20, 109 (1974b).

BELPOMME, D., LELARGE, N., GRANDJON, D., DANTCHEV, D., FEUILHADE DE CHAUVIN, F., JOSEPH, R., MATHE, G.: in "Lymphocytes and macrophages in cancer patients" (G. Mathé , I. Florentin, M.C. Simmler, eds). Heidelberg: Springer-Verlag, in press 1976.

BERKELHAMMER, J., MASTRANGELO, M.J., PREHN, R.T.: Proc. Amer. Assoc. Cancer Res. 16, 94 (abstract No. 375) (1975).

BERNARD, J.: in Third Meeting European African Division. International Society of Hematology, London. August 1975.

BERNARD, J., WEIL, M., JACQUILLAT, C.: p. 97 in Workshop on prognostic factors in human acute leukaemia. 1 vol., Oxford: Pergamon press 1975.

BILLINGHAM, R.E., BRENT, L., MEDAWAR, P.B.: Phil. Trans.Roy.Soc. London 239, 357 (1956).

BLUMING, A.Z., VOGEL, C.L., ZIEGLER, J.L.: p. 415 in "Investigation and Stimulation of Immunity in Cancer Patients" (G. Mathé, R. Weiner, eds.). Vol. 47 R.R.C.R., Heidelberg: Springer-Verlag 1974.

BLUMING, A.Z., VOGEL, C.L., ZIEGLER, J.L., MODY, N., KAMYA, G.: Ann. Intern. Med., 76, 405 (1972).

BREDT, A.B., MARDINEY, M.R.: Transplantation 8, 763 (1969).

BROUET, J.C., PREUD'HOMME, J.L., SELIGMANN, M.: Blood Cells 1, 81 (1975).

BRUCE, W.R., MEEKER, B.E., VALERIOTE, F.A.: J. Nat. Cancer Inst. 37, 233 (1966).

BUCHANAN, J.D., FAIRLEY, K.F., BARRIE, J.U.: Lancet 2, 156 (1975).
CANELLOS, G.P., DE VITA, V.T., ARSENEAU, J.C., JOHNSON, R.C.: p. 108 in "Complications of Cancer Chemotherapy" (G. Mathé, R.K. Oldham, eds.). Vol. 49 R.R.C.R., Heidelberg: Springer-Verlag 1974.
CHIHARA, G., HAMURO, J., MAEDA, Y., ARAI, Y., FUKUOKA, F.: Cancer Res. 30, 2776 (1970)
CHIHARA, G., MAEDA, Y., HAMURO, J., SASAKA, T., FUKUOKA, F.: Nature 222, 687 (1969).
CIKES, M.: Transplant. Proc. 3, 1161 (1971).
CURRIE, G.A., BAGSHAWE, K.D.: Brit. Med. J. 1, 541 (1970).
CUTTNER, J., HOLLAND, J.F., BEKESI, J.G., RAMACHANDAR, K., DONOVAN, P.: Proc. Amer. Assoc. Cancer Res. 16, 264 (abstract No. 1175) (1975).
CZAJKOWSKI, N.P., ROSENBLATT, M., CUSHING, F.R., VAZQUEZ, J., WOLF, P.L.: Cancer 19, 739 (1966).
DANTCHEV, D.: in "Lymphocytes and Macrophages in Cancer Patients" (G. Mathé, I. Florentin, M.C. Simmler, eds.).Heidelberg: Springer-Verlag, in press 1976.
DONALDSON, R.C.: An. J. Surg. 124, 527 (1972).
DORE, J.F., AJURIA, E., MATHÉ. G.: Europ. J. Clin. Biol. Res. 15, 81 (1970).
DORE, J.F., HADJIYANNAKIS, M.J., COUDERT, A., GUIBOUT, C., MARHOLEV, L., IMAI, Y.: p. 387 in "Investigation and Stimulation of Immunity in Cancer Patients" (G. Mathé, R. Weiner eds.). Vol. 47 R.R.C.R. Heidelberg: Springer-Verlag 1974.
EKERT, H., JOSE, D.G., WILSON, F.C., MATTHEWS, R.N., LAY, H.: Int. J. Cancer 16, 103 (1975).
EKERT, H., JOSE, D.G.: Lancet 2, 713 (1975).
European Organization for Research on the Treatment of Cancer (EORTC). Leukemia and Hematosarcoma Co-Operative Group. Biomedicine 18, 192 (1973).
European Organization for Research on the Treatment of Cancer (EORTC). Leukemia and Hematosarcoma Co-Operative Group. Europ. J. Cancer, 10, 413 (1974).
European Organization for Research on the Treatment of Cancer (EORTC). Hemopathies Working Party. Third Meeting of International Society of Haematology European and African Division, London, 24-28th August, 1975 (abstract 21:06).
FRANCHI, G., GARATTINI, S., KRAM, L.J.K., VAN PUTTEN, L.M., EORTC.: Europ. J. Cancer 9, 383 (1973).
FREEMAN, C.B., HARRIS, R., GERRY, C.G., LEYLAND, M.J., MacIVER, J.E., DELAMORE, I.W.: Brit. Med. J. 4, 571 (1973).
GALBRAITH, A.W., OXFORD, J.S., SCHILD, G.C., WATSON, G.I.: Lancet 2, 1026 (1969).
GEE, T.S., HAGHBIN, M., TAN, C., MURPHY, M.L., DOWLING, M.D., CLARKSON, B.D.: Proc, Amer. Assoc. Cancer Res. 15, 164 (abstract No. 720) (1974).
GIULIANI, F., CASAZZA, A.M., DI MARCO, A.: Biomedicine 21, 435 (1974).
GRANT, R.M., COCHRAN, A.J., HOYLE, D., MACKIE, R., MURRAY, E.M., ROSS, C.: Lancet 2, 1096 (1974).
GREAGER, J.A., CHAUDHURI, P.K., BOVAR, M., DAS GUPTA, T.K.: Proc. Amer. Assoc. Cancer Res. 16, 143 (abstract No. 571) (1975).
GROOMS, G.A., MORTON, D.L.: Proc. Amer. Assoc. Cancer Res. 16, 261 (1975).
GROSS, L.: Oncogenic viruses. 2nd ed., Oxford: Pergamon Press (1970).

Groupe Coopérateur Leucémies et Hématosarcomes de l'Organisation Européenne de Recherche sur le Traitement des Cancers (O.E.R.T.C.). Nouv. Presse Med., 4, 1553 (1975).

GUNVEN, P., KLEIN, G., HENLE, W., HENLE, G., ROCCHI, G., HEWETSON, J.F., GUERRA, A., CLIFFORD, P., SINGH, S., DEMISSIE, A., SVEDMYR, A.: Int. J. Cancer 12, 115 (1973).

GUTTERMAN, J.U., MacBRIDE, C., FREIREICH, E.J., MAVLIGIT, G., FREI, E. III, HERSH, E.M.: Lancet 1, 1208 (1973).

GUTTERMAN, J.U., MAVLIGIT, G., GOTTLIEB, J.A., BURGESS, M.A., McBRIDE, C.M., EINHORN, L., FREIREICH, E.J., HERSH, E.M.: New England J. Med. 291, 592 (1974b).

GUTTERMAN, J.U., MAVLIGIT, G., KENNEDY, A., HERSH, E.: Proc. Amer. Assoc. Cancer Res. 16, 245 (abstract No. 1096) (1975).

GUTTERMAN, J.U., MAVLIGIT, G.M., REED, R.C., HERSH, E.M.: Seminars in Oncology, 1, 409 (1974c).

GUTTERMAN, J.U., RODRIGUEZ, V., MAVLIGIT, G., BURGESS, M.A., GEHAN, E., HERSH, e.M., MacCREDIE, K.B., REED, R., SMITH, T., BODEY, G.P. fr., FREIREICH, E.J.: Lancet 2, 1405 (1974a).

HALLE-PANNENKO, O., BOURUT, C., KAMEL, M.: Cancer Immunol.Immunoth. 1, 17 (1976).

HAUSS, G.: Application des méthodes statistiques à l'étude pronostique rétrospective de 161 cas de leucémies aigues lymphoides. Thesis, Université de Paris-Sud, Faculté de Médecine du Kremlin-Bicêtre, 1975.

HAYHOE, F.G.J., FLEMANS, R.J.: Atlas de cytologie hématologique. Paris: Editions Médicales Flammarion (1970).

HERSH, E.M., GUTTERMAN, J.U., MAVLIGIT, G.: 1 Vol., Springfield: C.C.Thomas publ. 1973.

HOERNI, B., CHAUVERGNE, J., HOERNI-SIMON, G., DURAND, M., BRUNET, R., LAGARDE, C.: Cancer Immunol. Immunoth. 1, 109 (1976).

HOLLAND, J.F.: p. 217 in "Perspectives in Leukemia" (W. Dameshek, R.M. Dutcher, eds.). 1 Vol., New-York: Grune & Stratton 1968.

HOLMES, E.C., EILBER, F.R., MORTON, D.L.: J. Amer. Med. Ass. 232, 1052 (1975).

HUCHET, R., LHERITIER, J.: in "Lymphocytes and Macrophages in Cancer Patients" (G. Mathé, I. Florentin, M.C. Simmler, eds). Heidelberg: Springer-Verlag, in press (1976).

IKONOPISOV, R.L.: Tumori, 58,121 (1972).

ISRAEL, L.: p. 486 in "Investigation and Stimulation of Immunity in Cancer Patients" (G. Mathé, R. Weiner, eds.). Vol. 47 R.R.C.R., Heidelberg: Springer-Verlag 1974.

ISRAEL, L., HALPERN, B.: Nouv. Presse Med. 1, 19 (1972).

ISRAEL, L., SORS, C., REBOUL, R.: Presse Med. 73, 701 (1965).

JACQUILLAT, Cl., WEIL, M., GEMON, M.F., BOIRON, M., BERNARD, J.: p. 113 in "Therapy of Acute Leukemias" (F. Mandelli, S. Amadori, G. Mariani, eds.). Rome: Minerva Medica 1975.

JOSEPH, R.R., BELPOMME, D., MATHÉ, G.: Brit.J.Cancer 33, 567 (1976).

KALISS, N.: Cancer Res. 12, 379 (1952).

KALISS, N.: Proc. Nat. Acad. Sc. 42, 269 (1956).

KALISS, N.: p. 225 in "La biologie des homogreffes". 1 Vol., Paris: C.N.R.S., ed. 1957.

KALISS, N.: Cancer Res. 18, 992 (1958).

KHALIL, A., BOURUT, C., HALLE-PANNENKO, O., MATHÉ, G., RAPPAPORT, H.: Biomedicine 22, 112 (1975).

LACOUR, F., SPIRA, A., LACOUR, J., PRADE, M.: Cancer Res. 32, 648 (1972).
LACOUR, F., LACOUR, J., SPIRA, A.: p. 352 in "Investigation and Stimulation of Immunity in Cancer Patients" (G. Mathé, R. Weiner, eds.). Vol 47 R.R.C.R. Heidelberg: Springer-Verlag 1974.
LAMPERT, F., HEINZE, G., WUNDISCH, G.F., OLISCHLAGER, A., KLOSE, K., USENER, M., NEIDHART, M.: p. 595 in "Therapy of Acute Leukemias" (F. Mandelli, S. Amadori, G. Mariani, eds.). Rome: Minerva Medica 1975.
LEMONDE, P.: Union Med. Canada 103, 1716 (1974).
LEVY, N.L., MAHALEY, M.S. fr., DAY, E.D.: Int. J. Cancer 10, 244 (1972).
LEVY, J.G., WHITNEY, R.B., SMITH, A.G., PANNO, L.: Brit. J. Cancer 30, 289 (1974).
MacCULLOCH, P., DENT, P., LUI, V., BLAJCHMAN, M., MUIRHEAD, W.: Proc. Amer. Assoc. Cancer Res. 16, 174 (abstract No. 693) (1975).
MacLENNAN, I.C.M.: Proc. Royal Soc. Med. 68, 216 (1975).
MAEDA, Y.Y., CHIHARA, G.: Nature 229, 634 (1971).
MAEDA, Y.Y., HAMURO, J., CHIHARA, G.: Int. J. Cancer 8, 41 (1971).
MAGRATH, I.J., ZIEGLER, J.L.: Brit. Med. J. 1, 615 (1976).
MAGRATH, I.T., ZIEGLER, J.L., BLUMING, A.Z.: p. 461 in "Investigation and Stimulation of Immunity in Cancer Patients" (G. Mathé, R. Weiner, eds.). Vol. 47 R.R.C.R., Heidelberg: Springer-Verlag 1974.
MAK, T.W., MANASTER, J., HOWATSON, A.F., McCULLOCH, E.A., TILL, J.R.: Proc. Nat. Acad. Sci. 71, 4336 (1974).
MANDELLI, F., AMADORI, S., ANSELMO, M.P., DEL PRINCIPE, D., DERIU, L., DIGLIO, G., ISACCHI, G., MULTARI, G.: p. 609 in "Therapy of Acute Leukemias" (F. Mandelli, S. Amadori, G. Mariani, eds.). Rome: Minerva Medica 1975.
MARCOVE, R.C., SOUTHAM, C.M., LEVIN, A.G., HUVORS, A.G., MIKE, V.: p. 488 in "Investigation and Stimulation of Immunity in Cancer Patients" (G. Mathé, R. Weiner, eds.). Vol. 47 R.R.C.R., Heidelberg: Springer-Verlag 1974.
MARTIN, M., BOURUT, C., HALLE-PANNENKO, O., MATHÉ, G.: Biomedicine 23, 337 (1975).
MATHÉ, G.: Presse Med. 75, 2591 (1967).
MATHÉ, G., Rev. Fr. Et. Clin. Biol. 13, 881 (1968).
MATHÉ, G.: Adv. Cancer Res. 14, 1 (1971).
MATHÉ, G.: Biomedicine 18, 169 (1973).
MATHÉ, G.: Bull. Cancer 61, 191 (1974b).
MATHÉ, G.: p. 124 in "Complications of Cancer Chemotherapy" (G. Mathé, R.K., Oldham, eds.). Vol. 49 R.R.C.R., Heidelberg: Springer-Verlag 1974a.
MATHÉ, G.: p. 207 in "Clinical tumor immunology" (J. Wynbran, ed.). Oxford: Pergamon press 1976.
MATHÉ, G., AMIEL, J.L.: La greffe. Aspects biologiques et cliniques. 1 Vol., Paris: Masson et Cie, ed. 1962.
MATHÉ, G., AMIEL, J.L., HAYAT, M., VASSAL, F. DE, SCHWARZENBERG, L., SCHNEIDER, M., JASMIN, C., ROSENFELD, C.: p. 168 in "International Symposium on Adriamycin" Heidelberg Springer-Verlag 1972.
MATHÉ, G., AMIEL, J.L., SCHWARZENBERG, L., HAYAT, M., POUILLART, P., SCHNEIDER, M.,

CATTAN, A., JASMIN, C., BELPOMME, D., SCHLUMBERGER, J.R., VASSAL, F. DE, MUSSET, M., MISSET, J.L.: Nouv. Presse Med. 4, 1337 (1975b).

MATHÉ, G., AMIEL, J.L., SCHWARZENBERG, L., SCHNEIDER, M., CATTAN, A., SCHLUMBERGER, J.R., HAYAT, M., VASSAL, F. DE: Rev. Fr. Et. Clin. Biol. 13, 454 (1968).

MATHÉ, G., AMIEL, J.L., SCHWARZENBERG, L., SCHNEIDER, M., CATTAN, A., SCHLUMBERGER, J.R., HAYAT, M., VASSAL, F. DE: Lancet 1, 697 (1969a).

MATHÉ, G., AMIEL, J.L., SCHWARZENBERG, L., SCHNEIDER, M., HAYAT, M., VASSAL, F. DE, JASMIN, C., ROSENFELD, C., POUILLART, P.: Europ. J. Clin. Biol. Res. 16, 216(1971b)

MATHÉ, G., AMIEL, J.L., SCHWARZENBERG, L., SCHNEIDER, M., HAYAT, M., VASSAL, G. DE, JASMIN, C., ROSENFELD, C., SAKOUHI, M., CHOAY, J.: Europ. J. Clin. Biol. Res. 15, 671 (1970a).

MATHÉ, G., BELPOMME, D., DANTCHEV, D., POUILLART, P., JASMIN, C., MISSET, J.L., MUSSET, M., AMIEL, J.L., SCHLUMBERGER, J.R., HAYAT, M., VASSAL, F. DE, LAFLEUR, M.: Biomedicine 20, 333 (1974c).

MATHÉ, G., BELPOMME, D., DANTCHEV, D., POUILLART, P., SCHLUMBERGER, J.R., LAFLEUR, M.: Blood Cells 1, 25 (1975f).

MATHÉ, G., BELPOMME, D., POUILLART, P., SCHWARZENBERG, L., HAYAT, M., VASSAL, F. DE, JASMIN, C., AMIEL, J.L., MISSET, J.L., MUSSET, M.: Biomedicine 23, 465 (1975h),

MATHÉ, G., DAUSSET, J., HERVET, E., AMIEL, J.L., COLOMBANI, J., BRULE, G.: J. Nat. Cancer Inst. 33, 193 (1964).

MATHÉ, G., DELGADO, M., VASSAL, F. DE, SCHWARZENBERG, L., POUILLART, P., HAYAT, M., AMIEL, J.L., JASMIN, C., BELPOMME, D., MISSET, J.L., MUSSET, M.: in preparation (1976b).

MATHÉ, G., HALLE-PANNENKO, O., AMIEL, J.L.: p. 48 in "Applications in Experimental and Clinical Oncology". New-York: Plenum press, 1975d.

MATHÉ, G., HALLE-PANNENKO, O., BOURUT, C.: Exp. Hematol. 1, 110 (1973a).

MATHÉ, G., HALLE-PANNENKO, O., BOURUT, C.: Nat. Cancer Inst. Monogr. 39 107 (1973c).

MATHÉ, G., HALLE-PANNENKO, O., BOURUT, C.: Europ. J. Cancer 10, 661 (1974a).

MATHÉ, G., HALLE-PANNENKO, O., BOURUT, C.: Transplant. Proc. 6, 431 (1974b).

MATHÉ, G., HALLE-PANNENKO, O., BOURUT, C.: Fourth Annual Conference of the International Society for Experimental Hematology, Yugoslavia, September 1975a.

MATHÉ, G., HAYAT, M., SAKOUHI, M., CHOAY, J.: C.R. Acad. Sci., Paris 272, 170 (1971d).

MATHÉ, G., KENIS, Y.: La chimiothérapie des cancers (leucémies, hématosarcomes et tumeurs solides). 3rd ed., Paris: Expansion Scientifique 1975.

MATHÉ, G., MUSSET, M., SCHWARZENBERG, L., HAYAT, M., VASSAL, F. DE, AMIEL, J.L., POUILLART, P., MISSET, J.L.: Biomedicine 23, 291 (1975g).

MATHÉ, G., POUILLART, P., LAPEYRAQUE, F.: Brit. J. Cancer 23, 814 (1969b).

MATHÉ, G., POUILLART, P., LAPEYRAQUE, F.: Experientia 27, 446 (1971a).

MATHÉ, G., POUILLART, P., RAPPAPORT, H., HAYAT, M., STERESCU, M., LAFLEUR, M.: p. 6 in "Nomenclature, Methodology and Results of Clinical Trials in Acute Leukemias" (G. Mathé, P. Pouillart, L. Schwarzenberg, eds.). Vol. 43 R.R.C.R., Heidelberg: Springer-Verlag 1973b.

MATHÉ, G., POUILLART, P., SCHWARZENBERG, L.: Acta Neuropath. 6, 235 (1975c).

MATHÉ, G., POUILLART, P., SCHWARZENBERG, L., HAYAT, M., VASSAL, F. DE, LAFLEUR, M.: p. 145 in "Prognosis Factors in Human Acute Leukemias. 1 Vol., Oxford: Pergamon press 1975e.

MATHÉ, G., POUILLART, P., STERESCU, M., AMIEL, J.L., SCHWARZENBERG, L., SCHNEIDER, M., HAYAT, M., VASSAL, F. DE, JASMIN, C., LAFLEUR, M.: Europ. J. Clin. Biol. Res., 16, 554 (1971c).

MATHÉ, G., RAPPAPORT, H.: Histocytological typing on the neoplastic diseases of the haematopoietic and lymphoid tissues. Geneva: World Health Organization (1976).

MATHÉ, G., SCHNEIDER, M., SCHWARZENBERG, L.: Europ. J. Cancer 6, 23 (1970b).

MATHÉ, G., VASSAL, F. DE, AMIEL, J.L., SCHWARZENBERG, L., JASMIN, C., HAYAT, M., POUILLART, P., MISSET, J.L., MUSSET, M.: New-York: Plenum press, in press 1976c

MATHÉ, G., VASSAL, F. DE, DELGADO, M., POUILLART, P., BELPOMME, D., HAYAT, M., SCHWARZENBERG, L., AMIEL, J.L., MISSET, J.L., MUSSET, M., JASMIN, C.: Cancer Immunol. Immunoth. 1, 77 (1976a).

MEDICAL RESEARCH COUNCIL: Brit. Med. J. 4, 189 (1971).

MILAS, L., HUNTER, N., BASIC, I., MASON, K., GRDINA, D.J., WITHERS, H.R.: J. Nat. Cancer Inst. 54, 895 (1975).

MILAS, L., HUNTER, N., BASIC, I., WITHERS, H.R.: Cancer Res. 34, 2470 (1974).

MOERTEL, C.G., RITTS, R.E., SCHUTT, A.J., HAHN, R.G.: Proc. Amer. Assoc. Cancer Res. 16, 143 (abstract No. 569) (1975).

MOROSON, H., SCHECHTER, M.: Biomedicine in press (1976).

MORRIS, H.P.: Cancer Res. 20, 1252 (1960).

MORTON, D.L., EILBER, F.R., HOLMES, E.C., SPARKS, F.C., RAMMING, K.P.: Cancer Immunol. Immunoth. 1, 93 (1976).

OLDHAM, R.K., WEINER, R.S., MATHÉ, G., BREARD, J., SIMMLER, M.C., CARDE, P., HERBERMAN, R.: Int. J. Cancer, 17, 326 (1976).

OLSSON, L., FLORENTIN, I., KIGER, N., MATHÉ, G.: J. Nat. Cancer Inst., submitted to publication (1976).

ORMAI, S., DE CLERCQ, E.: Science 163, 471 (1969).

PARR, I.: Brit. J. Cancer 26, 174 (1972).

PEARSON, J.W., CHAPARAS, S.D., CHIRIGOS, M.A.: Cancer Res. 33, 1845 (1973).

PEARSON, J.W., PEARSON, G.R., GIBSON, W.T., CHERMANN, J.C., CHIRIGOS, M.A.: Cancer Res. 32, 904 (1972).

PIESSENS, W.F., LACHAPELLE, F.L., LEGROS, N., HEUSON, J.C.: Nature 228, 1210 (1970).

PIESSENS, W.F., HEIMANN, R., LEGROS, N., HEUSON, J.C.: Europ. J. Cancer 7, 377 (1971).

PIKES, A.: p. 466 in "Investigation and Stimulation of Immunity in Cancer Patients" (G. Mathé, R. Weiner, eds.). Vol. 47 R.R.C.R., Heidelberg: Springer-Verlag 1974.

POUILLART, P., PALANGIE, T., SCHWARZENBERG, L., BRUGERIE, E., LHERITIER, J., MATHÉ, G.: Biomedicine, 23, 469 (1975).

POUILLART, P., SCHWARZENBERG, L., SCHNEIDER, M., AMIEL, J.L., MATHÉ, G.: Presse Med. 1, 387 (1972).

POWLES, R.: Cancer 34, 1558 (1974).

POWLES, R.L., CROWTHER, D., BATEMAN, C.J.T., BEARD, M.E.J., MacELWAIN, T.J., RUSSELL, J., LISTER, T.A., WITHEHOUSE, J.M.A., WRIGLEY, P.F.M., PIKE, M., ALEXANDER, P., HAMILTON-FAIRLEY, G.: Brit. J. Cancer 28, 365 (1973).

PUTTEN, L.M. van: p. 115 in "The Design of Clinical Trials in Cancer Therapy" (M. Staquet, ed.). Bruxelles: Ed. Scientifiques Européennes 1972.

RENOUX, G., RENOUX, M.: Nature, New Biol. 240, 217 (1972b).

RENOUX, G., KASSEL, R.L., RENOUX, M., FIORE, N.C., GUILLAUMIN, J.M., PALAT, A.: in preparation (1976).

RENOUX, G., RENOUX, M.: C.R. Acad. Sci., Paris 274, 3034 (1972a).

ROJAS, A.F., FEIERSTEIN, J.N., MICHIEWICZ, E., GLAIT, H., OLIVARI, A.J.: Lancet 1, 7953 (1976).

ROSEN, R.B., NOWACK, J.: Cancer 34, 501 (1974).

RUMKE, P., BERNHEIM, J., VAN DER VORM, D.H., VAN PEPERZEEL, H.A., LOOYSEN, R.: p. 470 in "Investigation and Stimulation of Immunity in Cancer Patients" (G. Mathé, R. Weiner, eds.). Vol. 47 R.R.C.R., Heidelberg: Springer-Verlag 1974.

SCHNEIDER, M., MATHÉ, G., SCHWARZENBERG, L., POUILLART, P., WEINER, R., AMIEL, J.L., HAYAT, M., JASMIN, C., VASSAL, F. DE: p. 42 in "Investigation and Stimulation of Immunity in Cancer Patients" (G. Mathé, R. Weiner, eds.). Vol. 47 R.R.C.R., Heidelberg: Springer-Verlag 1974.

SCHWARZENBERG, L., MATHÉ, G., POUILLART, P., WEINER, R., LACOUR, J., GENIN, J., SCHNEIDER, M., VASSAL, F. DE, HAYAT, M., AMIEL, J.L., SCHLUMBERGER, J.R., JASMIN, C., ROSENFELD, C.: Brit. Med. J. 1, 700 (1973).

SCHWARZENBERG, L., SIMMLER, M.C., PICO, J.L.: Cancer Immunol. Immunoth. 1, 69 (1976).

SCOTT, M.T.: Seminars in Oncology 1, 367 (1974).

SEDLACEK, H.H., MEESMANN, H., SEILER, F.R.: Proc. Amer. Assoc. Cancer Res. 16, 141 (abstract No. 564) (1975).

SIMMONS, R.L., RIOS, A.: Science 174, 591 (1971).

SIMMONS, R.L., RIOS, A.: Ann. Surg. 176, 188 (1972).

SIMMONS, R.L., RIOS, A., RAY, P.K., LUNDGREN, G.: J.Nat.Cancer Inst. 47, 1087 (1971).

SIMONE, J.V.: p. 73 in "Therapy of Acute Leukemias" (F. Mandelli, S. Amadori, G. Mariani, eds.). Rome: Minerva Medica 1975.

SIMONE, J.V., HOLLAND, E., JOHNSON, W.: Blood 39, 759 (1972).

SJOGREN, H.O.: p. 51 in "The Role of Immunological Factors in Viral and Oncogenic Processes" (R.F. Beers jr, R.C. Tilghman, E.G. Bassett, eds.). Baltimore: John Hopkins Univ. Press 1974.

SKIPPER, H.E., SCHABEL, F.M., WILCOX, W.S.: Cancer Chemoth. Rep. 45, 5 (1965).

SKURKOVICH, S.V., KISLJAK, N.S., MACHONOVA, L.A., BEGUNENKO, S.A.: Nature 223, 509 (1969).

SMYTH, A.C., WIERNIK, P.H., SERPICK, A.A.: Proc. Amer. Assoc. Cnacer Res. 16, 236 (abstract 1062) (1975).

SOKAL, J.E., AUNGST, C.W.: p. 313 in XITH International Cancer Congress, Florence, October 1974. Vol. 1 Conferences, Symposia Workshops.

SONKIN, R., COUDEYRAS, M., BLONDON, J.: J. Gynec. Obstet. Biol. Reprod., Paris 1, 61 (1972).

SPARKS, F.C., O'CONNELL, T.X., LEE, Y.T.N.: Surg. Forum 24, 118 (1973).

SPIEGELMAN, S.: XVth Congress International of Hematology, Jerusalem, September 1974, (abstract No. 474).

SUIT, H., SEDLACEK, R., WAGNER, M., ORSI, L.: Proc. Amer. Assoc. Cancer Res. 16, 196 (abstract No. 781) (1975).

TAKAKURA, K.: Quoted by Medical World News, January 13, 1975.

TAKITA, H., BRUGAROLAS, A.: Cancer Chemoth. Rep. 4, 293 (1973).

THOMAS, J.W., PLENDERLEITH, I.H., CLEMENTS, D., LANDI, S.: p. 315 in XIth International Cancer Congress, Florence, October 1974. Vol. 1 Conferences, Symposia Workshops.

THOMPSON, R.B., ALBEROLA, V., MATHÉ, G.: Europ. J. Clin. Biol. Res.: 17, 900 (1972).

TOKUNAGA, T., YAMAMOTO, S., NAKAMURA, R., KATAOKA, T.: J. Nat. Cancer Inst. 53, 459 (1974).

TOWNSEND, C.M., EILBER, F.R.: Proc. Amer. Assoc. Cancer Res. 16, 261 (1975).

TURK, J.L., PARKER, D., POULTER, L.W.: Immunology 23, 493 (1972).

TURK, J.L., POULTER, L.W.: Clin. Exp. Immunol. 10, 285 (1972).

VASAREVIC, B., BORANIC, M., PAVELIC, Z.: Biomedicine in press (1976).

VILLASOR, R.: J. Philip. Med. Assoc. 4, 619 (1965).

VOGLER, W.R., CHAN, Y.K.: Lancet 2, 128 (1974).

WEISS, D.W.: Nat. Cancer Inst. Monogr. 35, 157 (1972).

WISSLER, R.W., CRAFT, K., KESDEN, D., POLISKY, B., DZOZA, K.: p. 539 in "Advance in transplantation" (J. Dausset, F.T. Rapaport, eds.). 1 Vol., Copenhagen: Munksgaard 1968.

WOODRUFF, M.F.A.: p. 272 in "Investigation and Stimulation of Immunity in Cancer Patients" (G. Mathé, R. Weiner, eds.). Vol. 47 R.R.C.R., Heidelberg: Springer-Verlag 1974.

YRON, I., COHEN, D., ROBINSON, E., HARBER, M., WEISS, D.W.: Cancer Res. 35, 17779 (1975).

YRON, I., WEISS, D.W., ROBINSON, E., COHEN, D., ADELBERG, M.G., MEKORI, T., HABER, M.: Nat. Cancer Inst. Monogr. 39, 33 (1973).

Chapter 4
Cancer Local Active Immunotherapy

As the mixture of BCG and irradiated tumour cells had been shown to be more effective than either one or the other administered alone (MATHÉ, 1968; MATHÉ *et al.*, 1969), it was tempting, in the case of a perceptible tumour, to inject BCG or another adjuvant into the neoplasia. This was tried on human tumours by MORTON *et al.* (1970a, 1970b), and on experimental models by ZBAR, RAPP and co-workers (ZBAR and TANAKA, 1971; ZBAR, 1972; ZBAR *et al.*, 1972; RAPP, 1973; ZBAR *et al.*, 1973).

4.1. Experimental

4.1.1. BCG and Other Mycobacteria

ZBAR, RAPP *et al.* (1971, 1972, 1973 immediately above) inoculated strain 2 guinea pigs ID with 10^6 isogeneic tumour cells from a diethyl-nitrosamine-induced hepatoma maintained in an ascitic form. After it had grown to a diameter of about 1 cm and (microscopic) metastases could be demonstrated in the regional lymph nodes, they injected living BCG (10^7 bacilli) into the primary neoplasia: the latter regressed in about 60% of the animals from its original site and from the lymph nodes.

In this guinea pig model, the injection of BCG into the primary tumour may inhibit not only its growth, but also that of distant metastases (ZBAR *et al.*, 1971).

More recently, similar results have been obtained in other species. A significant regression of a 3-methylcholanthrene-induced tumour in SWM/Ms mice has been obtained by injecting BCG into neoplasias, in 50% of animals (TOKUNAGA *et al.*, 1974).

SALOMON and LYNCH (1976) were recently able to reproduce these results not only in mice (carrying Moloney virus-induced sarcoma), but also in rats (grafted with an original methylcholanthrene sarcoma).

GREAGER *et al.* (1975) observed a slight, but significant, regression of Lewis tumour after local injection of BCG, but there was no effect on metastases.

Injecting this myobacteria into the tumour, the author observed, with MARTIN *et al.* (1976), neither regression, nor a decrease of lung metastases.

Mycobacterial extracts were also used with some success in local immunotherapy, specially by ZBAR *et al.* (1974) who injected BCG cell walls attached to mineral oil droplets into different established tumours (especially transplantable hepatomas). They caused local regressions and prevented the growth of lymph node metastases in more than 50% of the animals. The minimum effective dose of mycobacteria cell walls was 30 μg when injected in admixture with 1.4 μl of mineral oil and 0.8 μl Tween.

Cell walls from M. tuberculosis and M. kansasii possessed the same effect, but similar preparations from M. smegmatis, Nocardia asteroids, and C. parvum lacked these properties.

SALOMON and LYNCH (1976) used MER (WEISS, 1972) with favourable results, at approximately the same dose as that at which BCG acts.

4.1.2. Corynebacteria and Other Bacterias

The intratumoural injection of C. parvum on day 6 after grafting, significantly reduced in our experience, the number of perceptible lung heterotopias of Lewis tumour on day 20 (Table 4.I) (MARTIN *et al.*, 1976).

SALOMON and LYNCH (1976) also obtained regressions of several tumours with C. parvum (10^9 dead bacteria per injection of 0.010 ml).

ZBAR and RAPP (1974) did not observe such an effect on their guinea pig model with this adjuvant, but SCOTT (1974a and b) obtained a regression, in mice, of a mastocytoma after intra-tumour injection of C. parvum; the optimum dose was one-tenth of the intravenous optimal dose for systemic immunotherapy (SCOTT, 1974a, 1974b). The treatment was active for up to 12 days after tumour inoculation. Cell-mediated immunity was stimulated, while it was not by the i.v. injection.

This organism was also active in inducing rejection of two subcutaneous tumours isogeneically grafted in Fisher rats (LIKHITE, 1974a).

MILAS *et al.* (1975a) induced complete regression of a subcutaneously grafted isogeneic fibrosarcoma (originally 3-methylcholanthrene-induced) in 30% of the mice after intralesional injection of C. granulosum.

The injection of B. pertussis in mouse mammary tumours was also able to restrict tumour growth (LIKHITE, 1974b).

Table 4.I. Local intratumoural immunotherapy of solid Lewis tumour[a]: Efficacy of C. parvum in reducing the number of metastases versus inefficacy of BCG (MARTIN *et al.*, 1976).

	Average number of lung metastases at day +20	Statistics (Student-Fisher test)
Controls (inoculated i.m. with 5 x 10^5 tumour cells)	82 ± 18	
Saline (0.05 ml/tumour at day +6)	60 ± 20	t = 1.92 NS
B.C.G. (lyophilized 1 mg/tumour at day +6)	71 ± 22	t < 1 NS
C. parvum (0.1 mg/tumour at day +6)	48 ± 15	t = 3.4 S at 1%

[a] No effect was observed on the injected tumour.

4.1.3. Other Agents

Regressions were obtained by SIMMONS and RIOS (1972) on a methylcholanthrene fibrosarcoma and by SPARKS and BREEDING (1974) on MCA-2 tumours of mice into which they had injected neuraminidase (50 units). The latter authors obtained more striking results with the combination of neuraminidase and BCG.

4.1.4. Failures and Factors and Conditions of Efficacy

Many experiments with negative results have not been published. Initially, the question was raised as to whether or not the guinea pig model was the only one that was sensitive. Recent positive experiments on mice and rat models indicate that the *species*, if it is an influencing factor, is not the only one, and that the reason for failure may be sought in the quality and quantity of adjuvants, and the type and volume of tumours at the time of injection.

PIMM and HOPPER (1975) treating two transplanted methylcholanthrene-induced sarcomas in the same models, and using different adjuvants, BCG, C. parvum and B.pertussis, observed also that tumours susceptible to local infiltration with some agents may be resistant to others.

A third necessary condition for efficacy is the *quantity of the agent* used as an adjuvant; the number of viable bacilli in the case of BCG (ZBAR and RAPP, 1974), as will be described in chapter 7 on Pharmacology.

Other limiting factors that were demonstrated by the author for systemic immunotherapy are also found in its local form. The most important one, though it works on perceptible tumours, is the *volume* of the latter: the injection of BCG into i.d. hepatoma transplants induces a regression in 60% of the animals if the weight of the neoplasia is less than 95 mg; the regression appears in less than 20% if the weight is 425 mg or more (ZBAR *et al.*, 1972).

Another factor of success may be the *site of the implant of the tumour*, hence the immune reaction determined by the treatment. It is well known that the i.d. site is much more immunogenic than any other site of the organism: if one injects 20 SD[1]/50 Friend's Leukaemia virus i.d., one does not induce leukaemia, while the same dose administered i.p. or i.v. induces the disease in 100% of the animals (pers. observ.). LITTMAN *et al.* (1972) showed that the i.d. transplant of a given tumour (line 10) is much more susceptible than the intramuscular graft. Nevertheless, the skin site is not the only one which is sensitive, as SALOMON and LYNCH (1976) have successfully treated the i.m. inoculated moloney sarcoma in mice.

Particular Case(s) of Pleural and Peritoneal Neoplastic Localisations. PIMM and BALDWIN (1975) obtained remarkable regressions of intrapleural or peritoneal growths of a methylcholanthrene sarcoma or a hepatoma by local injections of (1×10^6 r) radiation sterilized BCG Glaxo vaccine.

MILAS *et al.* (1975b) obtained, after i.p. injections of C. granulosum, 58-80% cure in mice inoculated 3 days earlier with an isogeneic fibrosarcoma in the peritoneum.

---

[1] SD = Spleen doses.

4.1.5. Possibility of Neoplasia Growth Enhancement

Studying several experimental tumour systems, SPARKS and BREEDING (1974) obtained regressions of the neoplasias in some (MCA-2), and increased growth of both injected and uninjected tumours in others (BP and MC -43) after intratumour injections of BCG.

This remarkable demonstration of tumour growth facilitation risk needs more investigation, as it may also exist in some clinical trials.

4.2. Clinical

It was as early as 1911 that the direct inoculation of human sarcomata with extracts of streptococci and serratia were attempted and resulted in some regressions (COLEY, 1911).

4.2.1. BCG

MORTON *et al.* (1970a, 1970b, 1971, 1971, 1972) administered, on a single occasion, 0.05-0.2 ml of Glaxo BCG containing 4.9 10^6 viable units per ml into human melanomas: they observed 5 regressions at the injected sites, 2 of which were accompanied by regressions elsewhere. In one case, the remission lasted 3 years.

Many other clinicians have conducted similar trials with the same BCG (LEVY *et al.*, 1972a; SEIGLER *et al.*, 1972; PINSKY *et al.*, 1973; SPARKS *et al.*, 1973), or with the Tice preparations (NATHANSON, 1972; BORNSTEIN *et al.*, 1973; PINSKY *et al.*, 1973; SMITH *et al.*, 1973), or with Connaugh material (KLEIN and HOLTERMANN, 1972), lyophipized, while the author has treated, different tumours by similar local immunotherapy, using the lyophilized Pasteur Institute BCG (not published).

According to BAST *et al* .(1974a; 1974b), who gave a summary of the literature, the average rate of regressions of injected nodules (58%) varies from 5/15 (BORNSTEIN *et al.*, 1973), to 20/21 (SPARKS *et al.*, 1973), and that of regression of the noninjected ones (14%) in locally treated patients from 0/22 (SEIGLER *et al.*, 1972) to 5/21 (SPARKS *et al.*, 1973). The incidence of favourable results depends, at least in the experience of the author, in the respect of a few rules which will be discussed in chapter 9.

4.2.2. Other Agents

STEWARD and TOLNAI (1969) used a nontoxic extract of haemolytic streptococcus, streptokinase-streptodornase, and induced remarkable regressions of skin neoplasias, including myxosis fungoides or plasmocytoma.

HUNTER-CRAIG *et al.* (1970) obtained a complete regression of the treated lesions of malignant melanoma after injecting vaccinia virus.

KLEIN *et al.* (1972, 1973, 1973) and TISMAN *et al.* (1975) have treated different skin neoplasias by injection or topical applications of purified protein derivative (PPD) and have observed regressions. This effect of PPD was increased by oral or i.d. administration of BCG. The mechanism of action in local immunotherapy will be discussed but it is already evident that part of it is the immune reaction against PPD itself.

Hence, the idea of the replacement of BCG by a simple antigen, a hapten, by KLEIN (1968) and his coworkers (WILLIAMS and KLEIN, 1970). The patient is sensitized with DNCB (dinitrochlorobenzene) or TEIB (2,3,5-triethylene imino-benzoquinone) to induce delayed hypersensitivity. Afterwards, the tumour is treated with topical applications of the sensitizing agent. WILLIAMS and KLEIN (1970) have published accounts of the permanent cure of about 90% of squamous carcinomas and basal cell skin cancers. Similar results were obtained by the author (Fig. 4.1). A few local regressions of very superficial melanoma nodules with DNCB were obtained.

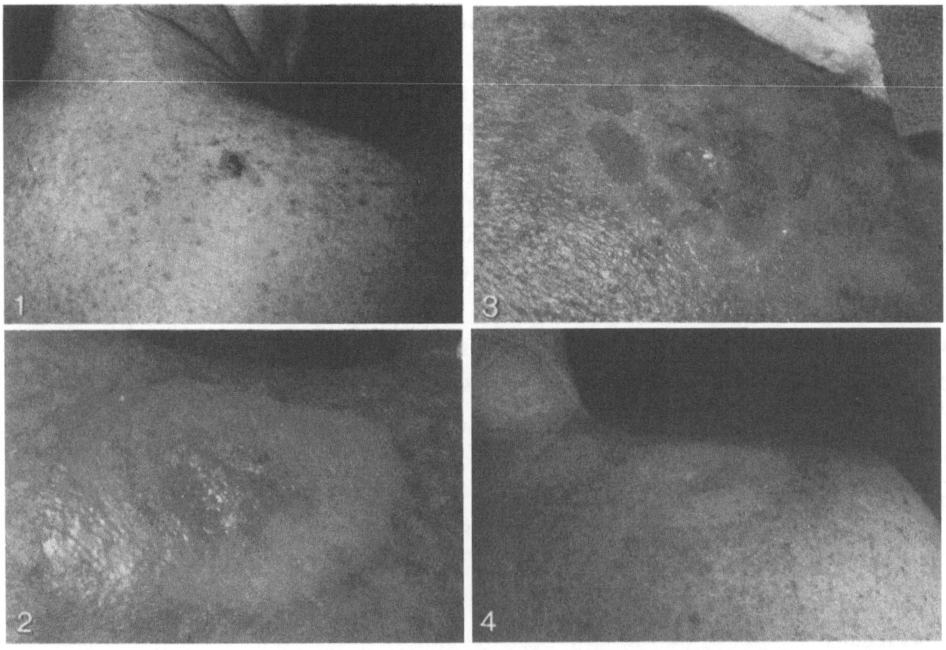

Fig. 4.1. Effect of local active immunotherapy with DNCB on a skin squamous cell carcinoma. 1. Before treatment; 2. one week after treatment; 3. 10 days after treatment; 4. 18 days after treatment (pers. exp.)

Mycosis fungoides have been similarly treated by RATNER *et al.* (1968), and by STIERNSJWARD and LEVIN (1971), who also treated skin reticulosarcomas or lymphosarcomas, melanomas and skin metastases of breast carcinoma. The author's personal results on the latter are poor.

4.2.3. Failures, and Factors and Conditions of Efficacy

No variation has been mentioned with the different preparations of BCG.

Complications, which are not rare, will be mentioned in chapter 8 on Toxicology.

Analysis of the above results suggests the importance of the following requirements for efficacy of local intratumoural immunotherapy.

(a) The *site of the tumour* is usually the skin or the subcutaneous tissue, and only the patients who had had their neoplasia localized in this tissue or satellite lymph node, became long-term survivors. Moreover, the most optimistic papers mention (almost) constant failures on visceral metastases.

Nevertheless, there are some interesting exceptions, such as those of hepatomas: with intratumoural injection of 37 mg living Pasteur BCG, Ton That TUNG *et al.* (1975) were able to prolong life significantly compared to controls. MERRIN *et al.* (1975) injected BCG into the prostate gland in cases of carcinoma of this organ, and they are quoted to have obtained some objective regressions.

(b) Usually, only *small tumours* have regressed.

(c) The last, but most important, condition for local immunotherapy by BCG to be successful, is that *the patient should be tuberculin positive:* when he is, or when he becomes so, there are 90% regressions, while the tumour invariably progresses in subjects who remain tuberculin negative (MORTON *et al.*, 1970a, 1970b; NATHANSON, 1972; SMITH *et al.*, 1973).

Hence the necessity of preceding local injections of BCG with a tuberculin skin test and, if this is negative, of submitting the patient to systemic BCG treatment in attempting to restore immunity (SIMMLER *et al.*, 1976).

The same condition is required for DNCB: the patient must be sensitized to this compound before its topical application on the tumour.

(d) A condition in which local immunotherapy most often failed in our experience is the treatment of tumour recurrences in the *irradiated* primary *site*, especially in the case of breast carcinoma.

4.2.4. Possibility of Neoplasia Growth Enhancement

Not only do metastatic visceral lesions generally not respond favourably to local immunotherapy applied to a perceptible tumour, but some authors have observed metastasis which seemed to start growing faster after local injection of an adjuvant (LEVY *et al.*, 1972; BORNSTEIN *et al.*, 1973; LEVY, 1973). Because the growth curve of a tumour is difficult to follow, such data are not easy to prove, but if confirmed, they could be explained by the general immune enhancement phenomenon.

References

BAST, R.C. jr., ZBAR, B., BORSOS, T., RAPP, H.J.: New England J. Med. 290, 1413 (1974a).

BAST, R.C. jr., ZBAR, B., BORSOS, T., RAPP, H.J.: New England J. Med. 290, 1458 (1974b).

BORNSTEIN, R.S., MASTRANGELO, M.J., SULIT, H., CHEE, D., YARBO, J.W., PREHN. L., PREHN, R.T.: Nat. Cancer Inst. Monogr. 39, 213 (1973).

COLEY, W.B.: Surg. Gynecol. Obstet. 13, 174 (1911).

GREAGER, J.A., CHAUDHURI, P.K., BOVAR, M., DAS GUPTA, T.K.: Proc. Amer. Assoc. Cancer Res. 16, 143 (abstract no. 571) (1975).

HUNTER-CRAIG, I., NEWTON, K.A., WESTBURY, G., LACEY, B.W.: Brit. Med. J. 2, 512 (1970).

KLEIN, E.: New-York State J. Med. 68, 900 (1968).

KLEIN, E.: Nat. Cancer Inst. Monogr. 39, 134 (1973).

KLEIN, E., HOLTERMANN, O.A.: Nat. Cancer Inst. Monogr. 35, 379 (1972).

KLEIN, E., HOLTERMANN, O.A., PAPERMASTER, B., MILGROM, H., ROSNER, D., KLEIN, L., WALKER, M.J., ZBAR, B.: Nat. Cancer Inst. Monogr. 39, 229 (1973).

LEVY, N.L.: Nat. Cancer Inst. Monogr. 37, 85 (1973).

LEVY, N.L., AMOS, D.B., SOLOVIEFF, G.V., DOS REIS, A.P.: Nat. Cancer Inst. Monogr. 35, 5 (1972a).

LEVY, N.L., MAHALEY, M.S. jr., DAY, E.D.: Int. J. Cáncer 10, 244 (1972b).

LIKHITE, V.V.: Cancer Res. 34, 2790 (1974b).

LIKHITE, V.V.: Int. J. Cancer 14, 684 (1974a).

LITTMAN, B., ZBAR, B., RAPP, H.J.: Proc. Amer. Assoc. Cancer Res. 13, 96 (abstract no. 381) (1972).

MARTIN, M., BOURUT, C., HALLE-PANNENKO, O., KHALIL, A., RAPPAPORT, H., MATHÉ, G.: Biomedicine 23, 337 (1975).

MARTIN, M., KHALIL, A., HALLE-PANNENKO, O., BOURUT, C., MATHÉ, G.: Biomedicine, in press (1976).

MATHÉ, G.: Rev. Fr. Et. Clin. Biol. 13, 881 (1968).

MATHÉ, G., POUILLART, P., LAPEYRAQUE, F.: Brit. J. Cancer 23, 814 (1969).

MERRIN, C., HAN, T., KLEIN, E., WAJSMAN, Z., MURPHY, G.P.: Cancer Chemoth. Rep. 59, 157 (1975).
MILAS, L., HUNTER, N., BASIC, I., MASON, K., GRDINA, D.J., WITHERS, H.R.: J. Nat. Cancer Inst. 54, 895 (1975b).
MILAS, L., WITHERS, H.R., HUNTER, N.: Proc. Amer. Assoc. Cancer Res., 16, 154 (abstract No. 614) (1975a).
MORTON, D.L.: J. Reticuloendothel. Soc. 10, 137 (1971).
MORTON, D.L.: Nat. Cancer Inst. Monogr. 35, 375 (1972).
MORTON, D.L., EILBER, F.R., JOSEPH, W.L., WOOD, W.C., TRAHAN, E., KETCHAM, A.S.: Ann. Surg. 172, 740 (1970a).
MORTON, D.L., EILBER, F.R., MALMGREN, R.A., WOOD, W.C.: Surgery 68, 158 (1970b).
MORTON, D.L., HOLMES, E.C., EILBER, F., WOOD, W.: Ann. Intern. Med. 74, 587 (1971).
NATHANSON, L.: Cancer Chemoth. Rep. 56, 659 (1972).
PIMM, M.V., BALDWIN, R.W.: Int. J. Cancer 15, 260 (1975).
PIMM, M.V., HOPPER, D.G.: Lancet 1, 806 (1975).
PINSKY, C., HIRSHAUT, Y., OETTGEN, H.: Nat. Cancer Inst. Monogr. 39, 225 (1973).
RAPP, H.J.: Israel J. Med. Sci. 9, 366 (1973).
RATNER, A.C., WALDORF, D., VAN SCOTT, E.: Cancer 21, 83 (1968).
SALOMON, J.C., LYNCH, N.: Cancer Immunol. Immunoth. in press (1976).
SCOTT, M.T.: J. Nat. Cancer Inst. 53, 855 (1974a).
SCOTT, M.T.: J. Nat. Cancer Inst. 53, 861 (1974b).
SEIGLER, H.F., SHINGLETON, W.W., METZGAR, R.S., BUCKLEY, C.E., BORGOC, P.M., MILLER, D.S., FETTER, B.F., PHAUP, M.B.: Surgery 72, 162 (1972).
SIMMLER, M.C., SCHWARZENBERG, L., MATHÉ, G.: Cancer Immunol. Immunoth. in press (1976).
SIMMONS, R.L., RIOS, A.: Surgery 71, 556 (1972).
SMITH, G.V., MORSE, P.A. jr., DERAPS, G.D., RAJU, S., HARDY, J.D.: Surgery 74, 59 (1973).
SPARKS, F.C., BREEDING, J.H.: Cancer Res. 34, 3262 (1974).
SPARKS, F.C., SILVERSTEIN, M.J., HUNT, J.S., HASKELL, C.M., PILCH, Y.H., MORTON, D.L.: New England J. Med. 389, 827 (1973).
STEWART, T., TOLNAI, G.: Cancer 24, 117 (1969).
STJERNSWARD, J., LEVIN, A.: Cancer 28, 628 (1971).
TISMAN, G., WU, S.-J.G., SAFIRE, G.E.: Proc. Amer. Assoc. Cancer Res. 16, 204 (abstract no. 813) (1975).
TOKUNAGA, T., YAMAMOTO, S., NAKAMURA, R.M., KATAOKA, T.: J. Nat. Cancer Inst. 53, 459 (1974).
TON THAT TUNG, NGUYEN NHU BANG, NGUYEN VAN VAN, TON THAT BACH, NGUYEN DANG TAM, SALOMON, J.C.: Lancet 1, 527 (1975).
WEISS, D.W.: Nat. Cancer Inst. Monogr. 35, 157 (1972).
WILLIAMS, A.C., KLEIN, E.: Cancer 25, 450 (1970).
ZBAR, B.: Nat. Cancer Inst. Monogr. 35, 341 (1972).

ZBAR, B., BERNSTEIN, I.D., BARTLETT, G.L., HANNA, M.G. jr., RAPP, H.J.: J. Nat. Cancer Inst. 49, 119 (1972).
ZBAR, B., BERNSTEIN, I.D., RAPP, H.J.: J. Nat. Cancer Inst. 46, 831 (1971).
ZBAR, B., RAPP, H.J.: Cancer 34, 1532 (1974).
ZBAR, B., RIBI, E., MEYER, T., AZUMA, I., RAPP, H.J.: J. Nat. Inst. 52, 1571 (1974).
ZBAR, B., RIBI, E., RAPP, H.J.: Nat. Cancer Inst. Monogr. 39, 3 (1973).
ZBAR, G., TANAKA, T.: Science 172, 271 (1971).

Chapter 5
Towards Immunoprophylaxis

It goes without saying that immunoprophylaxis of cancer, when it is no longer a risk, will be much more important than immunotherapy.

5.1. Experimental

There is a much more extensive and older literature on experimental immunoprophylaxis than on active immunotherapy.

5.1.1. Grafted Tumours

5.1.1.1. Use of Neoplastic Cells

There have been several attempts at using irradiated tumour cells administered before the graft of the same or isogeneic living tumour (GLYNN *et al.*, 1963; MATHÉ *et al.*, 1969; PARR, 1972) with a preventive effect on the neoplasia growth (Fig. 5.1).

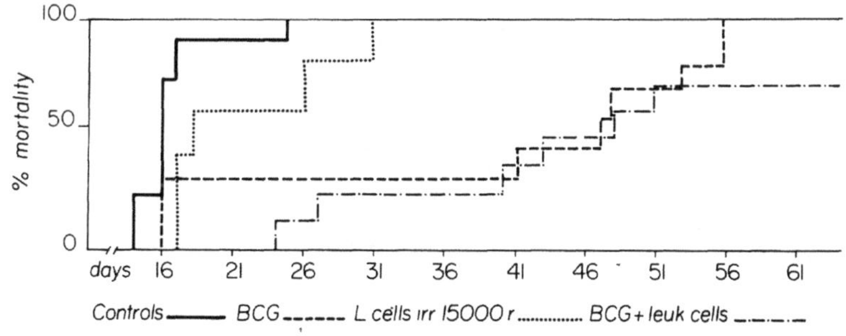

Fig. 5.1. Cumulative mortality of mice grafted with L1210 leukaemia, untreated or treated by BCG (1 mg intravenously) (first injection 14 days before graft and injections repeated every four days) or irradiated leukaemic cells (one injection 14 days before graft) or combination of both (MATHÉ et al., 1969)

KRONMAN *et al.* (1970) obtained remarkable results in pure strains of guinea pigs into which they had grafted a hepatoma i.m.. Intradermal injections of living hepatoma cells (3 injections per week on alternate weeks) induced an immunological reaction against the hepatoma.

The experiment with DORÉ *et al.* (1970) in which the author obtained a significant immunoprophylaxis against an isogeneic leukaemia grafted onto AkR mice was described earlier (Fig. 5.2), and the value of this result and of the antibodies found in the sera (T) against the concept of immunological tolerance of animals of this strain, as far as Gross antigens are concerned, was discussed in Chapter 2.

PRESCIA and BRAUN (cited by BRAUN, 1970) have tried to enhance immunogenicity of tumour cells (by complexing them with foreign basic protein, either methylated bovine serum or methylated gammaglobulin), and they observed that protective immunity elicited by small complexes in immunoprophylaxis experiments was stronger than that obtained in immunotherapy trials.

5.1.1.2. Use of Systemic Immunity Adjuvants

Since DUBOS and SCHAEDLER (1957) showed that *BCG* was able to prevent experimental infections, there have been many works devoted to the role of the "reticuloendothelial system in the host reaction to neoplasia" (OLD *et al.*, 1959, 1961). They were studying the role of these reactions in prophylaxis, as they administered the agents before the grafting of the tumours, and the first experiments were carried out on noncompatible neoplasias such as the Ehrlich ascitis tumour (BIOZZI *et al.*, 1959) and the T8 Guerin tumour (HALPERN *et al.*,1959). The preventive effect described

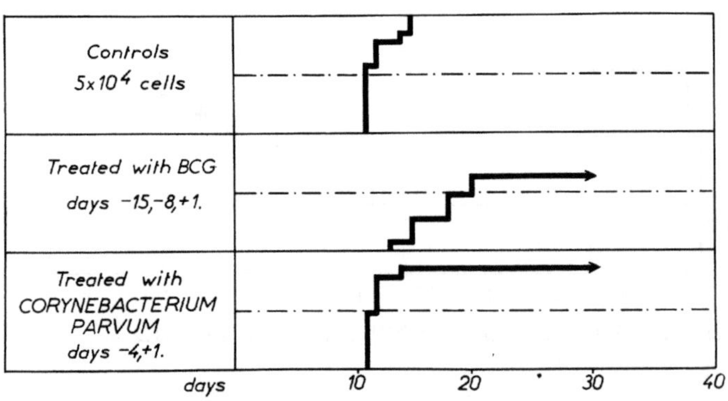

Fig. 5.2. Cumulative mortality of 2 months old normal AkR mice treated with BCG or Corynebacterium parvum, then given 10^3 *K36 cells i.p.. Significant prophylaxis by BCG and tendancy to growth enhancement by C. parvum (DORÉ et al., 1970)*

was also found on isogeneic neoplasias [BIOZZI *et al*, 1959; OLD *et al.*, 1961; AMIEL, 1967; MATHÉ *et al.*, 1969 (Fig. 5.1); DORÉ *et al*, 1970 (Fig. 5.2)].

The two last experiments were conducted to answer special questions. The first (MATHÉ *et al.*, 1969), on L1210 leukaemia, showed that, contrary to what was observed in immunotherapy experiments, BCG (1 mg i.v.) is much more effective than tumour cells (Fig. 5.1).

The second (DORÉ *et al.*, 1970) has shown that a significant immunoprophylaxis of a grafted isogeneic leukaemia can be obtained in AkR mice (Fig. 5.2), which, in the light of the information about specific immunization with leukaemic isogeneic cells quoted above, is a strong argument against the existence of immune tolerance for Gross leukaemia in the animals of this strain.

These favourable results were not found on the same leukaemia, with a different experimental model, by FINKLESTEIN *et al.* (1972) who showed that an i.p. injection of 1 mg (Chicago Research Foundation) BCG into AkR mice at 2 weeks of age did not decrease or delay the percentage of leukaemia and mortality in a group of animals inoculated at 9 weeks of age with Gross virus induced leukaemia isogeneic cells.

BCG (1 mg i.v. on day 14 before living neoplastic cell inoculation) can be strongly efficient on solid tumours as illustrated by its immunopreventive action on grafted Lewis tumour (MATHÉ *et al.*, 1973, 1975) (Fig. 5.3). (We will see later that the subcutaneous injection is not beneficial but, furthermore, that it has the contrary effect).

Different *extracts from Mycobacteria* have been shown to be active in immunoprophylaxis of different neoplasias: methanol extract residue (MER) administered before challenge with isogeneic grafts of different neoplasias may protect the recipients in certain circumstances (WEISS, 1972).

Extracts submitted to pronase digestion and delipidation obtained from BCG, Mycobacterium smegmatis and M. kansasii were shown to be active in the immunoprophylaxis of Ehrlich ascitic carcinoma (CHEDID *et al.*, 1973).

Corynebacterium parvum has been shown by HALPERN *et al.*, (1966) to prevent the growth of grafted tumours, and by WOODRUFF and BOAK (1966) to be a good prophylactic agent for the growth of an isogeneic tumour transplant.

In considering the results of the author's screening, we see that C. parvum and several preparations of C. Granulosum are effective in the prophylaxis of L1210 leukaemia and of the solid Lewis tumour (MATHÉ *et al.*, 1973, 1975).

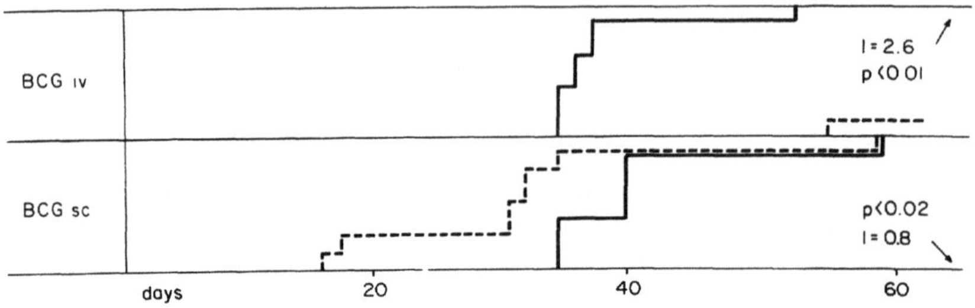

Fig. 5.3. Immunoprophylactic effect of i.v. BCG on Lewis lung tumour (LLT) and growth accelerating effect of subcutaneous injection of same dose (BCG i.v. or subcutaneous, day–14, 1 mg/mice) (LLT: 2.10⁶ cells SC, day 0)
——— controls; - - - - - - - treated animals;

$$I = \frac{\text{Median of survival of experimental animals}}{\text{Median of the controls}}$$

Statistics : nonparametric test of Wilcoxon

It was not effective against K36 leukaemia grafted into an isogeneic AkR host when used by the author (Fig. 5.2) (DORÉ *et al.*, 1970), while it was efficacious in the hands of LAMENSANS *et al.* (1968).

Bordetella pertussis has been active in grafted tumour immunoprophylaxis (BARTLETT and ZBAR, 1973; WISSLER *et al.*, 1968).

A lipopolysaccharide from *Proteus vulgaris* prepared according to TOUPLICHEVA *et al.* (1965), was shown to be able to prevent the growth of certain leukaemias, ascitic tumours, (MIZUNO *et al.*, 1968) and rhabdomyosarcoma TCH1 (DONNER and VAILLIER, 1974).

CHIHARA *et al.* (1969, 1970) claimed that *lentinan,* a polysaccharide extracted from Lentinus edodes, administered prior to grafting sarcoma 180, is able to delay mortality. The author has not observed such a result on L1210 leukaemia or on solid Lewis tumour prophylaxis (MATHÉ *et al.*, 1975).

Certain complexes of multistranded synthetic *polynucleotides*, which are known as interféron inducers (FIELD *et al.*, 1967), have been shown to exert an immunity systemic adjuvant action, and an immune antitumour effect, even on grafted neoplasias the origin of which was not viral. This action was demonstrated for poly I-poly C (BRAUN *et al.*, 1971; MATHÉ *et al.*, 1971) (Fig. 5.4) and for poly A-poly U; though the first prophylactic attempts concerned viral-induced neoplasias (YOUN *et al.*, 1968; LEVY *et al.*, 1969; SARMA *et al.*, 1969), it was shown (WEINSTEIN *et al.*, 1971) that there was no correlation between interferon induction of poly A-poly U and its antitumour effect.

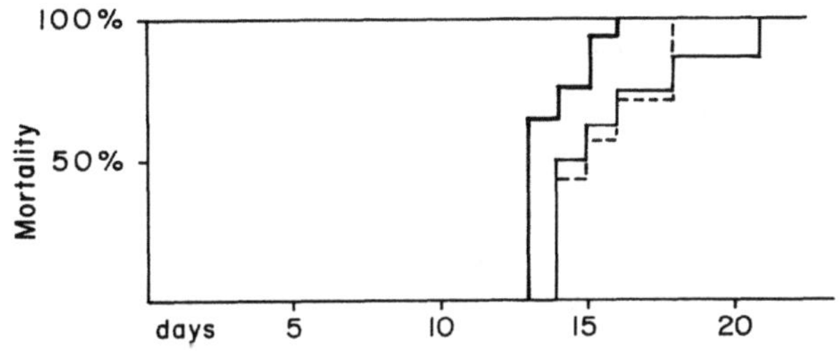

Fig. 5.4. Immunoprophylactic effect of poly I-poly C on L1210 leukaemia.
——— controls: 2.10⁴ L1210 cells subcutaneous (day 0) (m = 13d)
——— poly I-poly C, 10 mg/kg injected 3 days before leukaemia transplantation (m = 14.5 d)
-------- poly I-poly C, 1 mg/kg injected 3 days before leukaemia transplantation (MATHÉ et al., 1971)

5.1.1.3. Combination of an Adjuvant and of Tumour Cells

Comparing the preventive effect, on L1210 leukaemia, of the combination of BCG (1 mg i.v.) and irradiated neoplastic cells (10^7, s.c.) to the application of these agents applied alone, the author observed that the combination is much more active than BCG which is, itself, more active than the neoplastic cells (MATHÉ, 1968; MATHÉ *et al.*, 1969) (Fig. 5.1).

LIKHITE and HALPERN (1973) observed a significant delayed rejection of tumours by the previous administration of tumour cells mixed with killed C. parvum.

The following result of TANAKA and SASAKI (1974) must be repeated here: they administered BCG, then inoculated living tumour cells; they then gave a mixture of tumour cells and BCG: in this experiment, which can be defined as being between immunotherapy and immunoprophylaxis, they obtained significant tumour inhibition and tumour immunity.

5.1.2. Carcinogen Induced and Spontaneous Tumours

5.1.2.1. Virus-Induced Tumours

5.1.2.1.1. Use of Tumour Cells or Viruses

Many years ago, Charlotte FRIEND (1959) showed that the administration of a preparation of her virus treated with formalin, prevented the disease induced by this living agent in mice. A decade later, CHURCHILL *et al.* (1969), using an attenuated living vaccine, were able to prevent Marek's disease in chickens; this vaccine is now used throughout the world. Recently, JARRETT *et al.* (1974) prepared an efficient vaccine against feline leukaemia, and LAUFS and STEINKE (1975) an active one with Herpes-virus saimiri against lymphosarcoma in cotton-topped marmosets.

5.1.2.1.2. Use of Systemic Immunity Adjuvants

ALLISON (1964) observed that administration of *Freund's complete adjuvant* to hamsters during the latent period following neonatal inoculation with adenovirus type 12 affected tumour induction and the production of complement-fixing antibodies to adenovirus, virions and T-antigens.

LACOUR *et al.* (1975) injected newborn females with *poly A-poly U* (15 µg at days 1, 3, and 5 after birth), using C_3H/Hc mice, and observed a significant reduction of spontaneous mammary (viral) carcinoma.

The results were much more disappointing in attempts at immunoprophylaxis of spontaneous AkR leukaemia.

UNGARO *et al.* (1973) gave i.p. injections of *BCG* to AkR mice, from 8 to 42 weeks of age, every 2 weeks: this treatment increased the survival time, but did not prevent the eventual onset of leukaemia.

Treating AkR mice with 1 mg BCG injected i.v. 1 day after birth, we did not observe any difference between the mortality (from spontaneous leukaemia) curve of these animals and that of the controls (MATHÉ *et al.*, 1974).

However, LEMONDE (1973), injecting i.v. 1 mg of BCG into AkR mice when they were 10 weeks old, observed a significant decrease of spontaneous leukaemia incidence (45% instead of 91% in the controls).

5.1.2.1.3. Use of a Combination of an Adjuvant and of Tumour Cells or Viruses

EMERSON and EMERSON (1965) administered a preventive mixture of tumour homogenates and Freund's adjuvant to mice of a strain presenting a high incidence of spontaneous mammary tumours. They obtained a significant reduction of these neoplasias.

BERMAN *et al.* (1967) treated hamsters with virus and Freund's adjuvant during the latent period following their neonatal inoculation with adenovirus 12 and generally obtained a reduction in the frequency of the tumours.

5.1.2.2. Chemically Induced Tumours

5.1.2.2.1. Use of Cells

Although specific immunoprophylaxis has, theoretically, a very poor chance of being effective because the tumour-associated antigens are singular for each neoplasia, some researchers, interested in the exceptional cross reactions (REINER and SOUTHAM, 1969; OTH, 1975, 1976), have immunized animals with mixtures of chemical carcinogen-induced tumours: most results were negative (KLEIN *et al.*, 1960; PREHN, 1961); nevertheless, USUBUCHI (1956) published a successful attempt with an immunogenetic system in which cross reactions are frequent. MAISIN (1967) also obtained some prophylactic effect against a 20-methylcholanthrene-induced tumour with a cellular fraction of the mixture of different chemical carcinogen-induced tumours.

Because GOLD and FREEDMAN (1965), von KLEIST and BURTIN (1969), and many others (MARTIN *et al.*, 1975) described antigens common to embryonic tissues and several tumours, especially those of the digestive tract, and as PARMIANI and LEMBO (cited by OTH, 1975, 1976) observed that multiparous female mice treated with methylcholanthrene, develop less sarcomas than the controls, OTH (1975, 1976) tried to develop immunoprophylaxis of such tumours with anti-embryonic immunization;his results were negative.

5.1.2.2.2. Use of Systemic Immunity Adjuvants

BEKIERKUNST *et al.* (1971) prevented urethane-induced lung adenomas in mice by pretreatment with living *BCG* or *cord factor*. LAVRIN *et al.* (1973) observed some immunoprophylaxis of methylcholanthrene-induced tumours in mice submitted to BCG or MER.

LE GARREC *et al.* (1974) and OTH *et al.* (1973) obtained similar results with *Brucella abortus*.

PORT *et al.* (1975) were able to use a less toxic synthetic *vitamin A* analog, 13-cis-retinoic acid, to prevent lung cancer induced in hamsters by benzpyrene. The incidence of squamous metaplasia was also lowered.

5.1.2.3. Does Immunoprophylaxis Work before Carcinogenesis?

In the case of a virus-induced tumour, it may work by inhibiting the virus before it has transformed a cell.

It may be that Charlotte FRIEND (1959) accomplished this when she prevented her viral leukaemia by inoculation of formalin treated virus before administration of the active one.

BARSKI (1970) was able to obtain, in a cell culture chronically infected with the Rauscher virus, the $JLSV_5$ line, a progressive lowering of the leukaemogenic potential of this agent, which was, however, produced in the same quantity. The supernatant and cells from these cultures, injected into mice, immunized them strongly against the virulent virus. Similar observations were made by SINKOVICS *et al.* (1966), WRIGHT and LASFARGUES (1966), and, on Friend's virus, by YOSHIKURA *et al.* (1967) and DE TKACZEVSKI *et al.*, (1968).

BARSKI and YOUN (1971) inoculated newborn C3He B/Fe mice with virus material they had collected from two long-term in vitro clonal cell lines chronically infected with abundantly replicating C-type virus particles, which showed, by immunofluorescence, Gross virus type specific cell surface antigen. A few hours later, they inoculated the animals with virulent Gross virus: 11% became leukaemic, compared with 60% in the nonvaccinated ones. The possible effect of interferon has been eliminated.

It has been shown that the administration of homologous virus, during the latent period, to hamsters inoculated neonatally with SV40 or adenovirus, reduced the incidence of tumours as compared with controls receiving only a single inoculation of virus (ALLISON, 1964; DEICHMAN and KLUCHAVERA, 1964; EDDY *et al.*, 1964). Repeated doses of adenovirus 7 in the latent period had an inhibitory effect on adenovirus 12 tumour induction (EDDY, 1965; PERIES *et al.*, 1966).

KOUZNETZOVA *et al.* (1974) pretreated mice with C. granulosum 4 days before inoculation of the 200 SD_{50} or 20 SD_{50} of Friend's virus, and they obtained a significant prophylaxis against the leukaemia.

PILET *et al.* (1974) have obtained a significant protection with 500 µg of inactivated Brucella abortus injected 8 days after 0.1 mg methylcholanthrene (MCA) in DBA/2 mice against tumours induced by this carcinogen. TOKUNAGA *et al.* (1974) obtained an immunoprophylactic effect on a 3-methylcholanthrene-(MCA) induced tumour, injecting BCG in SWM/Ms mice 2 weeks before MCA administration.

WILSON *et al.* (1966) incubated normal tissues with 3-methylcholanthrene and administered them in order to prevent MCA carcinogenesis: they obtained significant protection. The specificity of this effect has to be demonstrated, as normal tissues may work as adjuvants.

PECK and PECK (1971), immunizing rats with a 2-anthrylamine protein conjugate, obtained a reduction in the treated animals, compared to controls, of the incidence of neoplasias induced by this carcinogen.

The case of bladder chemical carcinogen-induced tumours may be special, because HELLSTRÖM and HELLSTRÖM (1974a,b) have found an immunity to antigens common to methylcholanthrene induced bladder tumours which is effective in vivo. TARANGER *et al.*, (1972) have induced a significant immunoprophylaxis by tumours of the same tissue.

HUEBNER and TODARO (1969) developed a hypothesis according to which all tumours could be due to oncornaviruses, not only those which are induced by irradiation (LIEBERMAN and KAPLAN, 1959; LATARJET and DUPLAN, 1962), but also those induced by chemicals, which is not in agreement with the individual antigenic specificity of the latter. WHITHMIRE (1973) and WHITHMIRE and HUEBNER (1972) were able to reduce the incidence and delay the time of appearance of methylcholanthrene-induced sarcomas with pretreatment by radiation-leukaemia virus or other type C oncorna viruses. The specificity of the effect, if it is an immunoprophylactic one, has to be demonstrated.

5.1.2.4. Failures

Experiments with negative results have been mentioned, especially the failure to decrease the incidence of spontaneous AkR leukaemia (UNGARO *et al.*, 1973; MATHÉ *et al.*, 1974) and to decrease that of chemically induced tumours with embryonic antigens (OTH, 1975, 1976).

It is more important to underline the possibility of growth enhancement.

5.1.2.5. Possibility of Neoplasia Growth Enhancement

The risk of growth facilitation does, in fact, exist for immunoprophylaxis as for immunotherapy.

5.1.2.5.1. Grafted Tumours

An experiment in which 1 mg BCG was injected i.v. or subcutaneously 14 days before solid Lewis tumour was grafted has already been mentioned. While a remarkable beneficial effect was obtained with the i.v. route, acceleration of the tumour growth and of mortality of the animals was observed with the subcutaneous injection of mycobacteria, as seen in Fig. 5.3. (MATHÉ *et al.*, 1973, 1975).

A similar enhancement effect has been observed by other groups on other tumours, specially with BCG (OTH and BURG, 1966; PIESSENS *et al.*, 1970, 1971; DONNER *et al.*, 1972, 1974; WEPSIC *et al.*, 1975) and with C. parvum (BERD *et al.*, 1975).

Such an effect is dependent on the schedule of the treatment and, in some instances, on the sex of the animals (OTH and BURG, 1966; LEMONDE, 1974).

The enhancement of minimum deviation MORRIS (1960) hepatoma growth has been observed by WISSLER *et al.* (1968) after treatment with a mixture of hepatoma cells and pertussis vaccine before grafting.

5.1.2.5.2. Induced Tumours

Enhancement has also been reported for induced tumours: GOLDNER *et al.* (1965) wrote that, in hamsters inoculated neonatally with SV40, the administration of an adjuvant during the latent period appeared to favour tumour development. In hamsters inoculated with the Bryan strain of Rous sarcoma virus, the injection of Freund's complete adjuvant increased the incidence of tumours (ALLISON and BERMAN, cited by BERMAN *et al.*, 1967).

PILET *et al.* (1974) observed enhancement of MCA-induced tumours in CBA and C57Bl mice treated with Brucella abortus (500 µg) 8 days after the injection (0.1 mg) of this carcinogen.

Enhancement has been obtained after MER administration "under some circumstances" to autochthonous hosts exposed to carcinogenic stimuli (WEISS, 1972).

If the attempt at immunoprophylaxis of leukaemias should be favourable or nil, there is a third possibility for solid tumours: the enhancement of tumour growth.

5.2. Clinical

5.2.1. Prophylaxis of Blastic Crisis of Chronic Myeloid Leukaemia

SOKAL *et al.* (1969, 1974) have treated patients with chronic myeloid leukaemia (CML) with BCG and cells from a permanent line of an acute myeloid leukaemia (AML) culture: the authors write that they have delayed the blastic crisis; the survival rates for 39 patients are as follows: 1 year after immunomanipulations: 96%; two years: 91%; three years: 77%. The median survival is 7 years, which is considerably more than those of other statistics (SCHWARZENBERG *et al.*, 1973). As all patients with CML die from blastic crisis, and as the myelocytic proliferation was not suppressed by this treatment, it is reasonable to consider that the latter may have prevented the blastic crisis.

5.2.2. Prophylaxis of Childhood Leukaemia by BCG Vaccination

In 1970 and 1971, DAVIGNON *et al.* (1970, 1971a, 1971b) observed that the incidence of mortality due to leukaemia was half as common among children vaccinated with BCG against tuberculosis (between 21 to 31 per 100,000) as in nonvaccinated children (between 44 to 191 per 100,000).

HEMS and STUART (1971), tried to ascertain whether the introduction, during the early 1950's, of BCG vaccination for 13-year-old schoolchildren in the United Kingdom was followed by a decline in leukaemia mortality. Assuming that, in the absence of this vaccination, the leukaemia mortality rate would have continued to increase at the same rate as during the first half of the century, thus that a rate of about 40×10^{-6} could be predicted for males aged 15-19 years in 1967, they observed a rate of 24×10^{-6} in BCG vaccinated subjects, which suggests the same degree of protection against leukaemia as in Davignon's study.

In 1972, ROSENTHAL *et al.* (1972, 1974) revealed that leukaemia incidence in Chicago was considerably less in coloured children, who had been BCG-vaccinated, than those who had not been vaccinated. In comparing the mortality rates from this disease (all forms) in children from birth to 6 years of age, vaccinated at 2-3 days of age with BCG (2×10^8 organisms in 1 ml vials, administered by the multipuncture method), he found 21 deaths due to leukaemia between 1964 and 1969 in 172,986 nonvaccinated coloured children (2.02 per 100,000/y.) versus one in 54,414 subjects of the vaccinated group. The difference is significant (p = 0,04) (Table 5.I).

Table 5.I. BCG vaccination and leukaemia mortality rates. Chicago 1964 through 1969. (ROSENTHAL *et al.*, 1974)

Age (years)	Nonvaccinated			Vaccinated		
	Population[a]	Deaths	Rate 100,000/ year	Population	Deaths	Rate 100,000/ year
Under 1	21,901	0	-	7,820	0	-
1 - 3	68,824	8	1.94	26,482	0	-
4 - 6	82,261	13	2.64	20,112	1	0.83
Total	172,986	21	2.02	54,414	1	0.31

[a] All negro population.

These data are in favour of the antileukaemic effect of BCG used for vaccination against tuberculosis, but it does not compare populations taken at random, and other data exist which show no such difference between vaccinated and nonvaccinated subjects. One study from COMSTOCK *et al.* (1975) concerned a total of 191,827 children in Puerto Rico, who were enrolled in a controlled trial of BCG vaccination. The annual rates of cancer were 7.2 per 100,000 population in the vaccinated group and 10.3 in the nonvaccinated one. The first group had slightly fewer leukaemia cases and slightly more lymphosarcomas and cases of Hodgkin's disease.

Another study from KINLEN and PIKE (1971) did not show any significant reduction of leukaemia incidence in Canadian areas where BCG vaccination is used frequently.

The author has conducted a study of the frequency of specific systematic vaccination with BCG against tuberculosis in acute leukaemia patients of less than 20 years of age. No significant difference has been found when comparing a group of controls chosen for their similarity to the patients, regarding the place and date of birth (MATHÉ *et al.*, 1974). It is, nevertheless, possible that the patients from a specialized service may represent a socio-economic group.

The negative results do not have a more conclusive value than the positive ones. Hence, prospective trials should be organized, based on the possibility that BCG application at a given age may prevent leukaemia and, possibly, other cancers. This should lead one to reconsider its method of application, especially the age of application, and the dose.

The hypothesis according to which immunoprophylaxis may work on viral carcinogens or on cells already made antigenic but not yet transformed, may explain why BCG administered to very young children may give protection against a leukaemia which appears four to ten years later, although the doubling time is short (CLARKSON, 1969).

It may also explain the immunoprophylaxis of relapses. We know, from the work of SPIEGELMAN (1974) and MAK *et al.* (1974), that the DNA of lymphoid cells of patients with acute lymphoid leukaemia in remission maintained by chemotherapy, presents the same biochemical marker of viral presence, in the same quantity as the DNA of leukaemic cells in the perceptible phase of the disease. It would be interesting to know if this is also true of immunotherapy treated patients, for whom we know that late relapses are exceptional. It could be that immunotherapy works as immunoprophylaxis of relapses.

References

ALLISON A.C.: Life Science 3, 1415 (1964).

AMIEL, J.L.: Rev. Fr. Et. Clin. Biol. 12, 912 (1967).

BARSKI, G.: p. 323 in "Comparative Leukemia Research" (R.M. Durcher, ed.). 1 Vol., Basel: Karger 1970.

BARSKI, G., YOUN, J.K.: J. Nat. Cancer Inst. 47, 575 (1971).

BARTLETT, G.L., ZBAR, B.: J.Nat. Cancer Inst. 50, 1385 (1973).

BEKIERKUNST, A., LEVIJ, I.S., YARKONI, E., VILKAS, E., LEDERER, E.: Science 174, 1240 (1971).

BERD, D., MITCHELL, M.S., MARSH, J.C.: Proc. Amer. Assoc. Cancer Res. 16, 6 (abstract No. 23) (1975).

BERMAN, L.D., ALLISON, A.C., PEREIRA, H.G.: Int. J. Cancer 2, 539 (1967).

BIOZZI, G., STIFFEL, C., HALPERN, B.N., MOUTON, D.: C.R. Soc. Biol., Paris 153, 987 (1959).

BRAUN, W.: p. 17 in "Progress in antimicrobial and Anticancer Chemotherapy". Vol. 1, Tokyo: University of Tokyo 1970.

BRAUN, W., ISHIZUKA, M., YAJIMA, Y., WEBB, D., WINCHURCH, R.: p. 139 in "Biological Effects of Polynucleotides" (R.F. Beers, W. Braun, eds.). 1 Vol., Heidelberg: Springer-Verlag 1971.

CHEDID, L., LAMENSANS, A., PARANT, F., PARANT, M., ADAM, A., PETIT, J.F., LEDERER, E.: Cancer Res. 33, 2187 (1973).

CHIHARA, G., HAMURO, J., MAEDA, Y., ARAI, Y., FUKUOKA, F.: Cancer Res. 30, 2776 (1970).

CHIHARA, G., MAEDA, Y., HAMURO, J., SASAKA, T., FUKUOKA, F.: Nature 222, 687 (1969).

CHURCHILL, A.E., PAYNE, L.N., CHUBB, R.C.: Nature 221, 744 (1969).

CLARKSON, B.D.: Nat. Cancer Inst. Monogr. 30, 81 (1969).

COMSTOCK, G.W., MARTINEZ, I., LIVESAY, V.T.: J. Nat. Cancer Inst. 54, 835 (1975).

DAVIGNON, L., LEMONDE, P., ROBILLARD, P., FRAPPIER, A.: Lancet 2, 638 (1970).

DAVIGNON, L., LEMONDE, P., ROBILLARD, P., FRAPPIER, A.: Lancet 1, 80 (1971a).

DAVIGNON. L., LEMONDE, P., ROBILLARD, P., FRAPPIER, A.: Lancet 1, 799 (1971b).

DEICHMAN, E.I., KLUCHAVERA, T.E.: Nature 202, 1126 (1964).

DONNER, M., HOTTIER, D., BURG, C.: Europ. J. Cancer 8, 141 (1972).

DONNER, M., VAILLIER, D.: p. 308 in "Investigation and Stimulation of Immunity in Cancer Patients" (G. Mathé, R. Weiner, eds.). Vol. 47 R.R.C.R., Heidelberg: Springer-Verlag 1974.

DORE, J.F., AJURIA, E., MATHÉ, G.: Europ. J. Clin. Biol. Res. 15, 81 (1970).

DUBOS, R.J., SCHAEDLER, R.W.: J. Exp. Med. 106, 703 (1957).

EDDY, B.E.: p. 369 in "Slow, Latent and Temperate Virus Infections" (D.C. Gajdusek, C.J. Gibbs jr., M. Alpers, eds.). Vol. 1, Washington: N.I.N.D.B. 1965.

EDDY, B.E., GRUBAS, G.E., YOUNG, R.D.: Proc. Soc. Exp. Biol. Med. 117, 575 (1964).
EMERSON, J.D., EMERSON, G.M.: Fed. Proc. 24, 495 (1965).
FIELD, A.K., TYTELL, A.A., LAMPSON, G.P., HILLEMAN, M.R.: Proc. Nat. Acad. Sci. 58, 1008 (1967).
FINKLESTEIN, J.Z., TITTLE, K.L., IMAGAWA, D.T.: Lancet 2, 875 (1972).
FRIEND, C.: J.Exp.Med. 109, 217 (1959).
GLYNN, J.P., HUMPHREYS, S.R., TRIVERS, G., BIANCO, A.R., GOLDIN, A.: Cancer Res. 23, 1008 (1963).
GOLD, P., FREEDMAN, S.O.: J. Exp. Med. 122, 467 (1965).
GOLDNER, H., GIRARDI, A.J., HILLEMAN, M.R.: Virology 27, 225 (1965).
HALPERN, B.N., BIOZZI, G., STIFFEL, C., MOUTON, D.: C.R. Soc. Biol., Paris 153, 919 (1959).
HALPERN, B.N., BIOZZI, G., STIFFEL, C., MOUTON, D.: Nature 212, 853 (1966).
HELLSTROM, I., HELLSTROM, K.E.: p. 37 in "The Role of Immunological Factors in Viral and Oncogenic Processes" (R.F. Beers, R.C. Tilghman, E.G. Bassett, eds.). 1 Vol., Baltimore: Johns Hopkins University press 1974a.
HELLSTROM, K.E., HELLSTROM, I.: Adv. Immunol. 18, 209 (1974b).
HEMS, G., STUART, A.: Lancet 1, 183 (1971).
HUEBNER, R.J., TODARO, G.J.: Proc. Nat. Acad. Sci., USA 64, 1087 (1969).
JARRETT, W., MACKEY, L., JARRETT, O., LAIRD, J., HOOD, C.: Nature 248, 230 (1974).
KINLEN, L.J., PIKE, M.C.: Lancet 2, 399 (1971).
KLEIN, G., SJOGREN, H.O., KLEIN, E., HELLSTROM, K.E.: Cancer Res. 20, 1561 (1960).
KLEIST, S. von, BURTIN, P.: Int. J. Cancer 4, 874 (1969).
KOUZNETZOVA, B., BIZZINI, B., CHERMANN, J.C., DEGRAND, F., PREVOT, A.R., RAYNAUD, M.: p. 275 in "Investigation and Stimulation of Immunity in Cancer Patients" (G. Mathé, R. Weiner, eds.). Vol. 47 R.R.C.R., Heidelberg: Springer-Verlag 1974.
KRONMAN, B.S., WEPSIC, H.T., CHURCHILL, W.H. jr., ZBAR, B., BORSOS, T., RAPP, H.J.: Science 168, 257 (1970).
LACOUR, F., DELAGE, G., CHIANALE, C.: Science 187, 256 (1975).
LAMENSANS, A., STIFFEL, C., MOLLIER, M.F., LAURENT, M., MOUTON, D., BIOZZI, G.: Rev. Fr. Et. Clin. Biol. 13, 773 (1968).
LATARJET, R., DUPLAN, J.F.: Int. J. Rad. Biol. 5, 339 (1962).
LAUFS, R., STEINKE, H.: Nature 253, 71 (1975).
LAVRIN, D.H., ROSENBERG, S.A., CONNOR, R.J., TERRY, W.D.: Cancer Res. 33, 472 (1973).
LE GARREC, Y., SABOLOVIC, D., TOUJAS, L., DAZORD, L., GUELFI, J., PILET, C.: Biomedicine 21, 40 (1974).
LEMONDE, P.: Nat. Cancer Inst. Monogr. 39, 21 (1973).
LEMONDE, P.: p. 700 in XIth International Cancer Congress, Florence, October 1974. Vol. 4, Panels.
LEVY, H.B., LAW, L.W., RABSON, A.S.: Proc. Nat. Acad. Sci., USA 62, 357 (1969).
LIEBERMAN, M., KAPLAN, H.S.: Science 130, 387 (1959).
LIKHITE, V.V., HALPERN, B.N.: Int. J. Cancer 12, 699 (1973).

MAISIN, J.H.: Tumori 53, 55 (1967).

MAK, T.W., MANASTER, J., HOWATSON, A.F., McCULLOCH, E.A., TILL, J.E.: Proc. Nat. Acad. Sci., USA 71, 4336 (1974).

MARTIN, F., KNOBEL, S., MARTIN, M., BORDES, M.: Cancer Res. 35, 333 (1975).

MATHÉ, G.: Rev. Fr. Et. Clin. Biol. 13, 881 (1968).

MATHÉ. G., HAYAT, M., SAKOUHI, M., CHOAY, J.: C.R. Acad. Sci. Paris 272, 170 (1971).

MATHÉ, G., FACY, F., HATTON, F., HALLE-PANNENKO, O.: Biomedicine 21, 132 (1974).

MATHÉ, G., HALLE-PANNENKO, O., FLORENTIN, I., BRULEY-ROSSET, M., KAMEL, M., HIU, I.J., BOURUT, C.: Europ. J. Cancer 11, 801 (1975).

MATHÉ, G., KAMEL, M., DEZFULIAN, M., HALLE-PANNENKO, O., BOURUT, C.: Cancer Res. 33, 1987 (1973).

MATHÉ, G., POUILLART, P., LAPEYRAQUE, F.: Brit. J. Cancer 23, 814 (1969).

MIZUNO, D., YOSHIOKA, O., AKAMATU, M., KATAOKA, T.: Cancer Res., 28, 1531 (1868).

MORRIS, H.P.: Cancer Res. 20, 1252 (1960).

OLD, L.J., BENACERRAF, B., CLARKE, D.A., CARSWELL, E.A., STOCKERT, E.: Cancer Res. 21, 1281 (1961).

OLD, L.J., CLARKE, D.A., BENACERRAF, B.: Nature 184, 291 (1959).

OTH, D.: Biomedicine 22, 350 (1975).

OTH, D.: Biomedicine 24, 4 (1976).

OTH, D., BURG, C.: C.R. Soc. Biol. 160, 1644 (1966).

OTH, D., SABOLOVIC, D., ROBERT, F.: Beitr. Path. 149, 60 (1973).

PARR, I.: Brit. J. Cancer 26, 174 (1972).

PECK, R.M., PECK, E.B.: Cancer Res. 31, 1550 (1971).

PERIES, J., CHAUT, J.C., CANIVET, M., BOIRON, M.: Nature 209, 738 (1966).

PIESSENS, W.F., LACHAPELLE, F.L., LEGROS, N., HEUSON, J.C.: Nature 228, 1210 (1970).

PIESSENS, W.F., HEIMANN, R., LEGROS, N., HEUSON, J.C.: Europ. J. Cancer 7, 377 (1971).

PILET, Ch., LE GARREC, Y., TOUJAS, L., SABOLOVIC, D., MONTEIL, J.C., ROTHIER, F., MISHRA, U., GHEBREHIWET, B., GUELFI, J.: p. 294 in "Investigation and Stimulation of Immunity in Cancer Patients" (G. Mathé, R. Weiner, eds.). Vol. 47 R.R.C.R., Heidelberg: Springer-Verlag 1974.

PORT, C.D., SPORN, M.B., KAUFMAN, D.G.: Proc. Amer. Assoc. Cancer Res. 16, 21 (abstract No. 83)(1975).

PREHN, R.T.: J. Nat. Cancer Inst. 26, 223 (1961).

REINER, J., SOUTHAM, C.M.: Cancer Res. 29, 1814 (1969).

ROSENTHAL, S.R.: Bull. Inst. Pasteur 70, 29 (1972).

ROSENTHAL, S.R., CRISPEN, R.G., THORNE, M.G., PIEKARSKI, N., RAISYS, N., RETTIG, P.: p. 421 in "Investigation and Stimulation of Immunity in Cancer Patients" (G. Mathé, R. Weiner , eds.). Vol. 47 R.R.C.R., Heidelberg: Springer-Verlag 1974.

SARMA, P.S., SHIU, G., NEUBAUER, R.H., BARON, S., HUEBNER, R.J.: Proc. Nat. Acad. Sci., USA 62, 1046 (1969).

SCHWARZENBERG. L., MATHÉ, G., POUILLART, P., WEINER, R., LACOUR, J., GENIN, J., SCHNEIDER, M., VASSAL, F. DE, HAYAT, M., AMIEL, J.L., SCHLUMBERGER, J.R., JASMIN, C., ROSENFELD, C.: Brit. Med. J. 1, 700 (1973).

SINKOVICS, J.G., BERTIN, B.A., HOWE, C.D.: Nat. Cancer Inst. Monogr. 22, 349 (1966).

SOKAL, J.E., AUNGST, C.W.: p. 313 in XIth International Cancer Congress, Florence, October 1974. Vol. 1 Conferences, Symposia Workshops.

SOKAL, J.E., GRACE, J.T.: Proc. Amer. Assoc. Cancer Res. 10, 85 (1969).

SPIEGELMAN, S.: XVth Congress International of Hematology, Jerusalem, September 1974 (abstract p. 474).

TANAKA, T., SASAKI, N.: Gann 65, 395 (1974).

TARANGER, L.A., CHAPMAN, W.H., HELLSTROM, I., HELLSTROM, K.E.: Science 176, 1337 (1972).

TKACZEVSKI, L. DE, HARVEN, E. DE, FRIEND, C.: J. Virol. 2, 365 (1968).

TOKUNAGA, T., YAMAMOTO, S., NAKAMURA, R.M., KATAOKA, T.: J. Nat. Cancer Inst. 53, 459 (1974).

TOUPLICHEVA, A.P., IVANOV, K.K., SILINOVA, N.G.: Radiobiologia 5, 243 (1965).

UNGARO, P.C., DRAKE, W.P., MARDINEY, M.R.: J. Nat. Cancer Inst. 50, 125 (1973).

USUBUCHI, I.: Gann 47, 534 (1956).

WEINSTEIN, A.J., GAZDAR, A.F., SIMS, H.L., LEVY, H.B.: Nature 231, 53 (1971).

WEISS, D.W.: Nat. Cancer Inst. Monogr. 35, 157 (1972).

WEPSIC, H.T., HARRIS, S., SANDERS, J., ALAIMO, J.: Proc. Amer. Assoc. Cancer Res. 16, 183 (abstract No. 731) (1975).

WHITMIRE, C.E.: J. Nat. Cancer Inst. 51, 473 (1973).

WHITMIRE, C.E., HUEBNER, R.J.: Science 177, 60 (1972).

WILSON, R.H., JUREVICS, V., DELANEY, L.R., SCHRAM, A.C.: Cancer Res. 26, 1435 (1966).

WISSLER, R.W., CRAFT, K., KESDEN, D., POLISKY, B., DZOZA, K.: p. 539 in "Advance in Transplantation" (J. Dausset, F.T. Rapaport, eds.). 1 Vol., Copenhagen: Munksgaard 1968.

WOODRUFF, M.F.A., BOAK, J.L.: Brit. J. Cancer 20, 345 (1966).

WRIGHT, B.S., LASFARGUES, J.C.: Nat. Cancer Inst. Monogr. 22, 685 (1966).

YOSHIKURA, H., HIROKAWA, Y., YAMADA, M., SUGANO, H.: Jap. J. Med. Sci. Biol. 20, 225 (1967).

YOUN, J.K., BARSKI, G., HUPPERT, J.: C.R. Acad. Sci., Paris 267, 816 (1968).

Chapter 6
Towards Immune Restoration

Immunotherapy most often works by helping the organism to kill the last cell of residual disease, which still remains after other forms of treatment; alternatively, if the preliminary data on the use of immunotherapy in the treatment of perceptible tumours is confirmed, the aim is to assist the treatment in the destruction of more cells.

The object of immunoprophylaxis is to help the organism to kill the first cells of a tumour or, perhaps, a viral carcinogenic or cocarcinogenic agent, or even cells made antigenic by a chemical carcinogen (MATHÉ, 1967a), or by a carcinogenic virus (MATHÉ, 1967b), but not yet transformed.

Immunotherapy as well as immunoprophylaxis are expected to potentiate a normal immune machinery. But deficiency is often found in cancer patients. Moreover, it has been observed in other subjects where it is congenital or induced by age, diseases or therapeutics, that the risk of immunodeficient patients presenting a cancer is higher than in the general population.

Hence the concept of immunorestoration will be introduced with a few preliminary data. Before developing it, there will be a brief review of the problems of non-specific immunodeficiency in cancer patients and of immunodeficiency as a risk of cancer.

6.1. Immunodeficiency Associated with Cancer and Cancer Risk in Immunodeficiency Subjects

The immune deficiency specific toward tumour-associated antigens will not be described here: the present data are based on tests which are, as we saw in Chapter 2 and which will be discussed again in the appendix, open to criticism; their significance is far from being clear. Moreover, what is presented as cell-mediated immunity insufficiency may be finally interpreted as being due to serum factors in which the part of humoral-mediated immunity is important.

Hence, exclusively non specific immunity syndromes observed in cancer subjects will be described and a list of non-specific immunity diseases and syndromes, which seem to be associated with a high risk of cancer, will be given.

6.1.1. Experimental

6.1.1.1. Animals with Cancer

Several types of tumour *treatment*, such as chemotherapy (Fig. 6.1) (MATHÉ *et al.*, 1968), and local radiotherapy (Table 6.I) (MARTIN *et al.*, 1975), are capable of decreasing or suppressing immune responses in normal mice. Surgery, in our hands, has not induced a similar effect (MARTIN *et al.*, not published).

In animals carrying *tumours* and not submitted to any form of treatment, immunodeficiency is frequently found, if sought.

Among the possible factors is the volume of the tumour. LEVY *et al.* (1974) showed that mice with a 1 g transplanted methylcholanthrene-induced tumour have normal mitogen responsiveness, whereas mice bearing 2.5 g tumours have a markedly impaired immune system. Tumour volume may be responsible for the immune alterations which appear during the course of the tumour.

ROWLAND *et al.* (1971) showed, in fact, that throughout the lifespan of mice bearing isogeneic mammary tumours, there was an initial rise in the total number of immune

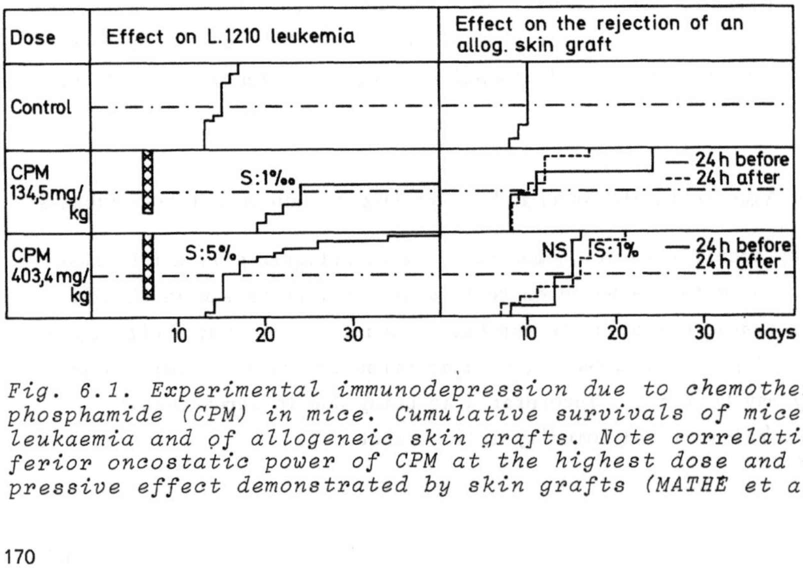

Fig. 6.1. Experimental immunodepression due to chemotherapy with cyclophosphamide (CPM) in mice. Cumulative survivals of mice carrying L1210 leukaemia and of allogeneic skin grafts. Note correlation between inferior oncostatic power of CPM at the highest dose and its immunosuppressive effect demonstrated by skin grafts (MATHÉ et al., 1968)

Table 6.I. Effect of repeated local irradiation of the hind leg[a] on systemic immune responsiveness in mice[b] (MARTIN *et al.*, 1975)

Cell type	Tests	Results		Student Fisher test
		Control	Treated	
T + B + macrophages	anti SRBC response: direct PFC/spleen at day 4	119,000 ± 51,000	11,000 ± 8,000	$p < 1\%$
T + B + macrophages	anti TNP response to TNP-KLH : indirect PFC/spleen at day 7	282,000 ± 47,000	136,000 ± 33,000	$p < 1\%$
B	anti TNP response to TNP-POL : direct PFC/spleen at day 4	51,000 ± 10,000	17,000 ± 3,000	$p < 1‰$
T + monocytes	DHS to picryl chloride : increase in ear thickness (1μ = 1/100 mm)	$14\mu \pm 1$	$6\mu \pm 1$	$p < 1‰$
T + macrophages	PHA responsiveness : c.p.m.	5,000 ± 1,900	1,600 ± 1,300	$1\% < p < 5\%$
T + B + macrophages	PWM responsiveness : c.p.m.	5,500 ± 1,700	5,400 ± 2,900	N.S.

[a] 450 rads, 3 times a week (10 sessions).

[b] Immune responsiveness was tested the day after the final irradiation.

SRBC: Sheep red blood cells; PFC: plaque forming cells; TNP-KLH: Trinitrophenyl keyhole limpet hemocyanine; TNP-POL: TNP - polymerised flagelline; DHS: Delayed hypersensitivity; PHA: phytohemagglutinin; PWM: pokeweed mitogen.

competent cells in the lymph nodes as well as in the spleen 15 days after tumour implantation, and that the number of competent cells was only reduced in the terminal stages of malignancy.

However, the volume of the neoplasia is not the only factor. *Carcinogenic agents*, themselves, can be immunosuppressive. These include chemical carcinogens (STJERNSWARD, 1965) as well as oncogenic viruses (CEGLOWSKI and FRIEDMAN, 1968); SCHNEIDER and DORE (1969) have shown, in our laboratory, that mice present a transitory but strong depression of hemolytic plaque forming cells a few days after they are inoculated with Friend's leukaemogenic virus (Table 6.II).

HUCHET and LHERITIER (1976), working in our institute, described a non-specific T-cell immunodepression in mice which received extracts (without living cells) of Lewis tumour (Table 2.IX) and of chemically-induced L20, L21, and L22 leukaemias. The *mechanisms* may be variable with the different tumour models.

ALEXANDER *et al.* (1969) who were quoted in Chapter 2, showed that a voluminous tumour blocks recirculation of sensitized lymphocytes (Fig. 2.22)

POUPON (1975) recently demonstrated another possible mechanism of immunodepression in cancer bearing animals: the increase of suppressor cells.

Host factors, such as genetic ones, may be important. LORING and SCHLESINGER (1970) showed that certain strains of mice, such as RIII, after being inoculated with an Ehrlich ascitis carcinoma, present an important fall in the number of θ positive lymphocytes in the lymphoid organs (thymus, nodes, and spleen); they did not find this phenomenon in $C_{57}Bl/6$ mice.

Table 6.II. Decrease of haemolytic plaque forming cell (HPFC) number by the injection of Friend's virus into mice (SCHNEIDER and DORÉ, 1969).

Mice	HPFC	5 days after Friend virus	20 days after Friend virus
(DBA/2 x $C_{57}Bl/6$)F1	HPFC/10^6 c.	185.7 ± 206.3[a]	567.3 ± 229
Controls	HPFC/spleen	28,382 ± 28,638	121,739 ± 74,000
(DBA/2 x $C_{57}Bl/6$)F1	HPFC/10^6 c.	22.59 ± 6.30	5.14 ± 7.21
Friend virus	HPFC/spleen	2,259 ± 6,447	5,857 ± 1,770

[a] Mean ± standard deviation ($p < 5\%$)

The exact mechanism of the induction of a general nonspecific immunodepression in animals carrying tumours, still remains unknown and is probably complex. ROWLAND *et al.* (1971) suggested that, during tumour growth, animals can develop an "inhibitory environment". A contaminant virus may be responsible in some conditions, as suspected by HUCHET and LHERITIER (1976).

6.1.1.2. Cancer Risk in Immunodeficient Animals

It was mentioned in Chapter 2, that immunosuppressive conditions, such as neonatal thymectomy (TRAININ *et al.*, 1967), treatment by antilymphocyte serum (ALLISON and LAW, 1968), or immunosuppressive drugs (SCHWARTZ, 1965), increase the incidence of chemically or virus-induced tumours.

6.1.2. Clinical

The main types and conditions of immune deficiency in cancer patients will be considered and a list prepared of congenital or acquired immunodeficiency syndromes and/or diseases which represent a high risk for cancer induction or promotion.

6.1.2.1. Cancer Patients

All cancer patients must be submitted, at any important stage of the evolution of their tumours and especially before and after treatment, to immune investigation (see monitoring battery in appendix).

6.1.2.1.1. Immunodeficiency Due to Treatments

Most anticancer drugs, including hormones, have been shown to be able to induce immunodepression (MATHÉ and KENIS, 1975).

Among *hormones*, corticosteroids are strongly immunosuppressive (Fig. 6.2) (MATHÉ *et al.*, 1967), while the sex hormones may not be. On the contrary, oestrogens may be immunopotentiators, as it is known that females of most species are immunologically more competent than males (BRENT and MEDAWAR, 1966) and as CASTRO *et al.* (1973) showed, orchidectomy is a powerful method of immunopotentiation.

Non hormonal cytostatic *chemotherapy* is immunosuppressive (Table 6.III). Nevertheless, intermittent modalities of administration with cycles and intervals are much less immunodepressive than continuous daily small dose chemotherapy (Table 6.III) (SCHNEIDER, 1968; MATHÉ *et al.*, 1970b).

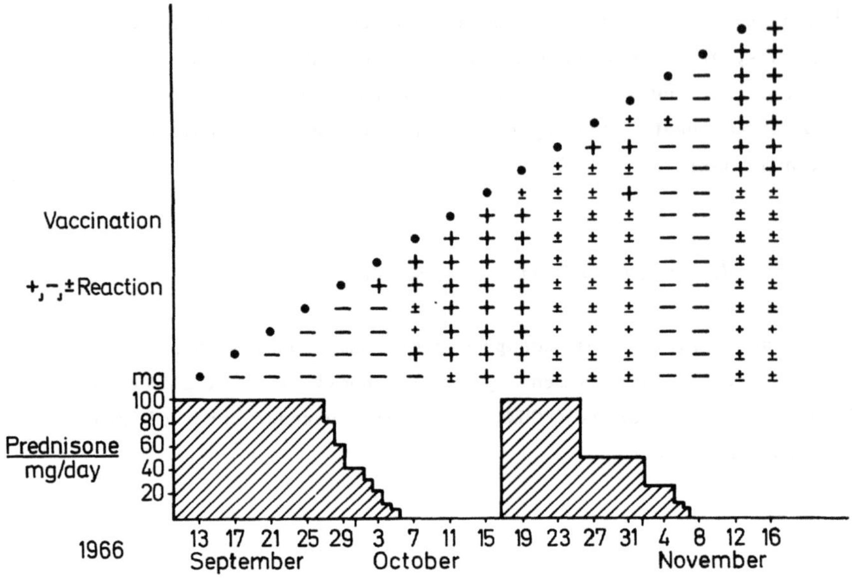

Fig. 6.2. Immunosuppression induced by corticosteroid treatment in a cancer patient (MATHÉ et al., 1967)

Table 6.III. Chemotherapy immunodepression according to the intermittent or continuous modality of administration[a] (SCHNEIDER, 1968; MATHÉ *et al.*, 1970b)

	Absence of visible immunosuppressive effect	Visible immunosuppressive effect
Discontinuous chemotherapies	51/57 (89.4 p.100)	6/57 (10.6 p.100)
Continuous chemotherapies	9/27 (33.3 p.100)	18/27 (66.6 p.100)

[a] The patients in whom the delayed hypersensitivity responses were negative before and after chemotherapy and in whom it was impossible to conclude that there had been an eventual immunosuppressive effect of the treatment do not appear in this Table.

A rebound is sometimes seen after spontaneous restoration, which occurs a few weeks after the end of a short chemotherapy (HERSH *et al.*, 1971).

According to LEVENTHAL *et al.* (1974), decrease of B-lymphocyte function is almost constant during chemotherapy and has no significant effect, while T-lymphocyte modification may be important: according to these authors, persistence of T-cell response in the phase of induction chemotherapy of acute leukaemia is associated with a good prognosis.

According to SIMMLER *et al.* (1976a), delayed skin hypersensitivity responses are more often decreased than in vitro mitogen lymphocyte transformation in patients submitted to chemotherapy. This seems to affect a peripheral event, rather than central proliferation of immune cells.

Radiotherapy even local, has been blamed by STJERNSWARD *et al.* (1972) as being long-term immunosuppressive and of inducing lymphopaenia mainly due to T-cell depletion (Fig. 6.3). According to McCREDIE (1973), local irradiation decreases lymphocyte stimulation by PHA.

In any case, it has been observed that the immune responses were worsened by postoperative radiotherapy (THOMAS *et al.*, 1971; STJERNSWARD *et al.*, 1972; JENKINS *et al.*, 1973; BRAEMAN and DEELEY, 1972).

Surgery itself can induce, in humans, a significant immunosuppression, but of limited duration (WINGARD *et al.*, 1967).

6.1.2.1.2. Immunodeficiency due to/or Associated with Cancer

A few years ago, it was conventionally considered that some neoplastic diseases were accompanied by: (a) humoral response deficiency, the typical example being chronic lymphoid leukaemia; (b) cell-mediated reaction alterations, the latter being consi-

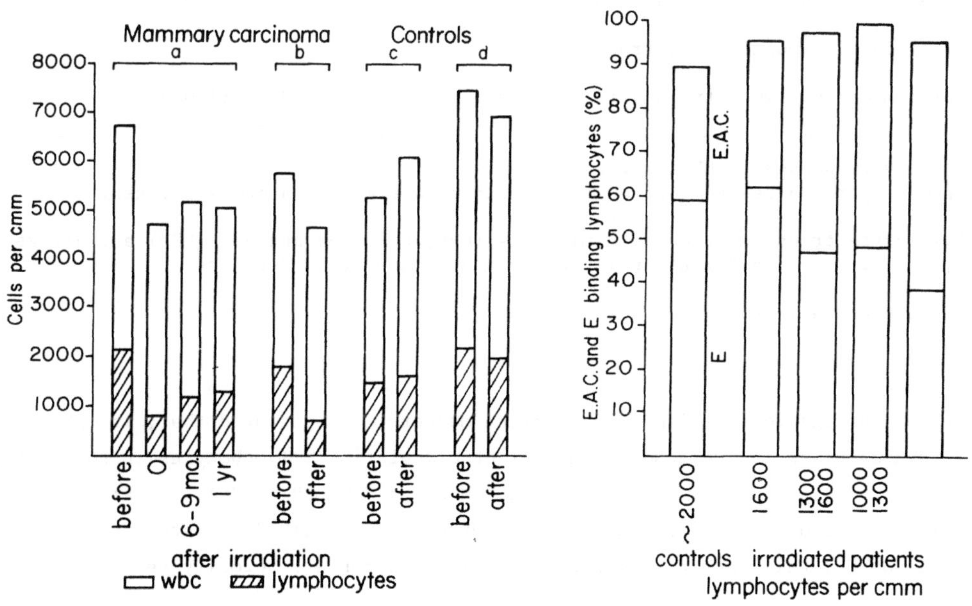

Fig. 6.3. Global and T-lymphopaenia secondary to radiotherapy (STJERNSWARD et al., 1972)

dered as a frequent manifestation of Hodgkin's disease; (c) a global immune insufficiency, mainly in the case of advanced terminal tumours.

Today, this may serve as a guide, but it appears to be far removed from reality, and the better the tests, and the more we use them for monitoring cancer patients (see appendix), the less valuable the distinction.

Humoral Immunodeficiency. Decreased antibody production (SCHNEIDER *et al.*, 1967), very often associated with hypogamma-globulinaemia (RUNDLES *et al.*, 1954; ULTMANN *et al.*, 1959; SCHNEIDER *et al.*, 1967), is considered a typical manifestation of chronic lymphoid leukaemia (CLL). Whether it is pure, as described by WEED (1965) and OPPENHEIM *et al.* (1965), or whether it may be associated with a defect of cell-mediated immunity (CMI) (SHARMAN *et al.*, 1966; SCHNEIDER *et al.*, 1967; BELPOMME and SIMMLER, 1976) (Table 6.IV) is still a matter for discussion (BLOCK *et al.*, 1969).

Table 6. IV. Immune deficiency in chronic lymphoid leukaemia patients (BELPOMME and SIMMLER, 1976)

	N° anergic/N° tested	
Delayed hypersensitivity reaction[a]	9/15	(60%)
Cell-mediated immunity in vitro[b]	9/15	(60%)
Humoral-mediated immunity in vitro[c]	8/8	(100%)

[a] Primary responses to DNCB and KLH.
Secondary responses to tuberculin, candidine antigen, and mumps antigen.

[b] Lymphocyte transformation with mitogens (PHA, PWM), an antigen (PPD) and allogeneic lymphocytes : transformation is more often delayed than decreased.

[c] Antibodies anti-KLH.

Recent investigations, using cell membrane markers, demonstrated that, in CLL, there may be a decrease in the absolute number of circulating T-lymphocytes in some cases, while there is an increase in others (BELPOMME and SIMMLER, 1976). Isolation of T-cells in chronic lymphoid leukaemia led WYBRAN *et al.* (1973) to study their response to PHA; this author suggests that T-lymphocytes are normally reactive in this disease.

Recently TAVADIA *et al.* (1974) showed that this normal T-lymphocyte stimulation can be inhibited by nonspecific plasma and B-lymphocyte factors. Thus impairment of cell-mediated immunity in chronic lymphoid leukaemia may be due to the release of factors by leukaemic B-lymphocytes.

Immunological insufficiency in myeloma patients is similar to that of CLL subjects: there is a decrease of antibody production and of normal immunoglobulin levels, though the global concentration of these globulins is extremely high, due to monoclonal production of one of these molecules alone (FAHEY *et al.*, 1963; SOLOMON *et al.*, 1963). But, as in CLL, cell-mediated immunity can be impaired, as shown by insufficient mitogen stimulation of lymphocytes (SALMON and FUDENBERG, 1969).

Cell-mediated Immunity Deficiency. Cell-mediated insufficiency was classically considered to be so frequent in patients with Hodgkin's disease that it was almost included in symptomatology (see MATHÉ and TUBIANA, 1974).

Delayed skin hypersensitivity is frequently missing as described by AISENBERG (1962, 1966), CHASE (1966), and SCHNEIDER *et al.* in our Institute (1970, 1974) (Tables 6.VI and 6.VII).

They have shown that: (a) immunoglobulin levels are normal whatever the circumstances; (b) the decrease of blood circulating lymphocytes and of the incidence of positive skin reactivity are diminished in perceptible phases compared to remissions, and in stages III and IV compared to stages I and II; (c) the incidence of skin responses is decreased in mixed cellularity and lymphocytic depletion types compared to lymphocytic predominance and nodular sclerosis types as well as controls (Tables 6.VI and 6.VII). Lymphopaenia is also correlated with the histological type (SCHNEIDER *et al.*, 1970, 1974; TUBIANA *et al.*, 1971).

It is also interesting to note that there is an increase of circulating immunoblasts during the perceptible phases (Tables 6.VI and 6.VII). (SCHNEIDER *et al.*, 1970, 1974) already mentioned by CROWTHER *et al.* (1969). Whether or not this abnormality indicates a certain specific immune reaction, has yet to be determined.

Impaired in vitro lymphocyte stimulation by PHA has been demonstrated in patients with anergy to delayed skin hypersensitivity antigens (HERSH and OPPENHEIM, 1965), suggesting that T-lymphocytes are intrinsically abnormal or damaged. In addition, serum inhibitory factors to PHA responsiveness have been detected, which may contribute to cell-mediated immunity deficiency (TRUBOWITZ *et al.*, 1966).

No correlation was found between PHA-induced lymphocyte transformation at the time of diagnosis and the clinical course of the disease (CORDER *et al.*, 1972; TRUBOWITZ, *et al.*, 1966).

Table 6.V. Routine "immune investigation" of patients with Hodgkin's disease. I. Perceptible phases versus "complete remission" (SCHNEIDER *et al.*, 1974)

	Number of patients	Cellular responses								Humoral responses		
		Positive responders to (%)			TL[a] PHA %	Circulating/mm^3 (means)				Immunoglobulins mg/100 ml (means)		
						Lymphocytes		Mono-cytes	"Immuno-blasts"			
		BCG	CD[b]	SK[c]		small	large			IgG	IgA	IgM
Normal controls	22	88	65	50	73	1560	410	300	4.2	1180	175	80
Patients in perceptible phase of the disease	149	33.1	31.2	20	37 ± 3[d]	1030 ± 56	260	250	18.4 ± 2.1	1276	227	75
Patients in remission	34	56.4	52.1	26	44 ± 4	1100 ± 110	270	170	11.3	1515	250	76

a TL PHA = mean of the percentage of transformed lymphocytes in vitro in the presence of phytohemagglutinin.

b CD = Candidin.

c SK = Streptokinase.

d Standard error of the means

Table 6.VI. Routine "immune investigation" of patients with Hodgkin's disease. II. Clinical Stages (SCHNEIDER *et al.*, 1974)

	Number of patients	Cellular responses								Humoral responses		
		Positive responders to (%)			TL[a] PHA %	Circulating/mm^3 (means)				Immunoglobulins mg/100 ml (means)		
						Lymphocytes		Mono-cytes	"Immuno-blasts"			
		BCG	CD[b]	SK[c]		small	large			IgG	IgA	IgM
Normal controls	22	88	65	50	73	1560	410	300	4.2	1180	175	80
Patients in perceptible phase of the disease	149	33.1	31.2	20	37	1030	260	250	18.4	1276	227	75
Stages I and II	42	30.8	44	30	43	1390	250	330	18.2	1307	238	106
Stages III and IV	107	33	18.4	10.5	32 ± 4[d]	803 ± 52	250	220	18.9	1150	189	59
Patients who died in the 6 months following the investigation	49	32.8	13.3	20	30 ± 3.2	610 ± 49	270	300	13.8	1170	182	55

[a] TL PHA = mean of the percentage of transformed lymphocytes in vitro in the presence of phytohemagglutinin.

[b] CD = Candidin.

[c] SK = Streptokinase.

[d] Standard error of the means.

Table 6.VII. Routine "immune investigation" of patients with Hodgkin's disease. III. Histological classification (SCHNEIDER *et al.*, 1974)

Histological type	Number of patients	Cellular responses								Humoral responses		
		Positive responders to (%)			TL[a] PHA %	Circulating/mm^3 (means)				Immunoglobulins mg/100 ml (means)		
						Lymphocytes		Monocytes	"Immunoblasts"			
		BCG	CD[b]	SK[c]		small	large			IgG	IgA	IgM
Normal controls	22	88	65	50	73	1560	410	300	4.2	1180	175	80
Lymphocyte predominance	43	40.5	55.5	37.5	42	1100	250	100	16.5	1417	134	66
Nodular slerosis	40	43.4	55.5	33.3	42	970	230	290	12.5	1445	256	87
Mixed cellularity	61	29.9	33.3	21.7	48	1280	310	220	14	1441	233	84
Lymphocyte depletion	21	33.3	25	7	38	910	290	230	11.8	1283	330	66

[a] TL PHA = mean of the percentage of transformed lymphocytes in vitro in the presence of phytohemagglutinin.

[b] CD = Candidin.

[c] SK = Streptokinase.

Recent investigations on Hodgkin's disease suggest the existence of an increase of T-lymphocytes in the organs involved (lymph nodes, spleen) (BELPOMME *et al.*, 1974; KAUR *et al.*, 1974), and that stimulation by PHA would be in the normal range in these organs.

This data favours the hypothesis of an immunological reaction in the early stage of the disease. Although close contact between T-lymphocytes and Reed-Sternberg cells has been observed (BELPOMME *et al.*, 1976; ARCHIBALD and FRENSTER, 1973), the nature of this relationship still remains unknown.

This study shows that the detection of cell-mediated immunity insufficiency is not constant in Hodgkin's disease, and hence it does not seem to be a symptom, but only a possible associated characteristic or complication.

Moreover, though this neoplasia has been classically considered as inducing a pure cell-mediated immunity deficiency, because immunoglobulin levels are usually normal or high [SCHNEIDER *et al.* 1970, 1974, Tables 6.VI and VII] and AISENBERG and LESKOWITZ (1963) have found normal antibody responses to pneumococcal polysaccharide, it is now known, after extensive studies of many patients, that they present normal secondary antibody responses, but impaired primary antibody production (CHASE, 1966; HERSH *et al.*, 1970).
Similar immunological abnormalities as in Hodgkin's disease have been observed in patients with non Hodgkin's haematosarcomas for which SCHNEIDER *et al.* (1970, 1974) demonstrated the correlation between cell-mediated immune deficiency and the evolution of the disease (Table 6.VIII).

In acute leukaemias, mild impairment of cell-mediated immunity, based on skin tests, have been observed (DUPUY *et al.*, 1970; CHAR *et al.*, 1973; SCHNEIDER *et al.*, 1974) (Tables 6.IX and 6.X). Although PHA response has been reported to correlate with therapeutic responsiveness (HERSH *et al.*, 1971), one wonders, however, whether reduced in vitro immunological reactions in poor prognosis patients may not be due to the presence of circulating leukaemic cells (WALKER *et al.*, 1973).

Haematosarcomas and leukaemias are not the only neoplastic diseases in which deficiency of cell-mediated immunity has been described. EILBER and MORTON (1970), on the one hand, and PINSKY *et al.* (1971, 1973) on the other, demonstrated the high frequency of negative responses to DNCB in patients with solid tumours. The correlation was poor with the presentation of the disease, but excellent with post-surgery prognosis (Table 6.XI).

The conclusion of the authors is that the ability to become sensitized to DNCB is more important as a prognostic factor than the ability to respond to previously encountered antigen(s).

Table 6.VIII. Routine "immune investigation" of patients with reticulosarcoma and/or immunoblastic lymphosarcoma, in perceptible phase of the disease. Clinical Stages III and IV (SCHNEIDER *et al.*, 1974)

	Number of patients	Cellular responses								Humoral responses		
		Positive responders to (%)			TL[a] PHA %	Circulating/mm³ (means)				Immunoglobulins mg/100 ml (means)		
						Lymphocytes		Mono-cytes	"Immuno-blasts"			
		BCG	CD[b]	SK[c]		small	large			IgG	IgA	IgM
Normal controls	22	88	65	50	73	1560	410	300	4.2	1180	175	80
Reticulosarcoma	12	30.7	33.3	10	38 ± 5[d]	990 ± 434	530 ± 220	450 ± 125	51.4 ± 19.6	854 ± 86	206 ± 56	62 ±5.88
Reticulosarcoma or Immunoblastic lymphosarcoma	60	44	42	17.8	41 ± 2	880 ± 98	220 ± 34	370 ± 48	33.3 ± 2.01	1020 ± 57	238 ± 21	77 ± 5
Patients who died in the 6 months following the investigation	14	24.2	21.4	0/14	41 ± 4	608 ± 206	170 ± 110	290 ± 116	8.9 ± 1.5	1090 ± 116	259 ± 56	72 ± 9

a TL PHA = mean of the percentage of transformed lymphocytes in vitro in the presence of phytohemagglutinin.

b CD = Candidin.

c SK = Streptokinase.

d Standard error of the means.

Table 6.X. Routine "immune investigation" of patients with acute lymphoid leukaemia. I. Perceptible phases versus "complete remissions" (SCHNEIDER *et al.*, 1974)

	Number of patients	Cellular responses								Humoral responses		
		Positive responders to (%)			TL[a] PHA %	Circulating/mm^3 (means)				Immunoglobulins mg/100 ml (means)		
						Lymphocytes		Mono-cytes	"Immuno-blasts"			
		BCG	CD[b]	SK[c]		small	large			IgG	IgA	IgM
Normal controls	22	88	65	50	73	1560	410	300	4.2	1180	175	80
Patients in perceptible phase of the disease	200	39.5	24	14.2	50	3382	717	339	40.2 ± 6.6[d]	978	154	83
Patients in remission	301	60	35	21	59 ±13	1650	290	270	18.9 ± 2.4	882	135	80

a TL PHA = mean of the percentage of transformed lymphocytes in vitro in the presence of phytohemagglutinin.

b CD = Candidin.

c SK = Streptokinase.

d Standard error of the means.

Table 6.IX. Routine "immune investigation" of patients with acute myeloid leukaemia (SCHNEIDER *et al.*, 1974)

	Number of patients	Cellular responses								Humoral responses		
		Positive responders to (%)			TL[a] PHA %	Circulating/mm³ (means)				Immunoglobulins mg/100 ml (means)		
						Lymphocytes		Mono-cytes	"Immuno-blasts"			
		BCG	CD[b]	SK[c]		small	large			IgG	IgA	IgM
Normal controls	22	88	65	50	73	1560	410	300	4.2	1180	175	80
Patients in perceptible phase of the disease	52	42.8	39	21.8	58 ±2.7[d]	2300 ± 396	750 ± 127	370 ± 7.9	26.2 ± 5	1317 ± 89	277 ± 55	116 ± 9
Patients in Remission	13	62	35	20	67 ±5.7	2750 ± 745	500 ± 190	290 ±133.2	12.4 ± 2.9	1338 ± 125	287 ± 36	168 ± 56

a TL PHA = mean of the percentage of transformed lymphocytes in vitro in the presence of phytohemagglutinin.

b CD = Candidin.

c SK = Streptokinase.

d Standard error of the means.

Table 6.XI. Solid tumour patients: correlation between responses to dinitrochlorobenzene (DNCB) and prognosis after surgery

	EILBER and MORTON (1970)		PINSKY *et al.* (1971, 1973)	
	Recurrence at 6 months after surgery	No recurrence	Recurrence at 6 months after surgery	No recurrence
DNCB +	2	50	3	54
DNCB -	4	27	7	9

It is interesting to note, however, that in the case of some tumours such as skeletal and soft tissue sarcomas, there is no correlation between the response to DNCB and prognosis, since almost all patients have a positive response, whatever the stage of the disease (EILBER and MORTON, 1970).

It has been demonstrated that the in vitro blastogenic transformation of lymphocytes by PHA is impaired in solid tumour patients, specially with advanced neoplasias GATTI *et al.*, 1970; AL-SARRAF *et al.*, 1972).

Although LEVIN *et al.* (1964) was not able to demonstrate any associated humoral immunity defect in such subjects, LEE *et al.* (1970) and WAGNER *et al.* (1972) observed an insufficient primary antibody production against a variety of protein antigens.

Macrophage functions (tested with the skin window technique) are also often modified in these patients (SOUTHAM and LEVIN, 1966).

Cell-mediated Immunity Dissociated Deficiencies. Both in vivo delayed skin hypersensitivity tests and in vitro mitogen transformation assays are commonly considered to inform the clinician about the cell-mediated immunity status.

With SIMMLER *et al.* (1976a), the author has conducted a systematic study on the correlation between these in vivo and in vitro tests.

Table 6.XII shows the respective incidences of normal skin and PHA responses in melanoma and sarcoma patients before or after treatment: the percentages of correlation are 73 % in nontreated patients, 42% in subjects submitted to chemotherapy, and 56% in those who were under immunotherapy.

Table 6.XII. Search for a correlation between in vitro and in vivo responses in melanoma and sarcoma patients (SIMMLER *et al.*, 1976a)

Treatment	Group[a]	Number of patients	Number of normal in vitro responses[b]	Number of allergic patients[c]	Percentage of correlation	
Before treatment	II	7	7	5	71	73%
	III	8	8	6	75	
After chemotherapy	II	9	9	4	44	42%
	III	3	3	1	33	
After immunotherapy	II	11	9	6	55	56%
	III	5	4	2	60	
After chemoradiotherapy	II	1	0	0		
	III	1	1	0		
Total		45	41	24		58%

[a]Group II: Melanóma (28 patients),
Group III: Sarcomas (17 patients).

[b]Transformation test by phytohemagglutinin.

[c]Allergy to at least one antigen (induration $\geqslant$ 5 mm).

Table 6.XIII shows that: (a) the in vitro mitogen nonresponders are slightly more often anergic (7/9) than the responders (20/31); (b) the anergic are slightly less often responsive to mitogens (20/27) than the allergic patients (11/13).

Table 6.XIII. Absence of correlation between skin reaction tests and in vitro lymphocyte transformation in melanoma and sarcoma patients (SIMMLER *et al.*, 1976a)

In vivo tests[a]	In vitro transformation test[b]	
	Nonresponders = 9 patients	Responders = 31 patients
Anergic = 27 patients	7	20
Allergic = 13 patients	2	11

[a] Delayed hypersensitivity to primary antigens (DNCB and KLH) and/or recall antigens (mumps, PPD, Candida).

[b] Lymphocyte transformation test with PHA and PWM.

Finally, Table 6.XIV indicates the correlation between the in vivo and in vitro responses to the same antigen PPD: 8/51 anergic patients respond to PPD in vitro, compared to 13/39 allergic patients; the difference is not significant.

In conclusion, there is no correlation between the delayed hypersensitivity responses to recall antigens and the in vitro lymphocyte transformation by mitogens or antigens, even those working mainly on T-cells, such as PHA.

Table 6.XIV. In vivo and in vitro responses to PPD (SIMMLER *et al.*, 1976a)

In vitro response to PPD[a]	In vivo response to PPD[b]	
0 - 2	(8/51)	16%
$>$ 2	(13/39)	33%
	N.S.	

[a] Stimulation Index.

[b] Induration $\geqslant$ 5 mm.

Another type of dissociation has been described by BRULEY-ROSSET *et al.*, (1976), also working in the author's laboratory, between skin test responses and leukocyte inhibition factor(s) (LIF) in the serum: while a good correlation has been found between both results in the case of epidermoid bronchus and breast carcinoma patients, they are totally dissociated in glioblastoma patients (Table 6.XV).

A tendency toward dissociation between skin tests and PHA response has also been found by ZIEGLER *et al.* (1975) in Hodgkin's disease.

Before describing such dissociations, one must take all precautions to avoid misinterpretations. For example, in some patients with probable immunodeficiency, one may detect positive skin tests 2 or 3 days later than in normal subjects (WEINER, pers. obs.). Hence, some patients may be considered negative while they do in fact respond, albeit slowly. The results of in vitro lymphocyte transformation tests by mitogen(s), such as PHA, should also be interpreted with care as will be described later (SIMMLER *et al.*, 1976a).

Among the different mechanisms likely to induce such dissociations, the hypothesis of the role of serum immunosuppressive factors should be considered. One example is given by STEWART (1973). In bronchogenic carcinoma patients, there was, after surgery, a shift from negative to positive responses to tuberculin, associated with the disappearance of serum factors which could suppress the in vitro lymphocyte stimulation by tuberculin.

Similar factors have now been described in different solid tumours (SILK, 1967; GATTI *et al* ., 1971). Their role still remains unknown. The most striking feature is that they can occur in patients with small primary tumours (HERSH *et al.*, 1974).

On the other hand, it should be stated that protease can render patients with negative skin tests positive, as will be seen later (THORNES, 1974).

Global Immunodeficiency. In conclusion, even neoplasias classically considered as associated with the deficiency of only one branch of immunity (humoral in chronic lymphoid leukaemia, cellular in Hodgkin's disease) are, in fact, very often complicated by both deficiencies, at least, later on in their evolution.

6.1.2.2. Immunodeficiency Diseases or Syndromes with High Cancer Risk

Several conditions characterized by, or associated with, immune deficiency have been quoted in Chapter 2 as being able to increase cancer incidence: ageing (DOLL, 1970;

Table 6.XV. Correlation between LIF[a] assay in the serum and skin test reactivity according to the type of neoplasia BRULEY-ROSSET *et al.* (1976)

	Epidermoid br. carcinoma		Breat Cancer		Glioblastoma	
	Skin Test +	Skin Test -	Skin Test +	Skin Test -	Skin Test +	Skin Test -
LIF +	30	0	8	2	10	0
LIF -	4	27	0	19	15	8
Total	34	27	8	21	25	8
Statistical analysis:	$\chi^2 = 49.7$ $p < 0.001$		$\chi^2 = 16,8$ $p < 0.001$		$\chi^2 = 2,83$ N.S.	

[a] Leukocyte inhibiting factor

Table 6.XVI. Classification of congenital immunodeficiency status (after SELIGMANN *et al.*, 1968)

		Lymphocytes	Plasma-cells	Germinal centres	Para-cortical lymphoid tissue	Immuno-globulins	Humoral antibodies	Delayed hyper-sensitivity	Thymus	Sensitivity to V: virus F: fungi B: bacteria
I. Lymphocytic disorders without plasmocytic disorders										
DI GEORGE (1965)	Thymic aplasia and Absence of Parathyroids (by dysembryo-genesis) (1)	M.D. (2)	N (3)	N	I.D. (4)	N	M.D.	I.D.	Absent	V F B
WISCOTT-ALDRICH (1954)	Immunological deficiency with thrombocytopenia and eczema (sex-linked recessive)	I.D.	N	N	I.D.	Variable	For certain antigens	I.D.	N	V F
NEZELOFF *et al.* (1964)	Lymphopenia with normal immuno-globulins (autosomal. recessive)	I.D.	N	N	I.D.	N	N	I.D.	I.D.	V F
II. Plasmocytic Disorders without lymphocytic disorders										
BRUTON (1952)	Sex-linked agammaglobu-linemia (recessive)	N	I.D.	I.D.	N	I.D.	I.D.	N	N	B

ROCKEY *et al.* (1964)	Hypo-IgA (Autosomal. recessive)	N	M.D.		N	I.D. for IgA	I.D.for IgA antibodies	N	N	B
	Child transitory hypogammaglobulinemia (familial genetic)	N	I.D.	I.D.		I.D.	I.D.	N		B
III. Plasmocytic and lymphocytic disorders										
GOOD and ZAK (1956)	Agammaglobulinemia with thymoma (genetic?)	M.D.	I.D.	I.D.	I.D.	I.D.	I.D.	I.D.	Thymome	B V F
LOUIS-BAR (1941)	Telangiectasia ataxia (autosomal recessive)	M.D.	M.D.	M.D.	I.D.	I.D.	M.D.	I.D.	No hassal corpuscles	
GITLIN and GRAIG (1963)	Primary lymphopenic immune deficiency (sex-linked recessive or autosomal)	M.D.	M.D.		I.D.	M.D.	I.D.for certain antigens	I.D.	Hypoplasia	V F
GLANZMAN and RINIKER (1950)	Alymphocytic agammaglobulinemia (autosomal recessive) (Swiss type)	I.D.	I.D.	I.D.	I.D.	I.D.	I.D.	I.D.	I.D.	B V F

1) Agenesia of 3rd and 4th bronchial arcs; 2) mild or inconstant decrease; 3) normal; 4) important decrease

N.C.I., 1971; MAKINODAN *et al.*, 1971; PRICE and MAKINODAN, 1972); immunosuppressive treatment, specially for allogeneic transplant "tolerance" (MacKHANN, 1969; PENN 1975) (Table 2.VIII); cytostatic treatment for the cure of a first cancer (CANELLOS *et al.*, 1974; BELPOMME *et al.*, 1974); and finally congenital immune deficiency diseases, which are numerous; they have been the object of a tentative classification by SELIGMANN *et al.* (1968) summarized in Table 6.XVI; the incidence of cancer in these cases is higher than that of the general population as has been described by GATTI and GOOD (1971) and reviewed by WALDMANN (1972) (Table 2.VII). The excess of suppressor T-cells which block normal B-cell maturation into antibody producing plasma cells, has been found in common variable hypogammaglobulinaemia, which is sometimes complicated by cancer (WALDMANN *et al.*, 1974).

6.2. Attempts at Immunorestoration

The therapeutic aim for immunodepressed patients carrying cancer or a high risk of cancer is restoration of immunity.

6.2.1. Experimental

Several agents have been shown to be able to restore partially or totally, transitorily or permanently, an immunodeficient animal.

6.2.1.1. Use of Thymic Lymphocyte-Stimulating Factor (or Thymosin)

Neonatally thymus-deprived mice are immunologically restored not only by isogeneic thymus grafts (BARCLAY *et al.*, 1964), but also by histo-incompatible ones (DALMASSO *et al.*, 1963). Their immune disorder can be corrected by implantation of thymus in millipore chambers (OSOBA and MILLER, 1964). These last two facts suggest that a thymus graft works via a humoral factor, and not via the introduction of thymocytes.

Thymus extracts have been prepared by BOMSKOV and SLADOVIC (1940), by TRAININ *et al.* (1969), by GOLDSTEIN and WHITE (1970) and by BACH *et al.* (1971), which were able to alleviate immunodeficiencies in thymectomized newborn mice, which normally die from infections and wasting disease (GOLDSTEIN *et al.*, cited by MARX, 1975).

The known characteristics of the active agent, now generally called thymosin, will be described in Chapter 7 on Pharmacology.

This substance can accelerate the development of resistance to progressive tumour growth in neonatal mice (ZISBLATT *et al.*, 1970), and its administration to rats bearing tumours, following immunosuppression, prolonged their survival (HARDY *et al.*, 1971).

6.2.1.2. Use of Adjuvants

There are data which make the possible immunopotentiating, hence restorative, effect of some adjuvants in immunodepressed animals doubtful: they would need an intact functional lymphoid system to exert their effect, at least as far as T-cells are concerned (KERBEL, 1973). MAEDA *et al.* (1973) have observed that *lentinan* is ineffective as an antitumour agent in neonatally thymectomized mice treated with antilymphocyte serum.

Nevertheless, this is not the case with *C. parvum* (HOWARD *et al.*, 1973), but its main effect is considered to be due to its action on macrophages, and one must analyse the different branches of the immune machinery carefully before concluding that an adjuvant which alters the cell-mediated immunity "is able to restore immunity" (SCOTT, 1972a, 1972b).

In any case, some adjuvants are able to do so.

Poly A-poly U has been shown to improve the capacity of neonatally thymectomized mice for rejecting skin homografts (CONE and JOHNSON, 1971); the same agent is able to restore responsiveness in ageing $C_{57}Bl_6$ mice which display impaired antibody responses. Moreover, poly A-poly U has been mentioned as being capable (as thymosine), of making in vitro bone marrow lymphocytes able to induce rosettes with sheep red blood cells, a T-lymphocytic character (SCHEID *et al.*, 1973).

Poly I-poly C is presently under study in our laboratory.

Levamisole restores the impaired immunological system in aged mice (RENOUX and RENOUX, 1972).

MOORE *et al.* (1975) observed that *BCG* contact-induced inhibition of isogeneic tumour growth was normal in rats deprived of T-cells by thymectomy followed by repeated irradiation of the whole body. The authors conclude that the local effect of the mycobacterium was not dependent on thymus-derived lymphocytes. As the effect of the treatment was favourable in lung metastases, the mechanism of action may be more complex than that envisaged by the authors, and immune restoration should have been discussed. BCG is currently being investigated as an agent capable of restoring immunity, in our laboratory.

MER has been shown to be able to raise the levels of immunological capacity in mice where it is low due to age or nonspecific environmental factors (WEISS and YASHPHE, 1973). It increases the response against antigens; it protects the antigen reactive cells against reduction in number following treatment with antilymphocyte serum or cortisone; it reverses the immunosuppressive effect of x-irradiation or antithymocyte serum, and it prevents induction of tolerance to several antigens.

6.2.2. Clinical

6.2.2.1. Transfer Factor

The attempt at using transfer factor[1] cannot be left out of this chapter, though: (a) its aim is to repair specific antitumour immunity (while this chapter is purely devoted to nonspecific restoration), and (b) its use belongs to adoptive infracellular immunomanipulation and not to the active form which is the exclusive object of this monograph. Moreover, this topic is the object of so much controversy that either a few pages are enough, or a book based on an international inquiry is necessary: as a matter of fact, according to some, it would be efficient not only in cancer and in congenital immunodeficiency diseases, but in lupus erythematosis, chronic hepatitis, multiple sclerosis, tuberculosis, candidosis, etc.. (LAWRENCE, 1955, 1969; KIRKPATRICK *et al.*, 1972; GROB, 1974; FUDENBERG *et al.*, 1974).

As far as the optimistic summary of the world experience on cancer patients (GROB *et al.*, 1975) is concerned, its conclusions should be taken with caution: the 8 cases of melanoma with "favourable effects" and 9 "skin test immunoconversions", are found in 4 authors; the 3 cases of sarcomas, with one "favourable effect" and 2 "immunoconversions" of skin tests, in 3 authors.

A random well-controlled trial is necessary if one does not wish to leave the transfer factor in purgatory for ever.

It is when transfer factor has been injected into the neoplasias, specially melanomas, that the most convincing regressions have been described (BRANDES *et al.*, 1971; MORSE *et al.*, 1973).

Among the most interesting observations are those of patients with advanced sarcomas who, whilst their migration inhibition factor assay was negative, became responders (LO BUGLIO *et al.*, 1973; HILBERG *et al.*, 1973). However, their tumours did not regress.

[1] Transfer factor is considered a small molecular weight (10,000) dialysable product of lysed peripheral blood leucocytes. Its biological activity is not destroyed by DNAse; for LAWRENCE (1969), it is not destroyed by pancreatic RNAse, but for DRESSER and ROSENFELD (1974), it is destroyed by RNAse III and appears to consist, therefore, entirely, or partly, of double-stranded RNA.

Also interesting are the data published by LEVIN *et al.* (1975) according to which cell-mediated cytotoxicity of osteosarcoma patients, which was low, increased after specific transfer factor injection, while it declined when the subjects received "nonspecific transfer factor".

It should be underlined that, according to LAWRENCE (1971), recipients should have T-lymphocytes for the transfer factor to be active: in Hodgkin's disease patients fail to respond to transfer factor if their cell-mediated immunity is poor.

Thus it may be a means of restoring specific tumour immunity only if there is no general immune deficiency. FROST and LANCE (1973) and KATZ (1973) do not agree with this restriction, as conversions of skin tests have been observed in the case of immunodeficient syndromes, such as Wiskott-Aldrich disease (T- and B-defect).

6.2.2.2. Thymus Lymphocyte Stimulating Factor (or Thymosin)

Grafts of fresh human thymus implanted into patients with congenital immunodeficiency diseases and malignancies of the lymphoid system resulted in a partial recovery of their immunological balance (KAY, 1971; MARCOLONGO and DI PAOLO, 1975). Favourable results have been obtained by RZEPECKI *et al.* (1974) after grafts of thymus tissue from myasthenia gravis patients into subjects with leukaemia or Hodgkin's disease.

In the case of congenital immune deficiency patients (associated with a high risk of cancer), a simple thymus graft was able to correct the immune abnormalities of two patients with the embryonic thymus deficiency syndrome of DI GEORGE (AUGUST *et al.*, 1968; CLEVELAND *et al.*, 1968): the establishment of a long lived dividing population of T-cells of donor origin was not shown, while the peripheral blood lymphocyte level returned rapidly to normal: this observation suggests a humoral mechanism. The hypothesis was confirmed by a graft of a thymus in a millipore chamber (STEELE *et al.*, 1972), which corrected identically the immune disorder of another case of Di George disease.

Studies of the effect of *thymosin* in neoplasia patients are still exceptional.

SCHAFER *et al.* (1975) administered thymosin to 7 immunodeficient patients with advanced cancers in successive doses of 1 mg/m^2/i.d. per day, then 10 or 30 mg/m^2/day s.c. or i.m. for 7 days. Lymphocyte blastogenesis, the number of positive skin tests, active erythrocyte rosettes and total erythrocyte rosettes increased in 3, 4, 3 and 3 patients, respectively, while they decreased in 0,0, 2 and 1 subjects, respectively. In conclusion, immunological reactivity increased in 5 patients. One showed tumour progression, the others remained stable.

With all this data, clinicians are very much tempted to use thymosin for immunorestoration. GOLDSTEIN *et al.* (1975) showed that this hormone increases the percentage of T-cell rosettes in patients with primary or secondary immunodeficiency syndrome.

Given to a 5-year-old girl with thymic hypoplasia, thymosin was able to increase rosette forming cells to 50%, while her skin test reactivity to mumps and Candida antigens became positive. However, the in vitro lymphocyte response to PHA and to an allogeneic lymphocyte population remained unchanged.

WARA *et al.* (1975) have shown that the low number of T rosette forming cells of a thymic insufficiency patient increased by incubation in vitro with thymosin. They demonstrated that, after in vivo treatment of such a patient with this agent, the number of rosette forming cells increased almost to normal, while his immune response improved significantly. This laboratory test may be useful for indicating such a treatment.

6.2.2.3. Adjuvants

Levamisole has been declared to be able to induce delayed hypersensitivity to PPD, DNCB (HIRSHAUT *et al.*, 1973; TRIPODI *et al.*, 1973; VERHAEGEN *et al.*, 1973), mumps, Candida, and streptokinase-streptodornase (LEVO *et al.*, 1975) in patients who did not respond to these respective antigens.

LEVO *et al.* (1975) described an increase of PHA-lymphocyte transformation in subjects with low initial reactivity. He also described a discrete increase of T-(sheep red cell rosette forming) cells in the peripheral blood of lymphopaenic patients with Hodgkin's disease.

BINIAMINOV and RAMOT (1975) found that in vitro incubation of peripheral blood lymphocytes of Hodgkin's disease patients with 40 µg per ml of levamisol, increases the ability to form E-rosettes. VERHAEGEN *et al.* (1975) confirmed this phenomenon and observed that inhibition of E-rosette forming lymphocytes by azathioprine could be reversed by this agent.

RENOUX *et al.* (1976) treated patients suffering from advanced tumours, which were resistant to radio and chemotherapy, with levamisol. They had shown evidence of a deeply impaired T-cell reactivity. Two months and six months later, the authors observed an increase in the number of T-cells and of their capacity to respond to PHA. A preliminary report on the patient's survival correlates survival time and

T-cell reactivity, which should be accepted with caution because of the heterogeneity of the population.

HIRSHAUT *et al.* (1974) reported only a very limited effect. As far as the author's experience of chronic lymphoid leukaemia patients is concerned, they have not observed any effect, either on humoral or on cell-mediated immunities (BELPOMME and SIMLLER, 1976).

Treatment with *BCG* on scarifications of Burkitt's lymphosarcoma patients with negative nonspecific immune responses, gave BLUMING *et al.* (1974) positive responses to one or more recall antigens as seen in Table 6.XVII.

Table 6.XVII. Specific and nonspecific effect of repeated BCG vaccination on cutaneous reactivity for four recall antigens (BLUMING *et al.*, 1974)

Effect	Mode of BCG administration		
	Total	Intradermal	Scarification
Conversion from negative to positive	6	0	6[a]
Conversion from positive to negative	4	2[b]	2[c]

[a] Tests included: PPD (4), mumps (3), streptokinase-streptodornase (SKSD) (1), Candida (1).

[b] Tests included: SKSD (1), mumps, Candida (1).

[c] Tests included: SKSD (2).

Treating immune deficiency patients who had negative delayed hypersensitivity skin tests with BCG, allowed us to show that the application of the living fresh Pasteur Institute mycobacterium was frequently efficient in converting negative response patients to positive response patients. However, comparing the effect of different modalities of administration, SIMMLER *et al.* (1975b, 1976) showed that the most efficient method on skin delayed hypersensitivity and mitogen lymphocyte transformation, if not the only one, is the application of 75 mg, 3 times per week for 2 weeks on one scarified area, given according to the method described in Chapter 7. The application on 4 scarified areas, of 300 mg, per os ingestion, or i.v. injection, were rarely, or never, effective (Tables 6.XVIII, 6.XIX and 6.XX). The importance of the BCG modality of administration factor, which is probably a dose factor, will be dealt with later.

Table 6.XVIII. Phase I trial on BCG: dosage schedules (SIMMLER *et al.*, 1976b)

Route of administration		Dosage	Culturable units	Time schedule
Scarification	1 Member	75 mg	375 x 10^6	2 x week/2 weeks
	4 Members	100 mg	500 x 10^6	
Intradermal or Intratumoural		1,25 mg	8.75 x 10^6	once (2-3 times in 2 cases)
Per os		75 mg	375 x 10^6	3 x week/2 weeks
Aerosol		25 mg	125 x 10^6	once
Intravenous		0.6 - 1.4 mg	1-8 x 10^6	once (2-3 times in 3 cases)

Table 6.XIX. Effect of different modalities of BCG on skin test in anergic patients[a] (SIMMLER *et al.*, 1976b)

	B C G				
	On one scarified area	On four scarified areas	On four scarified areas + per os	Per os	I.V.
Number of patients	(7)	(8)	(5)	(8)	(5)
Conversion to positive reaction	5	2	1	2	0
Unchanged reaction	2	6	4	6	5

[a] Delayed hypersensitivity to primary antigens (DNCB and KLH) and/or recall antigens (mumps, PPD, Candida).

SOKAL and AUNGST (1969) found, in Hodgkin's disease patients treated with BCG, a correlation between PPD conversion after BCG and the prognosis of the disease.

Table 6.XX. Effect of different modalities of BCG in patients non responsive in vitro to mitogens[a] (SIMMLER *et al.*, 1976b)

	B C G				
	On one scarified area	On four scarified areas	On four scarified areas + per os	Per os	I.V.
Number of patients	(3)	(0)	(2)	(2)	(2)
Conversion to positive reaction	3	0	1	1	0
Unchanged reaction	0	0	1	1	2

[a] Lymphocyte transformation test with PHA and PWM.

Treatment of PPD nonresponding patients with prostate cancer by the administration of BCG by mouth failed to restore the skin reactions against this antigen (MERRIN *et al.*, 1975).

MER induced conversion of negative skin tests to positive in seven patients out of 27 in RICHMAN's experience (1975).

C. parvum seems, according to THORNES (1975), to "delay suppression or blockage" of the cellular immune system. ISRAEL (1974), using the same agent, reported conversion from negative to positive skin tests, but in most patients, C. parvum was administered in combination with chemotherapy. Conversion from negative skin tests to positive would be more frequent, according to Gutterman (pers. comm.), with the i.v. injection of this corynebacteria than with the subcutaneous injection.

Polynucleotides, poly I-poly C and poly A-poly U are currently being studied by our group in immunodepressed patients. In a preliminary trial we used poly I-poly C (MATHÉ *et al.*, 1970). We did not obtain any conversion of negative skin tests into positive ones. Only some in vitro responses have been increased (SIMMLER *et al.*1976c).

6.2.2.4. Immune "Restoration" by Proteolysis

THORNES (1974) published a paper in which he wrote that, after induction of enhanced proteolysis (fibrinolysis) by streptokinase or protease I[2], the skin reactions of delayed hypersensitivity occurred with antigens to which the host had been sensitized but was anergic. He suggests that lymphocytes' reactive sites may be unmasked by this treatment, as he observed the increased binding of sheep red blood cells at the surface of T-lymphocytes after exposure to protease I.

6.2.2.5. Immunorestoration by Surgery

One has to be careful before concluding that an immunomanipulation agent is able to restore immunity.

It has been found that radical surgery may increase the number of circulating lymphocytes, without any significant decrease of lymphocyte stimulation by PHA (McCREDIE, 1973). In some patients, a striking shift from negative to positive response of the delayed hypersensitivity has been observed after surgery (STEWART, 1973). Furthermore, this improvement of cell-mediated immunity after an operation, has been detected in cancer bearing patients, in association with a good prognosis (MORTON, 1972).

6.2.2.6. Immunorestoration by Cancer Chemotherapy

Three patients with Burkitt's lymphosarcoma, who had been treated with cyclophosphamide by MAGRATH *et al.* (1974) and kept as controls for an immunotherapy group, presented conversions from negative to positive reactions to one or more recall antigens.

The author has observed such conversions in many patients after chemotherapy with different drugs. An example is shown in Figure 6.4.

In certain diseases, like lupus erythematosus, STROBER (cited by MacDEVITT, 1973) observed one patient who had no T-cells, and who regained them after therapy with corticosteroid and immunosuppressors.

---

2 Brinase.

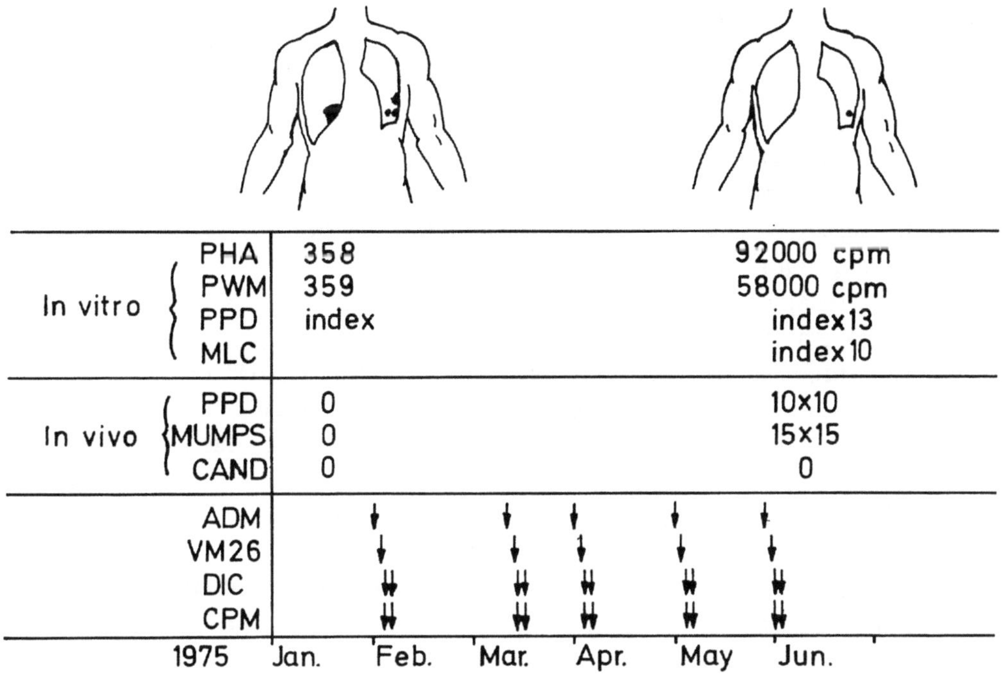

Fig. 6.4. Example of immunorestoration after chemotherapy

6. 2.2.7. Combined Methods

McILVANIE (cited by J. Amer. Med. Assoc., 1974) treated anergic Hodgkin's disease patients with lyophilized BCG (0.06 mg, i.d./week) and so-called "transfer factor", both applied just after chemotherapy. He obtained "objective improvement". It is regrettable that he did not divide the patients into four groups: controls, BCG, BCG + transfer factor, and chemotherapy.

ISRAEL (1974) converted more patients who were negative as far as delayed hypersensitivity to PPD, mumps, candidin, and DNCB is concerned, to positive with the combination of C. parvum and chemotherapy than with C. parvum alone. Hence, immunorestoration may often be due to complex mechanisms, including a general improvement of the patient.

References

AISENBERG, A.C.: J. Clin. Invest. 41, 1964 (1962).

AISENBERG, A.C.: Cancer Res. 26, 1152 (1966).

AISENBERG, A.C., LESKOWITZ, S.: New Engl. J. Med. 268, 1269 (1963).

ALDRICH, R.A., STEINBERG, A.G., CAMPBELL, D.C.: Pediatrics 13, 133 (1954).
ALEXANDER, P., BENSTED, J., DELORME, E.J., HALL, J.G., HODGETT, J.: Proc. Royal Soc. London B174, 237 (1969).
ALLISON, A.C., LAW, L.: Proc. Soc. Exp. Biol. Med. 127, 207 (1968).
AL-SARRAF, M., SARDESAI, S., VAITKEVICIUS, V.K.: Oncology 26, 357 (1972).
ARCHIBALD, R.B., FRENSTER, J.H.: Nat. Cancer Inst. Monogr. 36, 239 (1973).
AUGUST, C.S., ROSEN, F.S., FILLER, R.M., JANEWAY, C.A., MARKOWSKI, B., KAY, H.E.M.: Lancet 2, 1210 (1968).
BACH, J.R., DARDENNE, M., GOLDSTEIN, A.L., GUHA, A., WHITE, A.: Proc. Nat. Acad. Sci., U.S. 68, 273 (1971).
BARCLAY, T.J., WEISSMAN, I.L., KAPLAN, H.S.: p. 117 in "The Thymus" (V. Defendi, D. Metcalff, eds.). Philadelphia: Wistar Institute Press 1964.
BELPOMME, D., CARDE, P., OLDHAM, R.K., MATHÉ, G., JACQUILLAT, C., CHELLOUL, N., WEIL, M., AUCLERC, G., WEISBERGER, G., TANZER, T., BERNARD, J.: p. 115 in "Complications of Cancer Chemotherapy" (G. Mathé, R.K., Oldham, eds.) vol. 49 R.R.C.R., Heidelberg: Springer-Verlag 1974.
BELPOMME, D., DANTCHEV, D., JOSEPH, R., SANTORO, A., FEUILHADE DE CHAUVIN, F., LELARGE, N., GRANDJON, D., PONTVERT, D., MATHÉ, G.: Europ. J. Cancer in press (1976).
BELPOMME, D., JOSEPH, R., NEVARES, L., GERARD-MARCHANT, R., HUCHET, R., BOTTO, I., GRANDJON, D., MATHÉ, G.: New Engl. J. Med. 291, 1417 (1974).
BELPOMME, D., SIMMLER, M.C.: In preparation (1976).
BINIAMINOV, M., RAMOT, B.: Lancet 1, 464 (1975).
BLOCK, J., HAYNES, H., THOMPSON, W.L., NEIMAN, P.: J. Nat. Cancer Inst. 42, 973(1969).
BLUMING, A., VOGEL, L., ZIEGLER, J.L.: p. 415 in "Investigation and Stimulation of Immunity in Cancer patients" (G. Mathé, R. Weiner, eds.). Vol. 47 R.R.C.R., Heidelberg: Springer-Verlag 1974.
BOMSKOV, G., SLADOVIC, L.: Deutsch. Med. Wochschr. 66, 589 (1940).
BRAEMAN, J., DEELEY, T.J.: Ann. Clin. Res. 4, 355 (1972).
BRANDES, L.G., GALTON, D.A.G., WILTSHAW, E.: Lancet 2, 293 (1971).
BRENT, L., MEDAWAR, P.B.: Proc. Royal Soc. London 165, 281 (1966).
BRULEY-ROSSET, M., BOTTO, H.G., GAUTHIER, H., POUILLART, P., FLORENTIN, I., HUGUENIN, P.: Israel J. Med. Sci. in press (1976).
BRUTON, O.C.: Pediatrics 9, 722 (1952).
CANELLOS, G.P., DE VITA, V.T., ARSENEAU, J.C., JOHNSON, R.C.: p. 108 in "Complications of Cancer Chemotherapy" (G. Mathé, R.K., Oldham, eds.). Vol. 49 R.R.C.R. Heidelberg: Springer-Verlag 1974.
CASTRO, J.E., MEDAWAR, P.B., HAMILTON D.N.H.: p. 237 in "Immunopotentiation" (Ciba Found., Symp.no. 18). 1 Vol., Amsterdam: Elsevier Excerpta Medica 1973.
CEGLOWSKI, W.S., FRIEDMAN, H.: J. Immunol. 101, 594 (1968).
CHAR, D.H., LEPOURLHIET, A., LEVENTHAL, B.G., HERBERMAN, R.B.: Int. J. Cancer 12, 409 (1973).
CHASE, M.W.: Cancer Res. 26, 1097 (1966).

CLEVELAND, W.W., FOGEL, B.J., BROWN, W.T., KAY, H.E.M.: Lancet 2, 1211 (1968).
CONE, R.E., JOHNSON, A.G.: J. Exp. Med. 133, 665 (1971).
CORDER, M.P., YOUNG, R.C., BROWN, R.S., DE VITA, V.T.: Blood 39, 595 (1972).
CROWTHER, D., HAMILTON-FAIRLEY, G., SEWELL, R.L.: Brit. Med. J. 2, 473 (1969).
DALMASSO, A.P., MARTINEZ, C., SJODIN, K., GOOD, R.A.: J. Exp. Med. 118, 1089 (1963).
DI GEORGE, A.M.: J. Pediat. 67, 907 (1965).
DOLL, R.: in Tenth International Cancer Congress of Oncology. New-York: Yearbook Med. Publ. 1970.
DRESSER, D., ROSENFELD, S.: Proc. Nat. Acad. Sci. 71, 4429 (1974).
DUPUY, J.M., KOURILSKY, F.M., FRADELIZZI, D., FEINGOLD, N., JACQUILLAT, Cl., BERNARD, J., DAUSSET, J.: Cancer 27, 323 (1970).
EILBER, F.R., MORTON, D.L.: Cancer 25, 362 (1970).
FAHEY, J.L., SCOOGIN, R., UTZ, J., SZWEED, C.: Am. J. Med. 35, 698 (1963).
FROST, P., LANCE, E.M.: p. 28 in "Immunopotentiation" (Ciba Found. Symp. no. 18). 1 Vol., Amsterdam: Elsevier Excerpta Medica 1973.
FUDENBERG, H.H., LEVIN, A.S., SPITLER, L.E., WYBRAN, J., BYER, V.: Hospital Practice 95 (1974).
GATTI, R.A., GOOD, R.A.: Cancer 28, 89 (1971).
GATTI, R.A., HARRIOCH, D., GOOD, R.A.: p. 339 in "Proceedings of the Fifth Leukocyte Culture Conference" (J. Harris, ed.). New-York: Academic Press (1970).
GATTI, R.A., WHITTAKER, M.G., REES, K., CLARK, G.C.: Lancet 1, 1351 (1971).
GITLIN, D., GRAIG, J.M., Pediatrics 32, 517 (1963).
GLANZMANN, E., RINIKER, P.: Ann. Paediat. 175, 1 (1950).
GOLDSTEIN, A.L., ASANUMA, Y., WHITE, A.: Recent. Progr. Hormone Res. 26, 505 (1970).
GOLDSTEIN, A.L., WARA, D.W., AMMANN, A.J., SAKAI, H., HARRIS, H.S., THURMAN, G.B., HOOPER, J.A., COHEN, G.H., GOLDMAN, A.S., COSTANZI, J.J., McDANIEL, M.C.: Transplant. Proc. 7, 681 (1975).
GOLDSTEIN, A.L., WHITE, A.: p. 465 in "Biochemical actions of hormones" (G. Litwak, ed). 1 Vol., New-York: Academic Press, 1970.
GOOD, R.A., ZAK, S.J.: Pediatrics 18, 109 (1956).
GROB, P.J.: Med. Hyg. 32, 1961 (1974).
GROB, P.J., FRANKE, C.H., REYMOND, J.F., FREI-WETTSTEIN, M.: Europ. J. Clin. Invest. 5, 33 (1975).
HARDY, M.A., ZISBLATT, M., LEVIN, N., GOLDSTEIN, A.L., LILLY, F., WHITE, A.: Transplant. Proc. 3, 926 (1971).
HERSH, E.M., CURTIS, J., HARRIS, J.E.: p. 146 in "Leukemia-Lymphoma". Chicago: Yearbook Medical Publ. 1970.
HERSH, E.M., GUTTERMAN, J.U., MAVLIGIT, G.M.: Cancer Treatment Rev. 1, 65 (1974).
HERSH, E.M., OPPENHEIM, J.J.: New England J. Med. 273, 1006 (1965).
HERSH, E., WHITECAR, J.P., MacCREDIE, K.B., BODEY, G.P., FREIREICH, E.J.: New England J. Med. 285, 1211 (1971).
HILBERG, R.W., BALCERZAK, S.P., LO BUGLIO, A.F.: Cell. Immunol. 7, 152 (1973).

HIRSHAUT, Y., PINSKY, C., FRIED, J., OETTGEN, H.: Proc. Amer. Assoc. Cancer Res. 15, 126 (abstract No. 503) (1974).

HIRSHAUT, Y., PINSKY, C., MARQUARDT, H., OETTGEN, H.F.: Proc. Am. Assoc. Cancer Res. 14, 109 (abstract no. 436) (1973).

HOWARD, J.G., SCOTT, M.T., CHRISTIE, G.H.: p. 101 in "Immunopotentiation" (Ciba Found. Symp. 18). 1 Vol., Amsterdam: Elsevier Excerpta Medica 1973.

HUCHET, R., LHERITIER, J.: in "Lymphocytes and Macrophages in Cancer Patients" (G. Mathé, I. Florentin, M.C. Simmler, eds.). R.R.C.R., Heidelberg: Springer-Verlag, in press 1976.

ISRAEL, L.: p. 486 in "Investigation and Stimulation of Immunity in Cancer Patients" (G. Mathé, R. Weiner, eds.). Vol. 47 R.R.C.R., Heidelberg: Springer-Verlag 1974.

J. Amer. Med. Assoc. 227, 131 (1974).

JENKINS, V.K., OLSON, M.H., ELLIS, H.N.: Texas Rep. Biol. Med. 31, 19 (1973).

KATZ, D.H.: p. 140 in "Immunopotentiation" (Ciba Found. Symp. no. 18). 1 Vol., Amsterdam: Elsevier Excerpta Medica (1973).

KAUR, J., SPIERS, A.S., CATOVSKY, D., GLATON, D.A.: Lancet 2, 800 (1974).

KAY, H.E.M.: p. 249 in "Ontogeny of Acquired Immunity". Amsterdam: ASP 1971.

KERBEL, R.: in "Discussion of Immunopotentiation" (Ciba Found. Symp. 18). 1 Vol., Amsterdam: Elsevier Excerpta Medica 1973.

KIRKPATRICK, C.H., RICH, R.R., SMITH, T.K.: J. Clin. Invest. 51, 2948 (1972).

LAWRENCE, H.S.: J. Clin. Invest., 34, 219 (1955).

LAWRENCE, H.S.: Adv. Immunol. 2, 196 (1969).

LAWRENCE, H.S.: p.21 in "Immunologic Intervention" (J.W. Uhr, M. Landy, eds.). 1 Vol., New-York: Academic Press 1971.

LEE, A., ROWLEY, M., McKAY, I.R.: Brit. J. Cancer 24, 454 (1970).

LEVENTHAL, B.G., COHEN, P., TRIEM, S.C.: Israel J. Med. Sci. 10, 866 (1974).

LEVIN, A.S., BYERS, V.S., FUDENBERG, H.H., WYBRAN, J., HACKETT, A.J., JOHNSTON, J.O., SPITLER, L.E.: J. Clin. Invest. 55, 487 (1975).

LEVIN, A.G., McDONOUGH, E., MILLER, D.G., SOUTHAM, C.: Ann. New-York Acad. Sci. 120, 400 (1964).

LEVO, Y., ROTTER, V., RAMOT, B.: Biomedicine 23, 198 (1975).

LEVY, J.G., WHITNEY, R.B., SMITH, A.G., PANNO, L.: Brit. J. Cancer 30, 289 (1974).

LO BUGLIO, A.F., NEIDHART, J.A., HILBERG, R.W., METZ, E.N., BALCERZAK, S.P.: Cell. Immunol. 7, 159 (1973).

LORING, M., SCHLESINGER, M.: Cancer Res. 30, 2204 (1970).

LOUIS-BAR: Confinia Neurol., 4, 32 (1941).

MacCREDIE, J.A.: Arch. Surgery 107, 162 (1973).

MacDEVITT, H.O.: p. 332 in "Discussion of Immunopotentiation" (Ciba Found. Symp. no. 18). 1 Vol., Amsterdam: Elsevier Excerpta Medica 1973.

MacKHANN, C.F.: Transplantation 8, 209 (1969).

MAEDA, Y.Y., HAMURO, H.J., YAMADA, Y.O., ISHIMURA, K., CHIHARA, G.: p. 259 in "Immunopotentiation" (Ciba Found.Symp. no.18). 1 Vol., Amsterdam: Elsevier Excerpta Medica 1973.

MAGRATH, I.T., ZIEGLER, J.L., BLUMING, A.Z.: p. 461 in "Investigation and Stimulation of Immunity in Cancer Patients" (G. Mathé, R. Weiner, eds.). Vol. 47 R.R.C.R., Heidelberg: Springer-Verlag 1974.

MAKINODAN, T., CHINO, F., LEVER, W.E., BREWEN, B.S.: J. Gerontol. 26, 508 (1971).

MARCOLONGO, R., DI PAOLO, N.: Blood 41, 625 (1975).

MARTIN, M., HUCHET, R., FLORENTIN, I., HALLE-PANNENKO, O., JURCZYK-PROCYK, S., PRITCHARD, L., MATHÉ, G.: p. 59 Fourth Annual Conference of the International Society for Experimental Hematology, Yugoslavia, September (abstract No. 132) 1975.

MARX, J.L.: Science 187, 1183 (1975).

MATHÉ, G.: C.R. Acad. Sci., Paris 264, 2702 (1967b).

MATHÉ, G.: Rev. Fr. Et. Clin. Biol. 12, 380 (1967a).

MATHÉ, G., AMIEL, J.L. SCHWARZENBERG, L., SCHNEIDER, M., HAYAT, M., VASSAL, F. DE, JASMIN, C., ROSENFELD, C., SAKOUHI, M., CHOAY, J.: Europ.J.Clin.Biol.Res.15,671 (1970a).

MATHÉ, G., KENIS, Y.: La Chimiothérapie des Cancers (Leucémies, hématosarcomes et Tumeurs Solides) 3rd ed., Paris: Expansion Scientifique 1975.

MATHÉ, G., SCHNEIDER, M., SCHWARZENBERG, L.: Rev. Fr. Et. Clin. Biol. 13, 695 (1968).

MATHÉ, G., SCHNEIDER, M., SCHWARZENBERG, L.: Europ. J. Cancer 6, 23 (1970b).

MATHÉ, G., SCHWARZENBERG, L., AMIEL, J.L., SCHNEIDER, M., CATTAN, A., SCHLUMBERGER, J.R.: Europ. J. Cancer 3, 423 (1967).

MATHÉ, G., TUBIANA, M.: The natural history, diagnosis and treatment of Hodgkin's disease (Vol. I and II). Copenhagen: Munksgaard 1973.

MERRIN, C., HAN, T., KLEIN, E., WAJSMAN, Z., MURPHY, G.P.: Cancer Chemoth. Rep. 59, 157 (1975).

MOORE, M., LAWRENCE, N., NISBET, N.W.: Int. J. Cancer 15, 897 (1975).

MORSE, P.A., jr., DERAPS, G.D., SMITH, G.V., RAJU, S., HARDY, J.D.: Clin. Res. 21, 71 (abstract) (1973).

MORTON, D.L.: Nat. Cancer Inst. Monogr. 35, 375 (1972).

NEZELOF, C., JAMMET, M.L., LORTHOLARY, P., LABRUNE, B., LAMY, M.: Arch. Franç. Pediat. 21, 897 (1964).

OPPENHEIM, J., WHANG, H., FREI, E.: Blood 26, 131 (1965).

OSOBA, D., MILLER, J.F.A.P.: J. Exp. Med. 119, 177 (1964).

Patterns in cancer mortality in the United states: 1950-1967. Nat. Cancer Inst. Monograph 33 (1971).

PENN, I.: Transplant. Proc. 7, 323 (1975).

PINSKY, C., HIRSHAUT, Y., OETTGEN, H.: Nat. Cancer Inst. Monogr. 39, 225 (1973).

PINSKY, C.M., OETTGEN, H.F., EL DOMEIRI, A., OLD, L.J., BEATTIE, E.J., BURCHENAL, J.H.: Proc. Amer. Assoc. Cancer Res. 12, 100 (1971).

POUPON, M.F.: Cellules suppressives responsables de la suppression de la stimulation des lymphocytes par des mitogènes dans deux modèles expérimentaux: 1. Souris porteuses de tumeurs; 2. souris traitées par l'endoxan. Thesis, Université de Paris-Sud, Faculté de Médecine du Kremlin-Bicêtre, 1975.

PRICE, G.B., MAKINODAN, T.: J. Immunol. 108, 413 (1972).
Radiation, immunity and cancer. Lancet 2, 217 (1972).
RENOUX, G., RENOUX, M.: C.R. Acad. Sci., Paris 274, 3034 (1972).
RENOUX, G., RENOUX, M., PALAT, A.: in preparation (1976).
RICHMAN, S.P.: Proc. Amer. Ass. Cancer Res. 16, 227 (abstract no. 1025) (1975).
ROCKEY, J.H., HANSON, L.A., HEREMANS, J.F., HUNKEL, H.G.: J. Lab. Clin. Med. 63, 205 (1964).
ROWLAND, G.F., EDWARDS, A.J., HURD, C.M., SUMNER, M.R.: J. Nat. Cancer Inst. 47, 2 (1971).
RUNDLES, R.W., COONRAD, E.V., ARENDS, T.: Am. J. Med. 16, 842 (1954).
RZEPECKI, W.M., LUKASIENWICZ, M., ALEKSANDROWICZ, J., SZMIGIEL, Z., SKOTNICKI, A., LISIEWICZ, J.: Lancet 1, 990 (1974).
SALMON, S.E., FUDENBERG, H.H.: Blood 33, 300 (1969).
SCHAFER, L.A., WASHINGTON, M.L., GOLDSTEIN, A.L.: Proc. Amer. Ass. Cancer Res. 16, 233 (abstract no. 1051) (1975).
SCHEID, M.P., HOFFMANN, M.K., KONDO, K., HAMMERLING, U., ABBOTT, J., BOYSE, E., COHEN, G.H., HOOPER, J.A., SCHULOF, R.S., GOLDSTEIN, A.L.: J. Exp. Med. 138, 1027 (1973).
SCHNEIDER, M.: Rev. Fr. Et. Clin. Biol. 13, 877 (1968).
SCHNEIDER, M., CATTAN, A., HAYAT, M., TOULEMONDE, P.: Bull. Cancer 54, 381 (1967).
SCHNEIDER, M., DORE, J.F.: Rev. Fr. Et. Clin. Biol. 14, 1010 (1969).
SCHNEIDER, M., MATHÉ, G., SCHWARZENBERG, L., POUILLART, P., WEINER, R., AMIEL, J.L., HAYAT, M., JASMIN, C., VASSAL, F. DE: p. 42 in "Investigation and Stimulation of Immunity in Cancer Patients"(G. Mathé, R. Weiner, eds.). Vol. 47 R.R.C.R., Heidelberg: Springer-Verlag 1974.
SCHNEIDER, M., SCHWARZENBERG, L., AMIEL, J.L., HAYAT, M., MASCARO, G., OTMEZGUINE, Y., MATHÉ, G.: Presse Med. 78, 1969 (1970).
SCOTT, M.T.: Cell. Immunol., 5, 459 (1972a).
SCOTT, M.T.: Cell. Immunol., 5, 469 (1972b).
SCHWARTZ, R.S.: Prog. in Allergy 9, 246 (1965).
SELIGMANN, M., FUDENBERG, H.H., GOOD, R.A.: Amer. J. Med. 45, 817 (1968).
SHARMAN, C., CROSSEN, P.E., FITZGERALD, P.H.: Scand. J. Haematol. 3, 375 (1966).
SILK, M.: Cancer 20, 2088 (1967).
SIMMLER, M.C., RAMEAU, G., CHOU, M.J., MATHÉ, G.: Israel J. Med. Sci. in press (1976a).
SIMMLER, M.C., SCHWARZENBERG, L., MATHÉ, G.: Cancer Immunol. Immunoth. in press (1976b).
SIMMLER, M.C., BRULEY-ROSSET, M., BELPOMME, D., SCHWARZENBERG, L.: Europ. J. Cancer, in press (1976c).
SOKAL, J.E., AUNGST, L.W.: Cancer 24, 128 (1969).
SOLOMON, A., WALDMANN, T., FAHEY, J.L.: J. Lab. Clin. Med. 62, 1 (1963).
SOUTHAM, C.M., LEVIN, A.G.: Blood 27, 734 (1966).

STEELE, R.W., LIMAS, C., THURMAN, G.B., SCHUELEIN, M., BAUER, H., BELLANTI, J.A.: New Engl. J. Med. 287, 787 (1972).
STEWART, A.M.: J. Nat. Cancer Inst. 50, 625 (1973).
STJERNSWARD, J.: J. Nat. Cancer Inst. 35, 885 (1965).
STJERNSWARD, J., JONDAL, M., VANKY, F., WIGZELL, H., SEALLY, R.: Lancet 1, 1352 (1972).
TAVADIA, H.B., GOUDIE, R.B., NICOLL, W.D.: Clin. Exp. Immunol. 16, 177 (1974).
THOMAS, J.W., COY, P., LEWIS, H.S., YUEN, A.: Cancer 27, 1046 (1971).
THORNES, R.D.: Lancet 2, 382 (1974).
THORNES, R.D.: Lancet 1, 626 (1975).
TRAININ, N., LINKER-ISRAELI, M., SMALL, M., BOIATO-CHEN. L.: Int. J. Cancer 2, 326 (1967).
TRAININ, N., SMALL, M., GLOBERSON, A.: J. Exp. Med. 130, 765 (1969).
TRIPODI, D., PARKS, L.C., BRUGMANS, J.: New Engl. J. Med. 289, 354 (1973).
TRUBOWITZ, S., MASEK, B., DEL ROSARIO, A.: Cancer 19, 2019 (1966).
TUBIANA, M., ATTIE, E., FLAMANT, R., GERARD-MARCHANT, R., HAYAT, M.: Cancer Res. 31, 1801 (1971).
ULTMANN, J.E., FISH, W., OSSERMAN, F., GELLHORN, A.: Ann. Intern. Med. 51, 501 (1959).
VERHAEGEN, H., DE COCK, W., DE CREE, J., VERBRUGGEN, G., VERHAEGEN-DECLERCQ, M., BRUGMANS, J.: Lancet 1, 978 (1975).
VERHAEGEN, M., DECREE, J., DECOCK, W., VERBRUGGEN, F.: New Engl. J. Med. 289, 1148 (1973).
WAGNER, V., JANKU, O., WAGNEROVA, M., MATES, J., KONICKOVA, Z.: Neoplasma 19, 75 (1972).
WALDMANN, T.A.: Ann. Intern. Med., 77, 605, 1972.
WALDMANN, T.A., BRODER, S., BLAESE, R.M., DURM, M., BLACKMAN, M., STROBER, W.: Lancet 2, 609 (1974).
WALKER, J.S., DAVIS, D., DAVIES, P., FREEMAN, C.B., HARRIS, R.: Brit. J. Cancer 27, 203 (1973).
WARA, D.W., GOLDSTEIN, A.L., DOYLE, N.E., AMMANN, A.J.: New Engl. J. Med. 292, 70 (1975).
WEED, R.I.: Blood 26, 257 (1965).
WEISS, D.W., YASHPHE, D.J.: p. 163 in "Dynamic Aspects of Host-Parasite Relationships" (A. Zuckerman, D.W. Weiss, eds.). 1 Vol., New-York: Academic Press 1973.
WINGARD, D.W., LAIRY, R., HUMPHREY, L.J.: J. Surg. Res. 7, 430 (1967).
WYBRAN, J., CHANTLER, S., FUDENBERG, H.H.: Lancet 1, 126 (1973).
ZIEGLER, J.B., HANSEN, P., PENNY, R.: Clin. Immunol. Immunopath. 3, 451 (1975).
ZISBLATT, M., GOLDSTEIN, A.L., LILLY, F., WHITE, A.: Proc. Nat. Acad. Sci. U.S. 66, 1170 (1970).

Chapter 7
Developmental Pharmacology

7.1. Systemic Specific Active Immunotherapy or Immunoprophylaxis

Many experiments and some clinical trials have been conducted on specific stimulation for immunotherapy as well as for immunoprevention.

7.1.1. Tumour Cells and/or Antigens

There are several problems concerning the use of tumour cells, especially their histocompatibility with regard to the recipient, their preservation, their sterilization, and the possible increase of their immunogenicity.

7.1.1.1. Should Antigenicity be Restricted to Tumour-Associated Antigens or Not?

The question asks whether histocompatibility antigens can compete with tumour-associated antigens (TAA) and decrease the immune response against the latter, since it has been shown that antigenic competition may inhibit some immune responses to one or both antigens under certain conditions (LAWRENCE and SIMONSEN, 1967).

Preparing sera against TAA, MOTTA, in the author's laboratory (1970a, 1970b), showed that, by pretreating tumour cells with antisera against normal histocompatibility antigens, he was able to obtain humoral reactions almost purely directed against TAA, and which were very potent as seen in Table 7.I. This creates a method of obtaining cells with almost pure TAA for active immunization; one of the advantages of the method, if it could be developed to perfection, would be the absence of immunization against histocompatibility antigens, which could be very precious for an immune investigation later or, in case of relapse, for an allogeneic bone marrow graft (MATHÉ *et al.*, 1971a).

Table 7.I. Increase of the specificity of the immunological reaction against the C. Friend virus induced antigen by a differential "masking" of the normal mouse antigens in the various species (MOTTA, 1970a,b)

Immunized species	Type of injected cells	Masking by	Cytotoxic potency of the antisera obtained: Number of normal DBA/2 cells killed by 1 ml of antiserum	Number of C. Friend leukaemic cells killed by 1 ml of antiserum
C57Bl/6 mice	DBA/2 C. Friend leukaemic spleen cells	–	154×10^6	136×10^6
		C57Bl/6 anti-DBA/2 normal cells antiserum	4.6×10^6	92×10^6
CF rat	"	–	12×10^6	18×10^6
		CF rat anti-DBA/2 normal cells antiserum	0	25×10^6
"Fauve de Bourgogne" rabbit	"	–	157×10^6	240×10^6
		"Fauve de Bourgogne" rabbit anti-DBA/2 normal cell antiserum	40×10^6	210×10^6

7.1.1.2. *Allogeneic Versus Autologous Materials. Tumour Type Cells for Treated Tumour Types*

Nevertheless, in the author's clinical trials on neoplasias like leukaemias, which, on the basis of animal studies, are believed (GROSS, 1970) to be *virus-induced tumours* and, therefore, supposedly share the same TAA for a given variety (Chapter 2), we have only used allogeneic cells which were irradiated although a graft theoretically is not very feasible.

MOORE *et al.* (1973) used the same material in the treatment of osteosarcoma.

There are several reasons for this choice of allogeneic cells:

(a) When we start immunotherapy, the immune responses, which had been depressed by preimmunochemotherapy (SCHNEIDER *et al.*, 1974; OLDHAM *et al.*, 1976; SIMMLER *et al.*, 1976a), are theoretically restored because we leave a sufficient gap between the two treatments (15 days; see Chapter 6). There may, however, be some patients whose immune responses are not yet sufficient to react efficiently against TAA which are considered weak antigens.

(b) The types of leukaemia which seem to be immunotherapy resistant (MATHÉ *et al.*, 1971b), may be so, due to weak TAA. In their case, the risk taken would not be negligible.

(c) The author had observed, in the preliminary studies conducted before his first immunotherapy trial on murine leukaemia (MATHÉ, 1968), that allogeneic tumour cells are more efficient than isogeneic ones: some adjuvant effect may be achieved by histoincompatibility. KOBAYASHI *et al.* (1975) were able to prevent the growth of isogeneic tumours by a previous graft of normal allogeneic tissues, including skin. LIN and MURPHY (1969) attributed such an action to a lipoprotein which was present in the cells.

Until there is more known about the possible viruses which are respectively responsible for each variety in a group of diseases like acute leukaemias, acute lymphoid leukaemia cells are better used for ALL patients and acute myeloid leukaemia cells for AML patients, though there may be a cross reaction between both varieties (DORÉ and GUIBOUT, 1974).

Perhaps it would be a step forward to use cells of the cytological type in the case of ALL (prolymphoblastic, macrolymphoblastic, microlymphoblastic, prolymphocytic, or immunoblastic) (MATHÉ *et al.*, 1971b, 1974) which the patients present. We have, in fact, shown that some types (prolymphoblastic and macrolymphoblastic) are poorly immunotherapy-sensitive. This may be due to the restricted antigenicity or immunogenicity of their cells, which could be a weak source of antigenic stimulus for treating the immunotherapy-sensitive patients.

We saw in Chapter 2 that TAAs are supposed to be singular to each tumour in the case of *neoplasias of* very likely *chemical carcinogenic origin*. Therefore, if cells are included in immunotherapy, only autologous ones can be used. Few authors have done it; no complications have been mentioned (IKONOPISOV *et al.*, 1970; BLOOM *et al.*, 1970; MORTON *et al.*, 1973).

7.1.1.3. Antigenicity and/or Immunogenicity

One must remember, when taking cells from the peripheral blood or from a tumour, that their antigenicity may have been modulated by antibodies. The phenomenon of *antigenic modulation* has only been demonstrated experimentally (FENYO *et al.*, 1968). However, it may exist in clinical conditions and explain the difficulties met by those who are working on human tumour-associated antigens (DORÉ and GUIBOUT, 1974). It may also explain why the affinity of autologous serums for leukaemic cells increases after culture (BELPOMME *et al.*, 1972), which will be illustrated later. In any case, before using cells for immunotherapy, it is advisable to check that they are sensitive to a serum containing antibodies against TAA [such antibodies have been found in the case of Burkitt's lymphosarcoma, (HENLE *et al.*, 1969, 1970, 1971), and they may exist in acute lymphoid leukaemia (DORÉ and GUIBOUT, 1974), but they are far from being available for all human tumours], or can stimulate autologous lymphocytes, as shown in acute leukaemia (FRIDMAN and KOURILSKY, 1969; LEVENTHAL *et al.*, 1971).

7.1.1.4. Microbiologically Sterilized Cells

Before their cryopreservation, the cells are treated with 4p.1000 formaldehyde solution to inactivate any possible oncogenic or contaminant virus. This procedure, shown by FRIEND (1959), can suppress any leukaemia induction power of the virus, and maintain antigenicity and immunogenicity. This was confirmed by LIN *et al.* (1969) on Ib leukaemia cells.

7.1.1.5. Division-sterilized Cells

Before using irradiation of the cells administered for specific immunotherapy in humans (MATHÉ *et al.*, 1968), L 1210 leukaemia cells from this experimental neoplasia were checked to ensure that they do lose their division power and keep their antigenicity and immunogenicity (MATHÉ, 1968; MATHÉ *et al.*, 1969). This was also observed on other lines of cells used for immunotherapy or immunoprevention: Ib (LIN *et al.*, 1969), $6C_3$HED (PRAGER, 1973) and AkR (DORÉ *et al.*, 1970; MATHÉ *et al.*, 1973a) leukaemias.

For mice L 1210 leukaemia, the dose of in vitro irradiation was 14,000 rads (MATHÉ, 1968); for human leukaemia immunotherapy, in the author's experience, 4,000 rads. Such an irradiation, the other conditions of which are given on Table 7.II, is known to be unable to diminish antigenicity or immunogenicity (see BARTLETT and ZBAR, 1972).

It goes without saying that a higher dose of in vitro irradiation is necessary for human solid tumours where the cells are known to be poorly radiosensitive (10,000 rads), and it may be reasonable, for very radioresistant neoplasias, such as

Table 7.II. Irradiation of leukaemic cells for use in immunotherapy (author's experience)

	L1210 leukaemia	Acute lymphoid and myeloid human leukaemias
Apparatus:	Graviton	Neptune
Dose (rads):	15,000	4,000
Field:	6 x 6 cm	40 x 40 cm
Distance from source:	100 cm	100 cm
Debit:	180 rads/min.	200 rads/min.
Total duration:	84 min.	20 min.

melanomas, to replace irradiation by mitomycin C which, as mentioned earlier, was used by SIMMONS *et al.* (1971a) and SIMMONS and RIOS (1973) for inducing the production of the so-called deblocking factor. It is advisable, in such a case, to incubate the cells for 1 hour with mitomycin C (25 µg/ml/10^6 cells). They may then be washed, and mixed with M199 before being injected (SIMMONS and RIOS, 1972).

Hence, cell sterilization cannot be the same for all tumours, or even for the various types of a given neoplasia: irradiated P 1798 lymphoma cells are not protective in isogeneic hosts against the growth of isogeneic living tumour cells (PRAGER, 1973).

BALDWIN and PIMM (1973) noted, in an experiment on rat 3-methylcholanthrene-induced sarcoma, that 15,000 rads irradiated cells, in conjunction with BCG, were less effective than nonirradiated cells. This observation has not been confirmed. The dose of irradiation may have been too high.

Nevertheless, some authors have been afraid to lose antigenicity and/or immunogenicity by irradiation, and KRONMAN *et al.* (1970) developed an immunotherapy model in guinea pigs, in which the only stimulus was the intradermal injection of living tumour cells. Such a practice is not to be recommended in man. In treating 5 patients with living nonirradiated acute lymphoid leukaemia (ALL) cells, we observed their growth at the site of local injection in 4 subjects (see Chapter 8). The induced tumours then had to be removed surgically or irradiated. This observation is sufficient to make clear that the use of unirradiated or nonmitomycin C-treated cells is a dangerous, hence unethical, procedure.

Would such cells be more efficient than division-sterilized ones? POWLES (1974) treated a group of acute myeloid leukaemia patients during remission with unirradiated cells and BCG. At the time of publication (one and a half year after beginning the trial), 9 out of 17 patients had already relapsed and 3 had died, which is a much poorer result than that obtained by the same author with irradiated cells (POWLES, 1974).

Thus, in vitro irradiation or mitomycin C-treatment of the cells must be recommended for active specific immunotherapy, obviously when autologous cells are used, and even for allogeneic.

7.1.1.6. Cryopreserved Cells

We have used, in our clinical experience since 1963 (MATHÉ *et al.*, 1968, 1976a), cryopreserved allogeneic neoplastic cells (SCHWARZENBERG, 1973).

In the case of *leukaemia*, they were obtained with the IBM blood cell separator (Fig. 7.1) by several daily **leukophereses** from the circulating blood of patients in a perceptible phase of the disease, before any treatment was given (10^{12} can be collected from a patient).

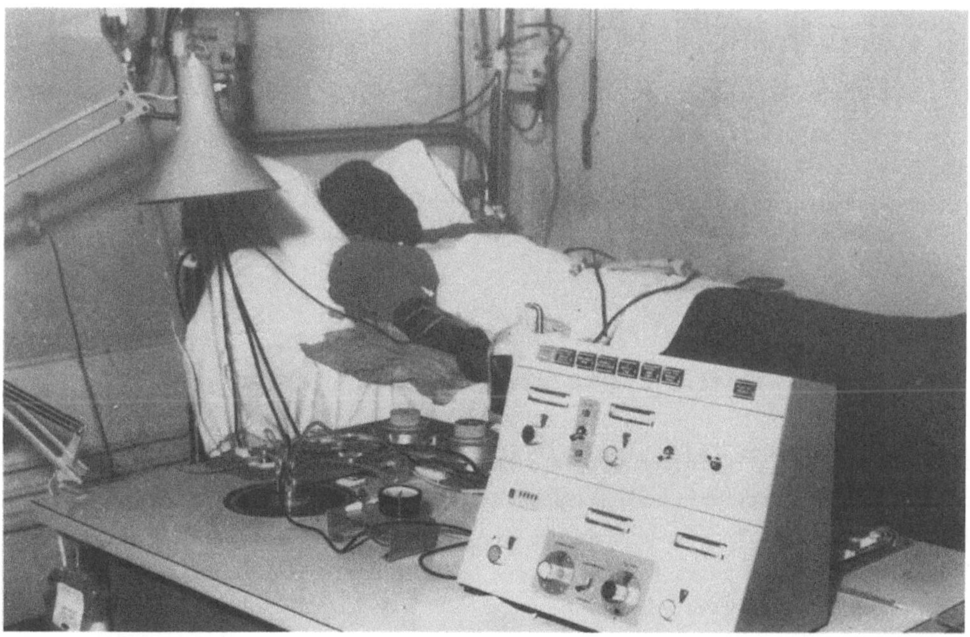

Fig. 7.1. The IBM blood cell in vivo separator used for obtaining leukaemic cells

The cells are sealed in glass ampoules with dimethylsulfoxide solution (10%), frozen slowly and regularly (1° C per min) to - 35° C, and stored at - 196° C in a liquid nitrogen container. A Planer Gas Phase Programmed Freezer is utilized.

When required for use, the cells are rapidly thawed to 37° C, washed and resuspended in Medium 199 at + 4° C. They are distributed into pools of samples coming from several donors so that a patient can never receive his own autologous cells. A trypan blue test applied to the cells from these pools checks their viability after preservation. The percentage of coloured cells, i.e. dead cells, is normally of the order of 5%.

These pooled cells are injected intradermally in the region of the BCG scarifications during the same treatment sessions.

As said above, for the first 6 weeks, the cells are treated with a 4/1000 solution of formaldehyde prior to injection, whereas, after the 6th week, the cells are only irradiated in vitro at a dose of 4000 rads before being injected.

In the case of *solid tumours*, the fragments, obtained by therapeutic surgery or biopsy, are lightly and rapidly minced in a blender and then pressed through a 60-mesh wire screen to obtain a single cell suspension.

7.1.1.7. Pooled Cells

The cells are maintained in ampoules containing 4 x 10^7 elements from a pool of 25 to 30 donors with the same disease (SCHWARZENBERG, 1973; POWLES *et al.*, 1974a).

These cells are pooled because of possible differences between various patients with respect to antigenicity and/or immunogenicity, especially those of different types within a given variety of cancer.

7.1.1.8. Subpopulations of Cells

Using cultured drug resistant L1210 leukaemia cells, KILLION *et al.* (1975) were able to isolate inherently antigenic subpopulations, using concanavalin A binding affinity. Their results indicated that the use of intact, untreated cells released by high affinity concanavalin A sugars were most effective in active immunotherapy. This method is worth developing.

7.1.1.9. Living Versus Killed Cells

Other physical treatments have been studied by PRAGER *et al.* (1974a): freeze-thaw, lyophilisation, heath (56° C); all of which kill the cells. They did not induce any immune protective effect in their model. LIKHITE and HALPERN (1973) observed enhancement of a mammary adenocarcinoma in transplanted DBA/2 mice, which received injections of killed tumour cells not mixed with adjuvant.

7.1.2.0. Modified Cells

Several workers have tried to make cells more antigenic and/or immunogenic. CZAJKOWSKI *et al.* (1966 , 1967) obtained encouraging results by chemically coupling a highly *antigenic foreign protein* carrier rabbit gammaglobulin to autologous cancer cells with bis-diazobenzidine. Plescia and Braun (quoted by BRAUN, 1970) have complexed the cells with foreign basic proteins, either methylated bovine serum albumin (BSA), or methylated bovine gammaglobulin: their therapeutic results were poor. Another trial using autologous cells coupled with BSA gave encouraging results (CZAJKOWSKI *et al.*, 1967), which were not confirmed by CUNNINGHAN *et al.* (1969). Similarly, BRUGAROLAS *et al.* (1972) obtained no effect with the treatment of MOPC 315 myeloma-carrying mice using isogeneic tumour cells coupled with human gammaglobulin.

Many other biochemical modifications of the immunizing tumour cells have been proposed and encouraging results have been reported. APFFEL *et al.* (1966) vaccinated C_{57}Bl/6 mice with *iodoacetate*- treated EL-4 cells. JASMIN *et al.* (1968) used, with moderate effect, iodoacetamine treatment of Rauscher virus-induced leukaemia cells. PRAGER (1973) and PRAGER *et al.* (1974a), utilized, with noticeable efficiency, neoplastic elements submitted to sulfhydril reagents iodoacetate, iodoacetamide, or N-ethyl-maleimide. More recently, the same author (PRAGER *et al.*, 1974a, 1974b) studied the efficacy of many modifying agents, listed in Table 7.III, and published the protective effect of 6 C_3HED cells treated by them against the graft of a living inoculum of the same intact living cells: the effective agents belong to two main groups, *sulfhydryl-blocking agents,* and *carbohydrate reagents.*

CAMPANILE *et al.* (1975), transferring L1210 leukaemia and LSTRA (Moloney leukaemia) in mice treated with *imidazol carboxamide,* observed that the cells were, after 16 transplant generations, more immunogenic than the original cells, and animals pretreated with them were highly resistant to parental tumour cells.

The in vitro treatment of tumour cells with Vibrio cholerea *neuraminidase* (VCN) has been shown by SANFORD (1967), BAGSHAWE and CURRIE (1968) (see also CURRIE and BAGSHAWE, 1969) to increase their immunogenicity.

Table 7.III. Summary of results of vaccination of C3H mice with modified 6C3HED lymphoma cells (PRAGER *et al.*, 1974a,b)

Modifying agent or process	Treatment of 6C3HED cells quantity/ml cells	Survivors/ total
Physical treatment		
X-ray	5,400 rad	5/5
Freeze-thaw	-	0/8
Lyophilization	-	0/10
Heat 56° C	-	0/10
Sulfhydryl-blocking agents		
Iodoacetamide	10-40 μ moles	16/16
Iodoacetate	10-40 μ moles	10/16
N-ethylmaleimide	10-40 μ moles	16/16
P-hydroxymercuribenzoate	10 μ moles	0/6
Iodoacetamide + ß-glucosidase	80 μ moles + 60 I.U[a]	6/6
Iodoacetamide + ß-glucuronidase	80 μ moles + 0.04 I.U	6/6
Carbohydrate reagents		
Concanavalin A	7.5 mg	9/12
Sodium periodate ($NaIO_4$)	3 mmoles	0/6
ß-Glucosidase	60 I.U.[a]	0/4
Other treatments		
pH 3.1, Earle's solution	-	10/12
Diazotized p-aminobenzoic acid	2.4 mmoles	0/6
Fluorodinitrobenzene	3-300 μ moles	0/18
Polypropyleneimine	10 mg	1/6

[a] From supplier's assay at pH 5.25; incubation was at pH 7.2.

This enzyme produces several modifications: enhanced phagocytosis (WEISS *et al.*, 1966a; LEE, 1968), inhibition of cell aggregation (KEMP, 1968), increased cell deformability (WEISS, 1965), alterations of cell migration (WOODRUFF and GESNER, 1969); but the most important is probably the intervention in the relationship between glycocalix and tumour associated antigens: CURRIE and BAGSHAWE (1968) postulated that the sialomucins on cells interfere because of their negative charges with the immunogenicity of the latter. APFFEL and PETERS (1970) broadened this concept, suggesting that sialoglycoproteins on the cell surface mask antigen (to some extent) and may be a system of immunoregulation. Neuraminidase removes sialic acid from the cell surface (Fig. 2.5), and therefore, may remove the steric hindrance to antigen recognition; by decreasing the negative charge (SIMON-REUSS *et al.*, 1964), it may increase deformability (WEISS, 1965) and susceptibility to phagocytosis (WEISS *et al.*, 1966b; LEE, 1968).

Several investigators have conducted experiments on the immunization of recipients with VCN-treated cells. Among them, BAGSHAWE and CURRIE (1968) and CURRIE and BAGSHAWE (1969) showed that they frequently fail to grow in isogeneic animals, and the recipients are rendered immune to a subsequent challenge with untreated viable tumour cells.

Moreover, inoculation of VCN-treated solid tumour cells for immunotherapy of animals grafted with cells from the same tumour is able, in the experiments of SIMMONS *et al.*(1971b, 1971c, 1974a), to impair the growth of the initial tumour cell inoculum of methylcholanthrene-induced sarcoma. The technique recommended by these authors (SIMMONS and RIOS, 1974a) is as follows: suspended tumour cells are incubated for 1 hour at 37° C, with 25 units of VCN[1]/ml/10^6 cells plus mitomycin C (25 µg/ml); the cells are then washed three times, and mixed with M 199, before being inoculated.

BEKESI *et al.* (1972) and BEKESI and HOLLAND (1974) have obtained a remarkable result on spontaneous AkR leukaemia carrying animals, submitting them firstly to chemotherapy to reduce the number of neoplastic cells, according to our general model, and then to VCN treated isogeneic neoplastic cells. The same group (BEKESI *et al.*, 1975) have treated acute myeloid leukaemia patients, after remission induction, with allogeneic neuraminidase treated myeoblasts. They inject every 4 weeks, at the same time as maintenance chemotherapy, 10^{10} cells in approximately 50 sites in different node draining areas. They have obtained encouraging preliminary results already mentioned above.

SEDLACEK *et al.* (1975) treated dogs with spontaneous tumours, mainly mammary, either with autologous neoplastic cells in vitro treated with mitomycin C, or with the same cells treated with this agent as well as with neuraminidase. Although they observed no regression in the first group, in the second group, 13 out of 15 animals had a significant regression.

The results which we have obtained with this enzyme have been, on the contrary, disappointing. Treating AkR mice at the age of 6 months (which is the age of the

---

[1] General Biochemicals, Chagrin Falls, Ohio, U.S.A.

beginning of the neoplasia) by immunotherapy using allogeneic tumour cells (from a C_{57}Bl/6 leukaemia induced, as AkR leukaemia, by Gross virus) and BCG, we obtained a significantly favourable result; the latter was not increased, but rather decreased if the cells were treated with VCN (MATHÉ *et al.*, 1973a). DORÉ *et al.* (1974) in our laboratory, using C_{57}Bl/6 mice grafted with the isogeneic E(AkR) leukaemia cells (this leukaemia had also been induced by Gross virus), and treating them at days 1, 5, 9, 12, and 17 after the transplantation with 5 x 10^6 irradiated isogeneic cells, either not modified or incubated with VCN (2 units/5 x 10^6 cells for 20 min. at 37° C), which increased antigenicity or immunogenicity in this experiment (COUDERT *et al.*, 1972), observed that, while a small percentage of mice treated only with irradiated cells were cured, those animals that received enzyme treated cells died as did the controls (Fig. 7.2).

A growth enhancement effect was observed by BRUM (1975) in the case of a hamster lymphosarcoma (initially induced by dimethylbezanthacene).

COUDERT *et al.* (1972) observed that the treatment of tumour cells with *papain* (0.1 mg of crude Papaya extract per ml per 10^7 cells for 10 min. incubated at room temperature) which, as neuraminidase, increases antigenicity and/or immunogenicity (COUDERT *et al.*, 1972), also diminishes the effect of irradiated tumour cells. Study of humoral immunity suggested that this may have been increased more than cell-mediated immunity.

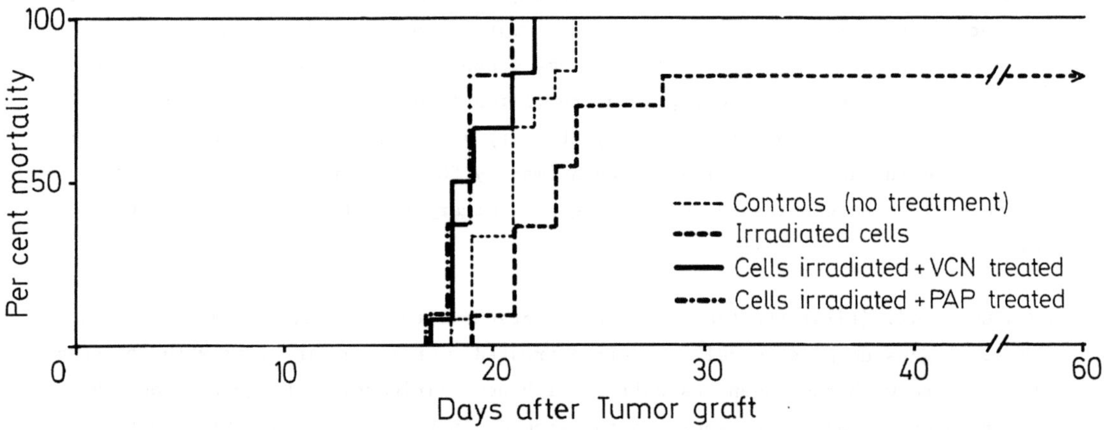

Fig. 7.2. Comparative effect on E (AkR) leukaemia immunotherapy by irradiated allogeneic leukaemic cells (significant favourable action) and by neuraminidase or papain treated cells (deterioration of the effect) (DORÉ et al., 1974)
VCN = Vibrio cholerae Neuraminidase
PAP = Papain

SPARKS and BREEDING (1974) used systemic injections of neuraminidase and observed some delayed tumour growth. It is not known if the mechanism of action is immunological.

Among workers who obtained the best results with neuraminidase are SIMMONS and RIOS (1974b) who also used *Concanavalin A* (Con A), which is known to increase the immunogenicity of tumour cells when bound to their surface (WUNDERLICH *et al.*, 1971). They dissolved it in M199 in various concentrations (50 µg/ml is the most efficient) with tumour cells at 37° C, and used the latter after washing and checking that they were viable before injection into tumour bearing hosts as challenging inocula. The degree of regression which was shown to be of immune nature was identical to the highest obtained with neuraminidase.

BRUGAROLAS *et al.* (1972) also obtained a moderate prolongation in survival of mice transplanted with MOPC 315 solid myeloma by treating them with single injections of 1.6×10^6 isogeneic tumour cells combined with Concanavallin A.

BUZZI and BUZZI (1974) have incubated in vitro Ehrlich cancer cells with *diphtheria toxin:* the cells injected into mice sensitive to this tumour did not induce any neoplasia, but made 80% of the animals resistant to a challenge with a number of virulent cells which usually kill 100% of them.

A *viral infection* of the cells may increase their antigenicity, as shown by STUCK *et al.* (1964), SVET-MOLDAVSKY and HAMBURG (1964), SJOGREN and HELLSTRÖM (1965), KOBAYASHI *et al.* (1969). An experiment to test the therapeutic antitumour effect of this method needs to be made.

BOONE *et al.* (1971, 1974), and LINDENMANN and KLEIN (1967), showed that tumour cells infected by certain viruses, could be used to render animals resistant to a later challenge with the same noninfected tumour cells. They used influenza and vesicular stomatitis viruses. They assumed that, as both are enveloped viruses, and as the envelope is derived from the cell membrane, virus antigens can serve as a "carrier" for tumour associated antigen(s). In the case of influenza virus, production of neuraminidase may play a complementary role (LINDENMANN and KLEIN, 1967).

SAUTER *et al.* (1974) infected an in vitro culture of human leukaemic "myeloblasts" with avian myxovirus and treated an acute myeloid leukaemia patient with his autologous virus infected leukaemic cells: they observed a spectacular regression of leucocytosis. Such a trial should be repeated, and the mechanism of the effect of the injected material, which is probably complex, analysed: direct oncolysis (LINDENMANN and KLEIN, 1967), interferon induction (YOUNG, 1971), postoncolytic immunity (LINDENMANN and KLEIN, 1967) should be considered.

YOUN *et al.* (1974) used a more specific experimental approach. They immunized mice, which were to be grafted with a Gross-virus induced lymphoma, with allogeneic cells from lines chronically infected with non-oncogeneic C-type viruses and presenting Gross virus-induced antigens. The best prevention of grafted lymphoma growth was obtained with living cells. Cell-free culture supernatant was inefficient.

7.1.2.1. Cultured Cells

We are very tempted to use cultured cells for leukaemia and lymphosarcoma specific immunotherapy, as permanent lines have been obtained in suspension from the peripheral blood of patients presenting circulating leukaemia or lymphosarcoma cells (MOORE and McLIMANS, 1968; ROSENFELD *et al.*, 1969; BELPOMME *et al.*, 1972). More than 50 lines are maintained in our laboratory (ROSENFELD *et al.*, 1969; BELPOMME *et al.*, 1972).

As similar lines can be obtained from normal donors, there is, for most of them, doubt as to their leukaemic nature. Nevertheless, ROSENFELD *et al.* (1972) believe they are able to recognize the lines from leukaemic patients on the following criteria: (a) a different thickness of their coat measured at electron microscopy after ruthenium red preparation (ROSENFELD *et al.*, 1974); (b) a different karyotypic evolution (VENUAT and ROSENFELD, 1973, 1974).

If it is proved that these cells are really the progeny of leukaemic cells, they could be a very convenient source of material for specific immunotherapy. They could also be good material, as BELPOMME *et al.* (1969) showed, for three reasons: (1) their antigenicity and/or immunogenicity increases with time (Table 7.IV); (2) it is possible, today, to produce them in bulk, and (3) it should be possible to synchronize them in order to take them during the G_1 phase of the cell cycle, the phase during which cells are the most antigenic and/or immunogenic (CIKES, 1971; ROSENFELD *et al.*, 1972), and the phase during which they are the most sensitive to immunotherapy (OLSSON and MATHÉ, 1976, not published).

A possible difficulty may rest in constant (or nearly constant) contamination of these cells by Epstein-Barr-virus, which may be necessary for the establishment of their culture. The administration of a huge quantity of virus may not be innocuous, especially in patients who are immunodepressed, either because of the disease, or because of preimmunochemotherapy. The measurement of the antibody titres of the patients before the use of such cells, and in case of absence of antibody, may necessitate vaccination with small doses of the virus or virus-containing cells as a precaution.

Table 7.IV. Increase of antigenicity shared by cells of permanent lines obtained from leukaemic donors (BELPOMME *et al.*, 1969)

Lines	Fresh cells		After 7 weeks culture		After 4 months culture		After 8 months culture	
	I.F	I.A	I.F	I.A	I.F	I.A	I.F	I.A
ICIG 101	0	1/64			+	1/1280	NT	1/3200
ICIG 102	0	0			0	1/320	NT	NT
ICIG 104B	0	1/10	0	1/80				

I.F: immunofluorescence; I.A: immune adherence; N.T: not tested.

ROSENFELD *et al.* (1975) recently described a line of cells, called Reh, which may be of greater interest for active immunotherapy than all others. It was established from the peripheral blood of a patient with acute lymphoid leukaemia. The cells from this line lack the Epstein-Barr-virus-determined nuclear antigen (EBNA) and no intra cellular EBV DNA was detected. They do not exhibit T-cell rosettes as described for another line called Molt 4 (MINOWADA *et al.*, 1972; WEST and HERBERMAN, 1974), but a small percentage of the cells carry receptors for complement. The kinetics of establishment of this line differs from that previously reported by the author for other cell lines (ROSENFELD *et al.*, 1969). The established cells possess chromosome markers which were present in the progenitor fresh leukemic cells. They show a metabolic block between glutamine and proline. The study of their membrane using TPCK-trypsin demonstrates two glycopeptides characteristic of this line which we have already used in immunotherapy trials. Such a line could have originated from the malignant cell population, as suggested by NILSSON and SUNDSTROM (1974) for two lymphoblastoid lines established from human lymphosarcomas.

There are, at present, very few clinicians who are using cultured cells. SPARKS *et al.* (1973) injected ID, every week, 5×10^7 irradiated allogeneic or autologous cells maintained in tissue culture.

In the case of solid tumours, the problem of the unknown nature of the cells obtained in tissue culture would be serious if one were to cultivate each tumour to treat the patients with their own autologous cells. The problem is solved satisfactorily if one uses well established lines such as those used in some virology or immunology laboratories (University of California and National Cancer Institute, 1970).

7.1.2.2. Modified Normal Cells

MADYASTHA and BARTH (1975) observed that DNP-tagged normal DBA/2 thymocytes treated with DNFB (at a ratio of 10^{11} mol:cells) were able, when injected 30 days before

L1210 leukaemia non-tagged cell inoculation, to protect significantly against the evolution of this tumour. DNP-tagged spleen cells were not effective. The author suggests that certain DNP-tagged normal cells may carry some antigenicity similar to their leukaemic counterparts.

7.1.2.3. Hybridized Cells

WATKINS and CHEN (1969) have attempted to make weak cellular antigen stronger by the fusion technique (HARRIS, 1970; BEREBBI and BARSKI, 1971; LE BORGNE DE KAOUEL *et al.*, 1972, 1973). A hybrid cell is formed containing strongly antigenic foreign components and the weak tumour associated antigens, so that the former might boost the latter. So far, the results are usually inconclusive (GOWANS, 1974). However, Meyer (personal communication) treated, with remarkable results, a polyoma tumour with (A9 x SEWA) hybrid cells.

7.1.2.4. Tumour-Associated Antigen Preparations

In our Laboratory, MARTYRÉ (1976), treating RC_{19} transplanted leukaemia (the disease was initially induced with the inoculation of 10^3 cells) with hydrosoluble tumour associated antigens obtained either by hypertonic lysis (3 MKCL) or by hypotonic sodium chloride lysis, was able to cure, by this specific immunotherapy, a high proportion of the animals, provided the treatment was started as early as possible, on the same day as the inoculation of the tumour. She obtained no prolongation of survival if the treatment was started 1 day later or more (Fig. 7.3).

Using extracts for immunoprophylaxis of the same leukaemia, MARTYRÉ *et al.* (1974a) obtained a significant reduction of mortality with the hydrosoluble antigen obtained with hypertonic lysis and, on the contrary, an enhancement of the tumour growth with the antigen obtained with the hypotonic lysis (Fig. 7.4).

RAI *et al.* (1975) obtained a similar facilitation of isogeneic HTD tumour growth in rats pretreated with soluble antigen extracted from this tumour, which was efficient in inhibition of complement dependent cytotoxicity.

These soluble tumour antigen extracts are submitted to physicochemical characterisation. MARTYRÉ *et al.* (1974b) have attempted that of the Rauscher leukaemia associated antigen, the result of which I quoted above (Fig. 2.4).

BENJAMINI and SCIBIENSKI (1974) based a hypothesis on the fact that certain degrees of aceto-acetylation of flagellin result in modification of the protein to the

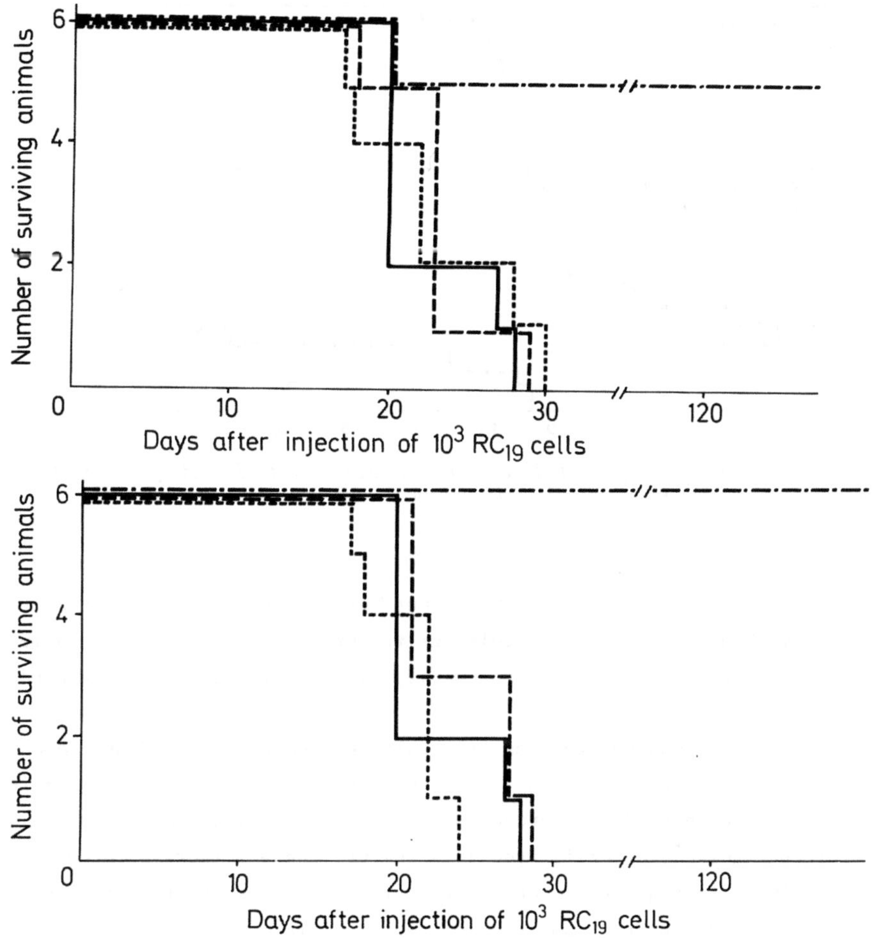

Fig. 7.3. Effect of soluble tumour associated antigen obtained by hypertonic lysis(above)or hypotonic lysis (below) on mortality from RC_{19} leukaemia (immunotherapy) (MARTYRE, 1976) (———) controls; (-----) first injection of antigen on day 0 (day of tumour cells injection); (-----) on day + 1; (---------) on day + 2

extent where the native and modified forms show little or no serological cross-reactivity, but still exhibit cross-reactivity with respect to delayed hypersensitivity (PARISH, 1971). They modified 3 Mkce tumour extracts by aceto-acetylating them with diketine, and they used this preparation for immunotherapy. Their experience with dog and human tumours was encouraging in the preliminary stages.

In man, HUMPHREY *et al.* (1971) prepared soluble antigens from tumours obtained at surgery or at autopsy, and used them without side effects for immunotherapy. Such antigens can be stored for prolonged periods of time.

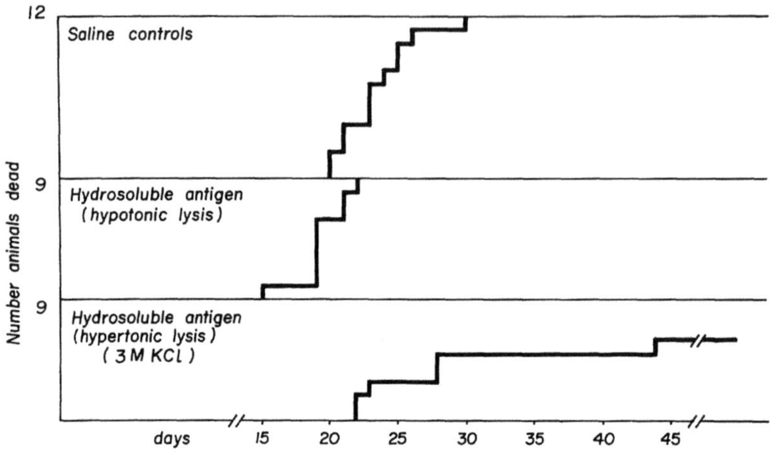

Fig. 7.4. Effect of soluble tumour associated antigen in the immunoprophylaxis of RC_{19} leukaemia (MARTYRE et al., 1974a)

KOKOSCHKA *et al.* (1975) treated malignant melanoma patients after resection with soluble membrane extracts of autologous tumour cells and BCG, and claimed they increased tumour specific immune reactivity, while relapses were prevented or delayed.

Soluble antigens can be prepared from cultured tumour cells: PELLIS and KAHAN (1975) obtained soluble TSTA extracts with 3 molar potassium chloride from an initially methylcholanthrene-induced sarcoma propagated by grafts, immuno-protective against suprelethal numbers of cells from this tumour. These authors also observed that trypsin treatment is required to demonstrate target cell susceptibility to the immune response, suggesting that antigens of target cells may be unavailable to immune recognition, unless modification in the cell membrane precedes introduction into the immunized host.

7.1.2. Dose Factor

The dose of the cells or of the antigen may be important; we have seen in Chapter 2 that the soluble antigen in the serum is supposed to play a role in the immune balance. Until more experiments are devoted to this important subject, one can only advise against giving a huge number of cells or a large quantity of antigen(s), for the purpose of avoiding some blocking phenomena.

At Villejuif (MATHÉ *et al.*, 1968; SCHWARZENBERG, 1973), we inject each week for 1 month, and each month for 5 years, both i.d. and s.c., 4×10^7 leukaemic cells obtained from a pool of allogeneic donors suffering from the same disease.

Studying, in acute myeloid leukaemia patients, the possibility of inducing a change in leukaemia cell recognition by autologous lymphocytes after immunization of the patients with their own irradiated neoplastic cells, POWLES *et al.* (1974) obtained a response with 1 x 10^9 cells.

This unique data is not sufficient to indicate that this number, provided the injections are repeated, is not too high.

SEDLACEK *et al.* (1975) injected dogs with spontaneous tumours (mainly mammary) with mitomycin and neuraminidase treated autologous tumour cells. In a group of 15 animals, each of which received 2.10^7 cells, there were 13 regressions; in another group in which the animals received 1.10^8 cells, there was acceleration of tumour growth in all the dogs.

GUTTERMAN *et al.* (1975a) compared adult acute leukaemia patients treated for remission maintenance with chemotherapy and BCG, or with both plus unirradiated autologous tumour cells: the median duration of remission was 40 weeks in the group receiving these cells, compared to 50 weeks in the other.

7.1.3. Sites of Administration

A few authors have used autologous neoplastic cells (BLOOM *et al.*, 1973; CURRIE, 1973; IKONOPISOV *et al.*, 1970; MORTON *et al.*, 1973) injected at many s.c. sites, in order to activate a larger portion of the immune machinery than injection into one site (McKHANN and GUNNARSSON, 1974). If it is demonstrated that the last hypothesis is valuable, it would also be more reasonable to use allogeneic cells.

As mentioned earlier, BEKESI *et al.* (1975) injected i.d. 10^{10} neuraminidase treated myeloblasts into about 50 sites.

The intradermal character of the injection is probably very important: GROSS and DREYFUSS (1974) showed that the i.d. injection of very small numbers of living cells of their guinea pig L_2C leukaemia resulted in spontaneous regressions and immunization against challenging reinoculations of tumour cells. Control animals receiving the first injection s.c.developed leukaemia.

7.1.4. Should Cells or Antigens be Mixed with the Adjuvant?

We initially showed in mice (MATHÉ, 1968; MATHÉ *et al.*, 1969) that, for immunotherapy as well as for immunoprophylaxis, the combination of BCG and irradiated tumour cells (not given at the same site) was considerably more efficient than each of the agents administered alone (Figs. 3.1, 3.3 and 5.1).

The question then arose as to whether the cells and the adjuvant should be mixed together: there are now many results of experiments conducted on guinea pigs (ZBAR *et al.*, 1970, 1971; BARTLETT and ZBAR, 1972; LITTMAN *et al.*, 1973), on mice (TANAKA and TOKUNAGA, 1971; BARTLETT *et al .*, 1972; CHUNG *et al.*, 1973; TOKUNAGA *et al.*, 1973), and on rats (BALDWIN and PIMM, 1971, 1973) which answer positively. In these systems, the simultaneous injection of cells and BCG at different sites were not effective, while the mixture suppressed tumour growth.

ZBAR (1974) has attempted to determine the conditions required for the activity of BCG and tumour cell mixtures. He limited his study to mycobacteria. Guinea pigs received simultaneously, on one side, BCG plus tumour cells i.d. and, on the other side, tumour cells as challenge. His conclusions are as follows: (a) more than 10^6 mycobacteria are necessary to produce systemic immunity; (b) 1.5×10^6 tumour cells seem to be an optimal number; (c) BCG cell walls in oil and water emulsion mixed with tumour cells are effective (75 µg is optimal); (d) living tumour cells are occasionally tumourogenic, while X-irradiated tumour cells are not, but are immunogenicity inducers; (e) no benefit is gained from multiple injections of the mixture of BCG plus cells; 300 µg of BCG cell walls is less effective than 75 µg; more than 1.5×10^6 tumour cells do not work any better than 1.5×10^6. Finally, he confirmed, in this study (BARTLETT and ZBAR, 1972; ZBAR *et al.*, 1972), our conclusions on the size of the tumour that can be eradicated (MATHÉ, 1968; MATHÉ *et al.*, 1969): in his experiment, only a tumour comprising 10^6 cells could be eradicated; the effect of immunotherapy is nil on a 10^7 cell neoplasia.

Living neoplastic cells have been used in admixture with living BCG: for two experimental rat sarcomata, they did not induce a tumour, and they were the site of intense inflammation; however, for a third sarcoma in the same species, though the local development was temporarily inhibited, metastasis appeared (MOORE *et al.*, 1974). Thus, even if they are mixed with BCG, the immunizing tumour cells must be sterilized.

WISSLER *et al.* (1968), treating minimal deviation hepatoma carrying rats with combined tumour cells and pertussis vaccine, obtained favourable effects in an immunotherapy experiment and enhancement in an immunoprophylaxis attempt.

Tumour cells emulsified in complete Freund's adjuvant were not active in an experiment conducted by BARTLETT and ZBAR (1972).

TANAKA and SASAKI (1974) have suggested that the mixture of heterospecific or nonallogeneic cells could be successfully mixed with tumour cells, as living neoplastic elements do not grow if they are inoculated together with cells non-histocompatible with the recipient.

7.2. Systemic Non-Specific Active Immunotherapy or Immunoprophylaxis

7.2.1. Systemic Immunity Adjuvants

I have quoted in the chapters devoted to experimental active immunotherapy and immunoprophylaxis, certain antitumour effects considered to be due to immunostimulation and obtained by the administration of certain agents. I have called them "immunity systemic adjuvants", and their main mechanism of action will be discussed below.

7.2.1.1. The Innumerable Adjuvants quoted in the Literature

Countless agents are mentioned in the literature, which have been described as "adjuvants" in the broad sense of the word. The main ones are listed in Table 7.V.

7.2.1.2. Experimental Screening and Pharmacology

It is impossible to compare the data of the literature usefully concerning these agents, because the tests to which they have been submitted differ greatly from one author to another.

Moreover, for each agent named, there may be many strains and/or preparations, and some may be efficient whilst others are not.

Hence, the need for a reference battery of tests that would permit comparison of different agents which have behaved as systemic immunity adjuvants in some experiments, or have been published as such by various authors (see Chapters 3, 4, 5 and 6).

On the other hand, so many microbiological or extracted or synthetized agents have been presented to us for clinical use, that we felt that it was necessary to set up a screening battery for systemic immunity adjuvants.

7.2.1.2.1. The ICIG[1]-EORTC[2] Experimental Screening for Systemic Immunity Adjuvants

The *first generation* of this screening (MATHÉ *et al.*, 1973b)comprised five tests: (a) the haemolytic plaque forming cell (PFC) test according to the JERNE (1963)

---

[1] Institut de Cancérologie et d'Immunogénétique.
[2] European Organization for Research on Treatment of Cancer.

Table 7.V. Agents which have been considered as immunity adjuvants[a]

Acetylated peptidoglycan from Mycobacterium tuberculosis	WERNER *et al.*, 1975 quoted by MARAL *et al.*, 1976
Aristolochic acid	NORPOTH and DEPPE, 1965 quoted by MARAL *et al.*, 1976
BCG	DUBOS and SCHAEDLER, 1957
Bis-diethyl amino-ethyl-fluorenone (see tirolone)	
Bordetella pertussis (endotoxins)	UNANUE *et al.*, 1969
Brucella abortus	RAMON *et al.*, 1949; TOUJAS *et al.*, 1972; OTH *et al.*, 1973; LE GARREC *et al.*, 1974
Carbon	RIOS and SIMMONS, 1972
Carrageenan	RIOS and SIMMONS, 1972
Cationic detergents	GALL, 1966
Cord factor[b] (6,6 dimycolate of trehalose) (secretion of tuberculosis bacilli)	BEKIERKUNST *et al.*, 1971
Corynebacterium granulosum	MATHÉ *et al.*, 1975b; RAYNAUD *et al.*, 1972; KOUTZNETZOVA *et al.*, 1974
Corynebacterium parvum	HALPERN *et al.*, 1963
Diethylamino-ethyl dextran	WITTMAN *et al.*, 1970, 1975 quoted by MARAL *et al.*, 1976
Diphteria toxin	BUZZI and BUZZI, 1974
Endotoxins	JANOSSY *et al.*, 1973; SAUTER and GIFFORD, 1966
Extrait bacterien phospholipidique (EBP) from Listeria monocytogenes or Salmonella typhimurium	FAUVE and HEVIN, 1974 quoted by MARAL *et al.*, 1976
Extracts of mycobacteria (see Mycobacteria)	
Freund's complete adjuvant[c]	FREUND, 1956
Freund's incomplete adjuvant[d]	JOLLÈS and PARAF, 1973

Hemophilus pertussis	WISSLER *et al.*, 1968
Hiu I non-hydrosoluble extract from BCG	HIU and AMIEL, 1971b
Hiu II hydrosoluble extract from BCG	HIU, 1972, 1974; MATHÉ *et al.*, 1975a
Interferon	GRESSER *et al.*, 1969; GRESSER, 1974
Lactic dehydrogenase virus	NOTKINS *et al.*, 1966
Latex	JACOTOT, 1947
Lectins (see phytohaemagglutinin and pokeweed mitogen)	
Lentinan (polysaccharide from Lentinus edodes)	CHIHARA *et al.*, 1970
Levamisole (-)-2,3,5,6-tetrahydro-6 phenylimidazo-(2,1-b)-thiazole monohydrochloride	RENOUX and RENOUX, 1972
Lipopolysaccharide B (LPS) (from Escherichia coli or Salmonella typhi)	ALLISON and DAVIES, 1971
Liposomes[e]	ALLISON and GREGORIADIS, 1974
Lysolecithin and synthetic analogues	WESTPHALL *et al.*, 1970
Methanol extraction residue (MER)	WEISS, 1972; BEN EFRAIM *et al.*, 1974
Methanol soluble fraction from M. butyricum	ESBER *et al.*, 1974
Methylxanthines (theophylline, caffeine)	WEBB *et al.*, 1972
M. bovis (see BCG)	
M. kansasii	CHEDID and LAMENSANS, 1974a
M. phlei	HALPERN, 1974
M. smegmatis	CHEDID and LAMENSANS, 1974b
M. tuberculosis (its wax D is rich in peptidoglycan amino acid)	HALPERN, 1974
M. wax D	WHITE, 1967
Neo WSA (water-soluble adjuvant extracted from whole cells of M. smegmatis or from Nocardia)	ADAM *et al.*, 1974; BONA *et al.*, 1974
Nocardia asteroides	ZBAR and RAPP, 1974

Table 7.V. (continuation)

Nucleic acids	MERRITT and JOHNSON, 1965
Pertussis vaccine	PARFENTJEV and GOODLINE, 1948; KIND, 1957; SLONECKER and RIEKE, 1965
Phytohaemagglutinin	NASPITZ *et al.*, 1968
Pokeweed mitogen	BASHAM and WAXDAL, 1975
Polyacrylic acids	DIAMANSTEIN *et al.*, 1971, quoted by MARAL *et al.*, 1976
Poly A-poly U[f]	SCHMIDTKE and JOHNSON, 1971
Poly I-poly C[f]	BRAUN 1970, 1971; MATHÉ *et al.*, 1971c
Polymethacrylic acid	ORMAI and DE CLERCQ, 1969
Polysaccharide peptidoglycan from M. tuberculosis, bovis, kansaii	MIGLIORE-SAMOUR and JOLLES, 1974
Polysaccharide of Proteus vulgaris	MIZUNO *et al.*, 1968
Pyran copolymer	BAIRD and KAPLAN, 1976; KAPLAN *et al.*, 1974
Saponin	GALL, 1966
Shigella lipopolysaccharide	ALEXANDER and EVANS, 1971
Shigella Minnesota lipid A	ALEXANDER and EVANS, 1971
Silica (SiO^2)	PERNIS and PARONETTO, 1962
Streptococcal preparation OK-432	HATTORI *et al.*, 1975
Tetramisole (levamisole isomer)	
Tetrasaccharide heptapeptide from M. tuberculosis	WERNER *et al.*, 1976
Tirolone (see bis-diethyl amino-ethyl-fluorenone)	MEGEL *et al.*, 1974, quoted by MARAL *et al.*, 1976
Viral double-stranded RNA	LARSON *et al.*, 1969
Vitamine A alcohol (retinol) palmitate	DRESSER, 1968
Vitamine A analog (13-cis-retinoic acid)	PORT *et al.*, 1964
Vitamin E (tocopherol)	TENGUDY *et al.*, 1972, 1973 quoted by MARAL *et al.*, 1976

Waxes D from M. tuberculosis, human strains (peptodoglycolipids)	WHITE *et al.*, 1964; ASSELINEAU and PORTELANCE, 1974
WSA (water-soluble adjuvant extracted from purified cells)	ADAM *et al.*, 1972; BONA *et al.*, 1974; LIACOPOULOS *et al.*, 1974

[a] See also JOLLÈS and PARAF, 1973.

[b] May be responsible for the special cordlike spatial in vitro growth of tuberculosis bacilli.

[c] FCA: Freund's "complete" adjuvant: water in oil emulsion containing suspended heat killed M. tuberculosis.

[d] FIA: Freund's "incomplete" adjuvant: emulsion containing mineral oil and an emulsifier (surface-active agent, poly-oxyethylene sorbitan mono-oleate).

[e] Concentric spheres consisting of phospholipid bilayers separated by aqueous compartments. Antigens can be entrapped into liposomes. The latter may be made non-immunogenic, biodegradable materials; their composition and properties such as electric charges can be varied at will.

[f] Complexing poly-L-lysine to polynucleotides makes them more resistant to serum nuclease; at the same dose as polynucleotides, they exert the same effect (LEVY, 1974, quoted by MARAL *et al.*, 1976).

method; (b) the graft versus host (GVH) reaction test according to our model (MATHÉ and AMIEL, 1962) (the adjuvant being administered to the donor and the same number of lymphoid cells being injected into the recipients); (c) the prophylaxis of L 1210 leukaemia (prophylaxis, because it is the condition in which BCG, given alone, worked in our operational experiment; L 1210 leukaemia, because it is on this leukaemia that we defined the conditions of immunotherapy (MATHÉ, 1968; MATHÉ *et al.*, 1969); (d) the prophylaxis of the Lewis solid tumour, which is a carcinoma, and on which we also conducted our studies on solid tumour immunotherapy (THOMPSON *et al.*, 1972; MARTIN *et al.*, 1975; (e) the prophylaxis of ICIGCI$_1$ solid tumour, which is a sarcoma. We published the result of the study with this battery of ten agents and of ten BCG preparations (MATHÉ *et al.*, 1973b, 1973c).

It became apparent to us that this battery was expensive, utilizing a large number of mice, that one of its tests was not of great use (the ICIGCI$_1$ solid tumour), and also that the results obtained with the GVH test could not be reproduced to perfection; in addition, we did not include a test studying the effect of the candidate-adjuvant on the macrophages.

We, therefore, modified the battery by dropping two tests (GVH and ICIG CI$_1$ tumour), and adding a macrophage activation study (MATHÉ *et al.*, 1975).

The *second generation of ICIG-EORTC screening* comprised 4 tests (Table 7.VI) (MATHÉ *et al.*, 1975):

(a) The first is the macrophage nonspecific cytolysis: it consists of studying the capacity of peritoneal macrophages of donors treated with the adjuvant-candidate, to lyse tumour cells against which the donors have not been sensitized. Chromium release is used to quantify lysis (Fig. 7.5)(BRULEY-ROSSET *et al.*, 1976a).

(b) The second is the counting of the haemolytic plaque-forming cells (HPFC) in the spleen of mice immunized with sheep red blood cells and treated with the adjuvant-candidate. This reaction requires macrophage, T-lymphocyte and B-lymphocyte co-operation (CLAMAN, 1966).

(c) The third test studies the prophylactic effect on mortality of L 1210 leukaemia of the adjuvant-candidate given alone.

(d) The fourth test studies the prophylactic effect on mortality from solid Lewis tumour of the adjuvant-candidate given alone.

For these two last tests, the dose and the time of injection of the adjuvant are only the optimal ones in the Jerne test, but four routes of administration, i.v., i.p., s.c., and i.d., are used.

Table 7.VI. Tests and methods used in second generation of ICIG-EORTC screening of immunity systemic adjuvants

Technique	Day 0	Injection of the adjuvant Before[a]	Results
(1) Macrophage activation	Sample taken of peritoneal macrophages	Days 3, 7 or 14	Tumour cell non-specific lysis (^{51}Cr release)
(2) HPFCT[b] (Jerne)	10^9 SRBC[c] i.p.	Day 14 (BCG) Day 2.5(others)	PFC[d] at day +4
(3) L1210	10^3 leukaemic cells i.v.	Day 14 (BCG) Day 2.5(others)	Mortality
(4) Lewis tumour	2 x 10^6 tumour cells s.c.	Day 14 (BCG) Day 2.5(others)	Mortality

[a] The administration of the antigen (Jerne test) (2); or the inoculation of the tumours [(3) and (4)]

[b] HPFCT = Haemolytic plaque-forming cells test

[c] SRBC = Sheep red blood cells

[d] PFC = Plaque forming cells

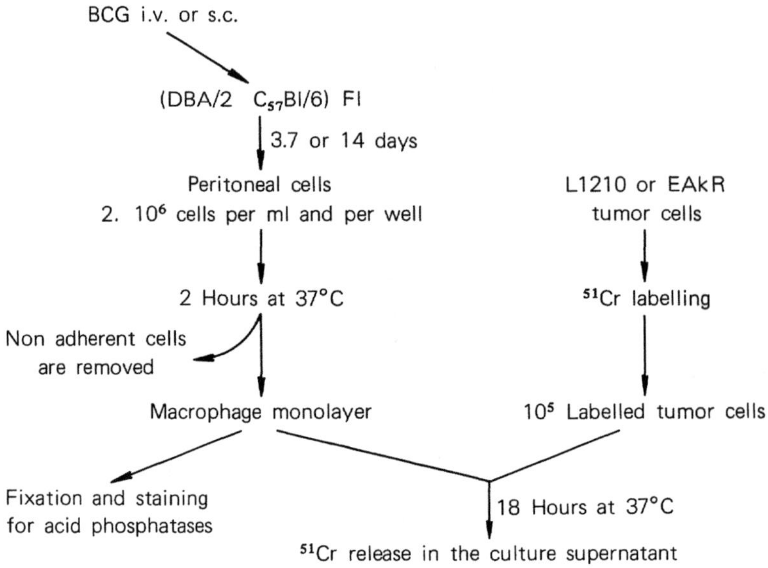

Fig. 7.5. The macrophage activation test. Scheme of the method

It goes without saying that a screening is nothing more than a systematic testing, similar to a school examination. As the latter tests the students under certain conditions, so does screening test the adjuvant-candidates in certain activities.

It means that an agent which had presented an adjuvant effect in the hands of certain authors, in certain experiments, may be negative in our hands, because of the conditions we have chosen. An absence of effect in the screening does not necessarily mean that the agent is not an adjuvant.

As will be illustrated later, this is rarely the case. We find that this is a quality of our screening, because our object is not to detect many new adjuvants in a few years, to be added to the already high number of agents which are known to have induced some stimulation of some immune responses (Table 7.V).

Our object is to select, among the numerous agents which increase these responses, those which may have a significant action on the two "severe operational" tests, L 1210 leukaemia and Lewis solid tumour, and/or a strong in vivo effect on macrophage nonspecific cytotoxicity.

Two examples, fresh living Pasteur Institute BCG and a hydrosoluble extract from mycobacteria called Hiu II, and prepared in our laboratory (HIU, 1972), will show the type of information we obtained on them with this simplified battery.

Table 7.VII gives a summary of the composition and/or the preparation and the doses determined by preliminary experiments which have been used for the screening (for the effect of the dose factor, see p. 224).

Table 7.VII. Two examples of agents submitted to the second generation of ICIG-EORTC screening of immunity systemic adjuvants (ISA)

Agents	Obtained from	Composition	Doses/mouse[a]
BCG (fresh)	Institut Pasteur	Suspension of living bacteria	1 mg = 5×10^6
HIU II	ICIG	Component of the BCG membrane (obtained by catalytic hydrogenolysis)	0.5 mg

[a] Determined by preliminary studies.

Figure 7.6 shows the results given by the macrophage nonspecific lysis. One sees that BCG induces a significant lysis when injected s.c. at days -7 or -14 before taking peritoneal cells for the test, and i.v. at days -3, -7, -14 (but day -7 is much less effective than the two others); Hiu II induces a much more important lysis when injected at days -3, -7, -14, whatever the route - s.c. or i.v. (Fig. 7.6) (BRULEY-ROSSET *et al.*, 1976a).

Table 7.VIII shows their effect according to the time of injection and the route of administration on the number of HPFC in the spleen: one sees that BCG only gives a significant result when injected i.v. at day -14, before the inoculation of the antigen. Table 7.IX shows that Hiu II gives a positive result only when administered i.v. at day -3.

Table 7.X indicates the results of BCG injected i.v. or i.p., i.d., or s.c. at day -14, and of Hiu II administered by the same routes at day -3, on immunoprophylaxis of L 1210 leukaemia. Only BCG gives a significant prevention, and only when administered by the i.v. route.

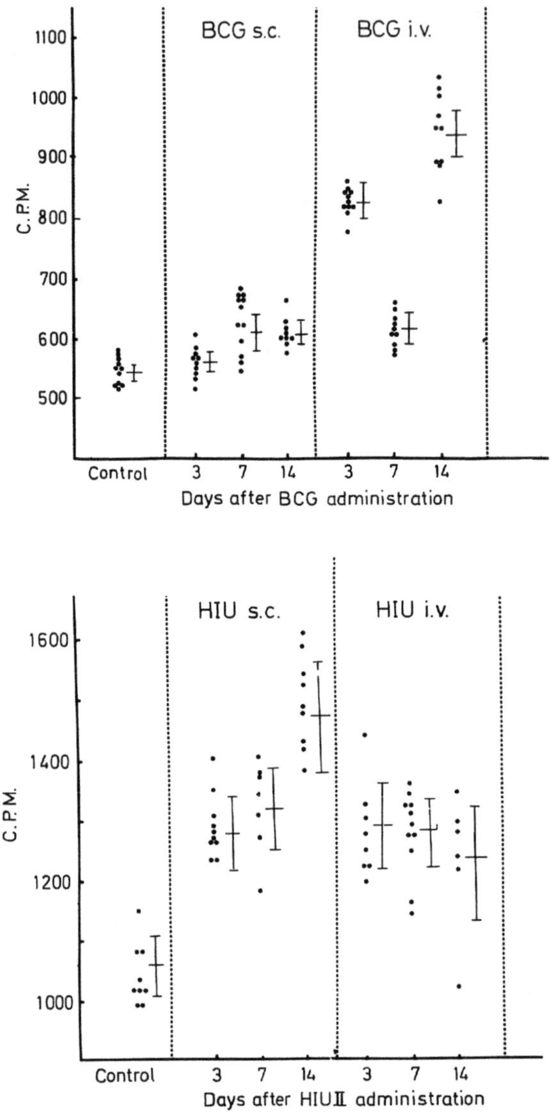

Fig. 7.6. In vitro cytotoxic activity of peritoneal mouse macrophages at different times after s.c. or i.v. administration of BCG (1 mg) (above). In vitro cytotoxic activity of peritoneal mouse macrophages at different times after s.c. or i.v. administration of HIU II, a soluble BCG extract (0.5 mg) (below) (MATHÉ et al., 1975a; BRULEY-ROSSET et al., 1976a)

Table 7.XI indicates the results obtained with BCG and HIU II on Lewis tumour: whereas HIU II is not efficient by any route, BCG induces a significantly favourable response when injected i.v., but a pronounced enhancement of the tumour growth when administered s.c.

Table 7.VIII. Results given by fresh Pasteur institute BCG (1 mg/mouse) on the haemolytic plaque-forming test (HPFCT) (time and route of administration factors)

	i.v.		i.p.		i.d.		s.c.	
Days[a]	I[b]	Significance	I	Significance	I	Significance	I	Significance
-14	2.8 ↗	S[c] at 2%	0.75	NS	0.75	NS	0.75	NS
- 5	1.1	NS	0.75	NS	0.75	NS	1.1	NS
- 3	1.1	NS	0.8	NS	0.75	NS	1.1	NS
0	1.1	NS	0.75	NS	1.1	NS	1	NS
+ 1	1.1	NS	1	NS	1.1	NS	1	NS
+ 2	1.1	NS	0.8	NS	1.1	NS	0.75	NS
+ 3	1	NS	1	NS	1.1	NS	1	NS

[a] - before, + after injection of SRBC.

[b] I = Mean number of PFC per spleen of the tested mice/mean number of PFC per spleen of controls.

[c] Statistics: Student-Fisher test. S, significant; NS, not significant.

Table 7.IX. Results given by HIU II (soluble) (0.5 mg/mouse) on the haemolytic plaque-forming test (HPFCT) (time and route of administration factors)

	i.v.		i.p.		i.d.		s.c.	
Days[a]	I[b]	Significance	I	Significance	I	Significance	I	Significance
-14	0.75	NS[c]	1.1	NS	0.75	NS	0.75	NS
- 5	0.75	NS	1.1	NS	0.75	NS	0.75	NS
- 3	1.6 ↗	S at 5%	1.1	NS	0.75	NS	0.8	NS
0	0.75	NS	1.1	NS	0.75	NS	0.75	NS
+ 1	0.9	NS	2 ↗	S at 5%	0.75	NS	0.9	NS
+ 2	0.75	NS	2.4 ↗	S at 2%	0.75	NS	1	NS
+ 3	0.75	NS	2.4 ↗	S at 5%	0.75	NS	0.75	NS

[a] - before, + after injection of SRBC.

[b] I = Mean number of PFC per spleen of the tested mice/mean number of PFC per spleen of controls.

[c] Statistics. Student-Fisher test. S, significant; NS, not significant.

Table 7.X. Effect of living Pasteur institute BCG and of HIU II on immunoprophylaxis of L 1210 leukaemia

	i.v.			i.p.		i.d.		s.c.	
"SAI"[a]	I[b]		Significance	I	Significance	I	Significance	I	Significance
Pasteur Institute BCG	1.6	↗	S[c] at 5%	1	NS	1	NS	1	NS
HIU II	1		NS	1	NS	1	NS	1	NS

[a] SAI = systemic adjuvant of immunity. BCG is injected at day -14, HIU II at day -3.

[b] I = Median of survival of experimental animals/median of the controls.

[c] Statistics: Student-Fisher test. S, significant; NS, not significant.

Table 7.XI. Effect of living Pasteur institute BCG and HIU II on immunoprophylaxis of Lewis tumour

	i.v.		i.p.		i.d.		s.c.	
"SAI"[a]	I[b]	Significance	I	Significance	I	Significance	I	Significance
Pasteur Institute BCG	2.6 ↗	S[c] at 1%	1	NS	1	NS	0.8 ↘	S at 2%
HIU II	1	NS	1	NS	1.10	NS	1.08	NS

[a] SAI = systemic adjuvant of immunity. BCG is injected at day -14, HIU II at day -3.

[b] I = Median of survival of experimental animals/median of the controls.

[c] Statistics: Nonparametric test of Wilcoxon. S, significant; NS, not significant.

These two examples show the extent of information to be obtained from this screening: both of these agents are efficient in the haemolytic plaque-forming cell test (in which macrophages and T and B lymphocytes intervene), but when given at different times before the antigen. Hiu II seems to be considerably more efficient than living BCG on macrophage non-specific cytotoxicity. However, this preparation does not prevent tumour growth, an action that only BCG is capable of doing. But it is only by the IV route that immunoprophylaxis of L 1210 leukaemia is obtained by this mycobacterium. By the s.c. route, it produces growth enhancement of solid Lewis tumour, a phenomenon not found on L 1210 leukaemia.

We realize that agents may not give substantial response in this screening, although they may induce very interesting responses in other tests, and in other hands. The aim of this screening is to select agents which can be candidates for clinical use, and we consider the agents which give positive results as interesting for preclinical study; the latter includes toxicology in monkeys (see JURCZYK-PROCYK *et al.*, 1976).

7.2.1.2. 2. BCG Preparations and Presentation

BCG (Fig. 7.7) was originally prepared by CALMETTE and GUERIN at the Pasteur Institute in Lille through progressive attenuations of a virulent strain of Mycobacterium bovis (GUERIN, 1957). These scientists spent 13 years and made 231 passages in culture mediums containing beef bile as an emulsifying agent.

The strain has been given to other laboratories and substrains have arisen due to selection of phenotypes which are now available in various preparations. While all of them are efficient as local adjuvants in subjects sensitized to tuberculosis, only some, as we shall see, are frequently active as systemic immunity adjuvants. Comparing our screening results with the ten preparations mentioned in Table 7.XII, we have only found a constantly immuno-adjuvant effect with the fresh living Pasteur Institute preparation as seen in Table 7.XIII. (MATHÉ *et al.*, 1973; HALLE-PANNENKO *et al.*, 1976).

The difference found between several strains and/or preparations of BCG may be explained by multiple factors: the mode of preparation, the mode of preservation and the selection of substrains. TURCOTTE and QUEVILLON (1974, 1976) have conducted some interesting work, showing that, in modifying the conditions of culture by successive subcultures of the Montreal strain, they could dissociate three types of bacilli according to the aspects of the colonies in their cultures. They suggest that, because of the reversibility of the phenomenon, they have obtained three phenotypic variations (and not three mutants) of the Montreal BCG. They do not know if the latter was originally constituted of several types, or if the three they identified have been induced de novo. Be that as it may, the latter differ in their chemical

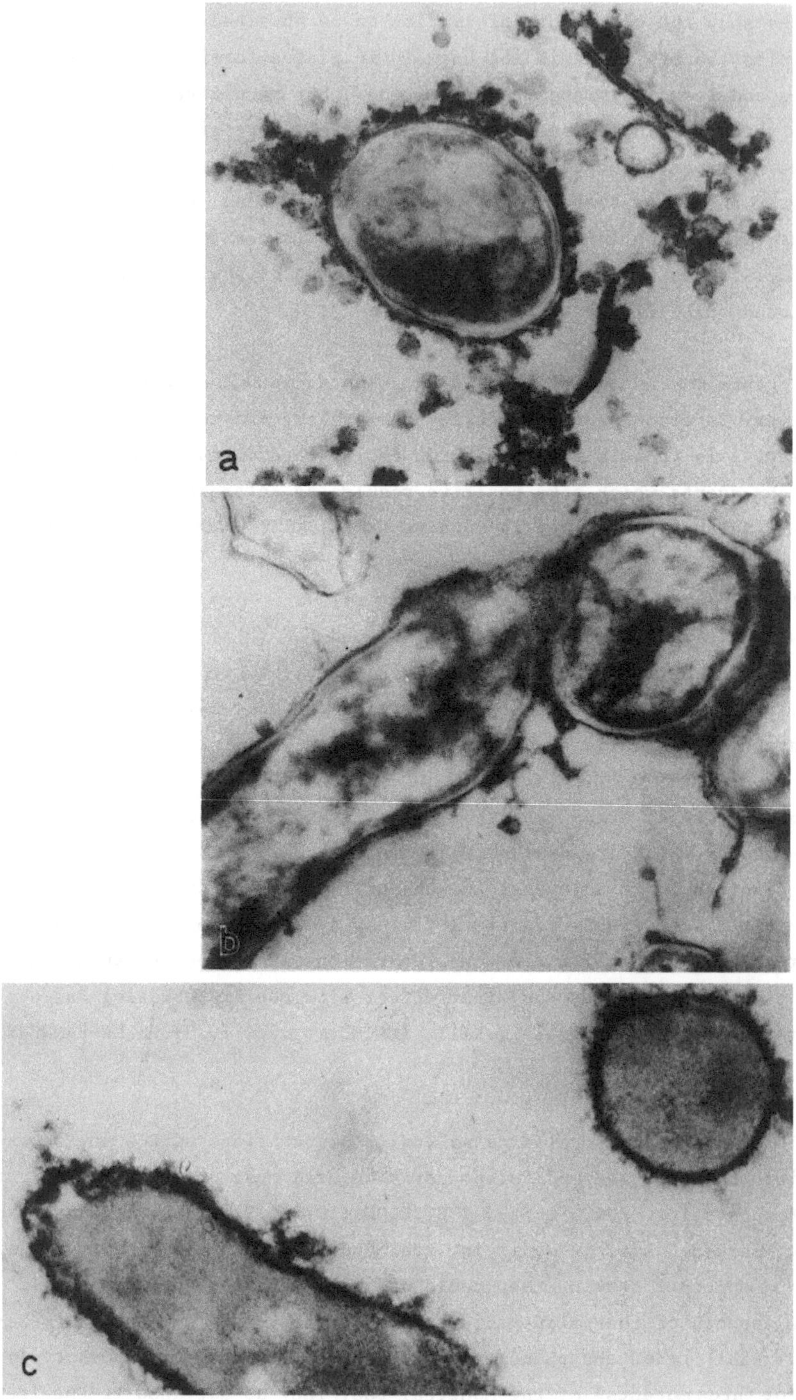

Fig. 7.7. Electron micrographs of : (a) untreated BCG; (b) solvent treated BCG; (c) diluted acid treated BCG. (Kindly provided by I.J. HIU and M. PAINTRAND)

Table 7.XII. Various preparations or strains of BCG submitted to the second part of our screening tests (MATHÉ *et al.*, 1973b)

Preparations or strains	Obtained from	"Viable units" per mg	Other characteristics[a]		
			Virulence for Golden hamster	Germination rate (% at 24 hours)	Opacity (suspension of 0.75 mg/ml)
French Pasteur Institure (fresh)	Pasteur Institute	7×10^6			
French Pasteur Institute 1173 P_5 (lyoph.)	Statens Serum Institute	6.2×10^6	++	75	0.35
Glaxo (lyoph.)	Glaxo Laboratories Ltd.	60×10^6	+	50-75	0.35
Chicago (lyoph.)	Research Foundation	4.2×10^6			
Brazilian (Moreau) (lyoph.)	Statens Serum Institute	9×10^6	++++	75-90	0.40
Madras 809 (lyoph.)	Statens Serum Institute	4×10^6	+	75-90	o.38
Swedish (lyoph.)	Statens Serum Institute	16×10^6	++	75	0.35
Prague 725 (lyoph.)	Statens Serum Institute	1.8×10^6	+	75	0.33
Danish 1331 (lyoph.)	Statens Serum Institute	6.3×10^6	+++	75	0.36
Japanese 172 (lyoph.)	Statens Serum Institute	66.5×10^6	++	75	0.38

[a] See BUNCH-CHRISTENSEN *et al.*, 1968, and LADEFOGED *et al.*, 1970.

Table 7.XIII. Summary of results obtained with ten preparations of BCG (MATHÉ *et al.*, 1973b)

Strains of BCG	Jerne test (i.v.)	Immunoprophylaxis of L 1210 (i.v.)	Lewis tumour (i.v.)
French Pasteur Institute (fresh)	I^a = 2.8 S^c 2% ↗	I^b = 1.6 S^c 2% ↗	I^b = 2.6 S^d 1% ↗
French Pasteur Institute 1173 P5 (lyoph.)	I = 1.1 NS	I = 1.15 S 1% ↗	I = 0.9 NS
Glaxo (lyoph.)	I = 0.3 S 1‰ ↘	I = 1.27 S 1% ↗	I = 0.9 NS
Chicago (lyoph.)	I = 1.1 NS	I = 1 NS	I 2 S 1‰ ↗
Brazilian (Moreau) (lyoph.)	I = 1.1 NS	I = 1.27 S 1‰ ↗	I = 1 NS
Madras 809 (lyoph.)	I = 1.1 NS	I = 1 NS	I = 1 NS
Swedish (lyoph.)	I = 1.1 NS	I = 1 NS	I = 0.9 NS
Prague 725 (lyoph.)	I = 1.08 NS	I = 1 NS	I = 0.9 NS
Danish 1331 (lyoph.)	I = 1.1 NS	I = 1 NS	I = 0.9 NS
Japanese 172 (lyoph.)	I = 1.09 NS	I = 1.15 S 1% ↗	I = 0.9 NS

[a] $I = \frac{\text{Mean No. of PFC/spleen of the tested mice}}{\text{Mean No. of PFC/spleen of controls}}$

[b] $I = \frac{\text{Median survival of experimental mice}}{\text{Median survival of controls}}$

[c] Statistics : Student Fisher test. S, significant; NS, not significant.

[d] Statistics : Wilcoxon's non-parametric test.

constitutions, and they exert different immuno-adjuvant effects: the type called PH1 is less antigenic than PH3 and than the original BCG; its tuberculin activity is inferior; it is less virulent and it often facilitates tumour growth by the SC route, while it possesses a certain anti-growth activity by the i.p. route. PH3 is a more regular and powerful antitumour agent than PH1 and also than the original BCG.

The phenotypic dissimilarities do not entirely explain the differences in activity between preparations.

Table 7.XIII shows that only the *fresh* living Pasteur BCG preparation was effective in all the tests of our screening. The *lyophilized* preparations gave inconstant results, including that of the Pasteur Institute strain (MATHÉ *et al.*, 1973b). This may be due to the power of multiplication of the bacilli, which is diminished by lyophilization (Table 7.XIV) despite the systematic control of quality to which the lyophilized preparations are submitted (GHEORGHIU *et al.*, 1971).

Tables 7.XV and 7.XVI show that several factors intervene in the effect of the different preparations: this effect is not simply correlated either to the virulence for golden hamsters or to the number of "viable units" per mg.

Freezing at -70° for 2 months does not suppress the activities of the Pasteur and of the Phipps[1] preparation on the HPFC test and on the L 1210 leukaemia, but suppresses their action on Lewis tumour. The Tice[2] preparation presented no activity on any of the three tests after this conservation, while the Glaxo preparation was not active on L 1210 leukaemia, was immunodepressive on the HPFC assay, and enhanced the growth of Lewis tumour (Table 7.XVII).

After 9 months at -70° C, the Pasteur preparation was only active on L 1210 leukaemia and lost its activity in the HPFC test (Table 7.XVII).

The Pasteur Institute asked us to compare, in a double blind study, the adjuvant effect of a preparation of *fresh living BCG* which was *maintained at 4° C in a special stabilizing solution* (Dubos' medium to which had been added

---

1 Transmission from Institut Pasteur: J. Aronson, BCG Laboratory, Henry Phipps Institute, Philadelphia, P.A. Dannenberg, jr., John Hopkins Univ., Baltimore, Md. (1965), Trudeau.

2 Transmission from University of Illinois, Institution for Tuberculosis Research, Tice Lab., Chicago, Ill. (1934), Trudeau. Deposited at Trudeau 1967 (lyophilized 9-8 (s) 104); special Handling at Trudeau; retained in deep freeze -70° C.

Table 7.XIV. Institute Pasteur BCG: various preparation used

	Number of mg per ml	Number of culturable units per mg	Number of culturable units per ml	Viability	Presence of aggregates
Fresh Pasteur "BCG - S"	50	$3\text{-}5 \times 10^6$	200×10^6	95% : 1st day 50% : 11th day	+++
Lyophilized Pasteur "BCG - ID"	1.25	7×10^6	5×10^6	50% : 1st day 30% : 30th day	++
New fresh Pasteur "Immuno-BCG"	75	15×10^6	$1,125 \times 10^6$	95% : 30th day 75% : 90th day	+++

Table 7.XV. Correlation between virulence for golden hamsters of the different preparations submitted to our screening and their effect in the three tests (HALLE-PANNENKO *et al.*, 1976)

Virulence for golden hamsters	Effect		
	Negative	Nil	Positive
++++		I = 2	I = 4
+++		I = 2	I = 1
++		I = 7	I = 2
+	I = 1	I = 7	I = 1

The Index (I) is the algebric sum of the values given by each test : + 1 = significant immunostimulation or antitumour effect; 0 = no significant effect; - 1 = significant immunodepression or tumour growth enhancing effect.

Table 7.XVI. Correlation between content in viable units of different preparations submitted to our screening and their effect in the three tests (HALLE-PANNENKO *et al.*, 1976)

"Viable units" per mg	Effect		
	Negative	Nil	Positive
60.10^6 – 66.10^6	I = 1	I = 3	I = 2
10.10^6 – 16.10^6		I = 3	
6.10^6 – 9.10^6		I = 3	I = 6
10^6 – 5.10^6		I = 8	I = 1

glycerol and human albumin); this form of preservation of living non-lyophilized BCG[1] maintains its colony-forming capacity without change (Table 7.XIV) (GHEORGHIU and CHAMBON, 1976) and its adjuvant activity in our screening for at least 3 months. This preparation will facilitate the transportation and the conservation of this living adjuvant agent, hence its use in any country.

[1] Immuno BCG Pasteur F

Table 7.XVII. Results obtained with four preparations of BCG frozen at -70° C for 2 months and for 9 months for the Pasteur preparation (1 mg i.v./mouse; day -14)

Conditions and Sources of Strains of BCG			Jerne test	Immunoprophylaxis of	
				L 1210	Lewis tumour
Fresh French Pasteur			I[a] = 2.80 S[c] 2% ↗	I[b] = 1.60 S 5% ↗	I = 2.60 S[d] 1% →
Frozen at -70° C	French	Pasteur 2 months	I = 1.94 S 5% ↗	I = 1.30 S 3% ↗	I = 0.80 NS
		Pasteur 9 months	I = 1.06 NS	I = 1.07 S 3% ↗	I = 1.10 NS
	Trudeau	Phipps 2 months	I = 3.10 S 5% ↗	I = 1.90 S 1% ↗	I = 0.79 NS
		Tice 2 months	I = 1.17 NS	I = 1.15 NS	I = 1.05 NS
		Glaxo 2 months	I = 0.11 S 1% ↘	I = 1.10 NS	I = 0.56 NS →

[a] $I = \frac{\text{Mean No. of PFC/spleen of treated mice}}{\text{Mean No. of PFC/ spleen of control mice}}$

[b] $I = \frac{\text{Median survival of treated mice}}{\text{Median survival of control mice}}$

[c] Statistics: Student Fisher test; S, significant; NS, not significant.

[d] Statistics: Wilcoxon's non-parametric test.

7.2.1.2.3. Extracts of BCG and other Mycobacterias

The differences between substrains and even between two phenotypes of the same substrain (TURCOTTE and QUEVILLON, 1974) are supposed to be due to the quantity or the "quality" of a molecule which would be responsible for the adjuvant effect.

Tuberculin was used by some clinicians who thought that it was the adjuvant factor in BCG: MOORE *et al.* (1974) showed that its effect is nil in a system where living BCG is completely suppressing tumour growth.

Hence the search for other compounds, especially in the bacillus *cell walls*: these walls, administered by WEPSIC *et al.* (1975) before, or at the same time as, transferable Morris hepatoma, induce inhibition or enhancement.

Crude preparations of mycobacteria may be very active. It is the case in: (a) *methanol extract residue (MER)* (WEISS, 1972; BEN EFRAIM *et al.*, 1973, 1974), the effects of which, observed in several experiments, were mentioned in Chapter 3 and which is a complex material; (b) an extract from M. smegmatis or M. kansasii prepared in the form of a gelatinous fraction collected between the hexane and water phases after lysis of bacterial cells, thus called *"interphase material" (IPM)*, with which CHEDID and LAMENSANS (1974a, 1974b) have obtained encouraging results; (c) *cord factor* (so-called because it may be responsible for special cord-like spatial in vitro growth of tuberculosis bacilli) which has been shown, according to BEKIERKUNST (1974), to be active at a dose of 10 µg in 0.2 ml of an emulsion and which is a simple molecule, 6,6-dimycolate of trehalose (NOLL *et al.*, 1956).

Extensive research has, therefore, been devoted to the search for active molecules.

(a) Several methods have been described for the extraction of liposoluble molecules. These molecules have been designated Pmko (CHOUCROUN, 1947), wax D (RAFFEL, 1948), bound wax D (KOTANI *et al.*, 1963), A fraction (HIU *et al.*, 1970) , and acetolysis extract AE (HIU and AMIEL, 1971). They are soluble in chloroform, but insoluble in hot acetone. They present a similar chemical constitution: they are peptidoglycolipids (mot. wt.~47.000) with various lipid contents.

Pmko and wax D were extracted from bacterial metabolic products. *Pmko* (CHOUCROUN, 1947) was isolated from heat-killed mycobacteria of all human strains by paraffin oil. It is a mixture of mycolic acids and of other material, among which is the adjuvant. *Wax D* (RAFFEL, 1948) was extracted, among the metabolic products of ether-ethanol killed mycobacteria of all human strains, by chloroform, according to Anderson's method as modified by ASSELINEAU and LEDERER (1953).

The lipid fractions similar to Pmko and wax D (i.e., fractions soluble in chloroform and insoluble in hot acetone) can be extracted from all strains of mycobacteria (human and non-human), but only those extracted from human strains were found to possess the adjuvant activity (WHITE *et al.*, 1958), even though whole cells from non-human strains are known to be active (FREUND and McDERMOTT, 1942).

Following this latter observation, intensive investigations in this field have been done to isolate the active material from the non-human strains of mycobacteria, and three non-human adjuvant molecules, called "bound wax D", A fractions, and AE have been successfully obtained by three different methods.

Bound wax D was extracted from bacterial separated BCG cell walls (KOTANI *et al.*, 1963): it was isolated after successive treatments with egg white lysozyme and L_{11} enzyme.

A fraction and AE were prepared from the whole cells of BCG.

A fraction (HIU *et al.*, 1970) was extracted from those cells treated with diluted acid in a suitable organic solvent mixture and then isolated by selective dissolution.

AE (HIU and AMIEL, 1971b) was extracted from the whole BCG cells by acetolysis. It was then isolated by successive precipitations with various neutral organic solvents.

Many structural studies have been conducted on mycobacterial liposoluble adjuvants. Alkaline hydrolysis, described by ASSELINEAU (1949), is a valuable tool for fragmenting the peptidoglycolipid molecules of mycobacterial liposoluble adjuvants. Cleavage of ester linkage (ASSELINEAU, 1949) by alkaline hydrolysis of Pmko, wax D, bound wax D, A fraction, or AE resulted in two distinct moieties: (a) a hydrophilic moiety, identified as a glycopeptide (Fig. 7.8) (proposed by MARKOVITS *et al.*, 1971); and (b) a lipophilic moiety mainly composed of very large mycolic acid molecules $\sim$ $C_{88}H_{176}O_4$ (STODOLA *et al.*, 1938). There are a variety of mycolic acid structures. Most are unsaturated, especially those from peptidoglycolipids of the human Peurois and Canetti strains of mycobacteria (HIU and AMIEL, 1971b). The structures shown in Fig. 7.9 of mycolic acids from various strains of mycobacteria have been proposed by ETEMADI *et al.* (1964a, 1964b) and MARKOVITS *et al.* (1966).

Much work has also been devoted to the relationship between structures and activity. WHITE *et al.* (1958) studied the role of the peptide portion. They showed that wax D fractions (i.e., fractions soluble in chloroform and insoluble in hot acetone) from non-human strains of mycobacteria failed to increase either cell-mediated or humoral

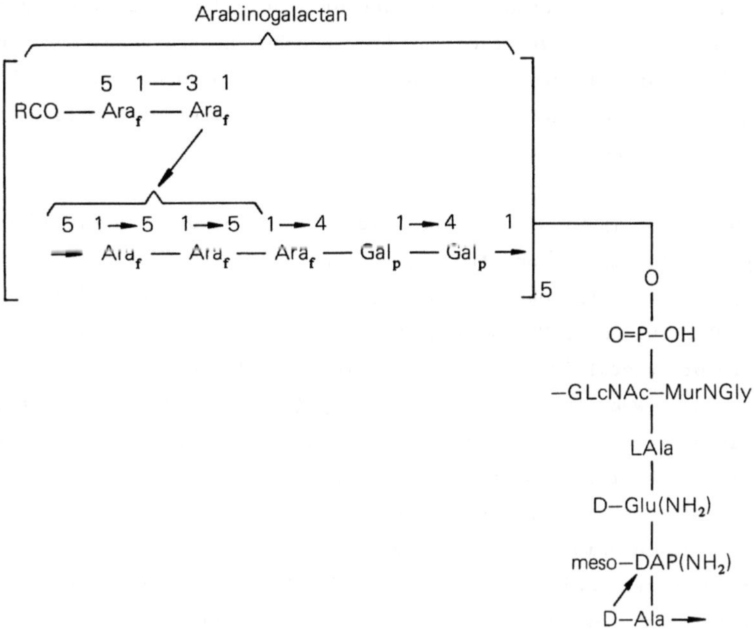

Fig. 7.8. Structure of a "monomer" of the hydrophilic moiety from peptodoglycolipids of M. tuberculosis human Peurois strains (MARKOVITS et al., 1971)

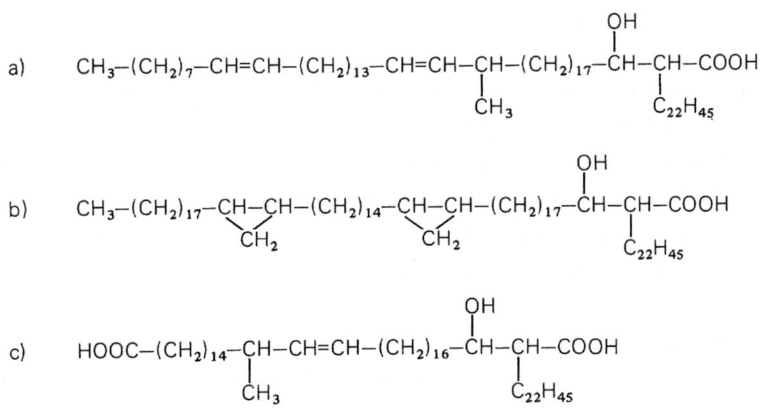

Fig. 7.9. Unsaturated mycolic acids from various strains of mycobacteria. (a) M. smegmatis (ETEMADI et al., 1964b); (b) M. kansasii (ETEMADI et al., 1964a); (c) M. phlei (MARKOVITS et al., 1966)

immune responses under conditions in which wax D from various human strains of mycobycteria exhibited a marked adjuvant effect. Inactivity of wax D from non-human strains was attributed to the lack of a peptide portion. The peptide-containing and wax D similar active fractions from non-human strains of mycobacteria were shown to be incorporated into the cellular structure in such a manner that they can only be extracted after the whole cells have been treated with diluted acid in a suitable organic solvent mixture (HIU *et al.*, 1970).

HIU *et al.* (1969) studied the relationship between the ratio "lipophilic moiety-hydrophilic moiety" and the adjuvant activity. They demonstrated that an increase of lipid content (lipophilic moiety) in mycobacterial peptidoglycolipid molecules (MPM) resulted in MPM derivatives highly active as adjuvants either for cell-mediated or humoral immune responses. This result suggests that in MPM, an optimal ratio exists between the hydrophilic moiety (glycopeptide part of MPM) and the lipophilic moiety (lipid part of MPM). The hydrophilic-lipophilic structure of MPM confers certain physical properties, such as adsorption and interfacial properties which are important in the mode of action of mycobacterial peptidoglycolipid adjuvants and related compounds in biological systems. The isolation of the natural high-lipid-content adjuvant, A fraction (high-lipid-content peptidoglycolipid) from BCG confirms the essential role of the lipophilic-hydrophilic ratio in the adjuvant activity of these substances (HIU *et al.*, 1970): (a) the adjuvant activity of A fraction is as intense as MPM derivatives; (b) the hydrophilic-lipophilic ratio found in the A fraction and in MPM derivatives is similar.

HIU and AMIEL (1971a) studied the relationship between the fatty acid nature and the adjuvant activity. They showed that hydrogenated MPM derivatives containing saturated mycolic acids are far more active than MPM in which the mycolic acids, constituting the lipophilic moiety of MPM, are not saturated. The increase in adjuvant activity of hydogenated MPM derivatives results from fatty acid saturation, possibly by giving a more stable compound with a longer-lasting immunostimulant activity. By catalytic hydrogenation and acylation with various long-chain fatty acid chlorides, HIU et al. have prepared various MPM derivatives with powerful adjuvant effects (HIU *et al.*, 1969; HIU and AMIEL, 1971a).

(b) Mycobacterial liposoluble adjuvants must be administered in water-in-oil emulsions which is a great inconvenience. In attempting to avoid the use of oily material, various watersoluble agents have recently been isolated.
However, most of them are only active when administered with Freund's incomplete adjuvant (FIA). They concern agents extracted from M. smegmatis and M. tuberculosis of human virulent strains.
An arabinogalactan-peptidoglycan , *WSA* or "water soluble adjuvant" (ADAM *et al.*, 1972; CHEDID *et al.*, 1972), was extracted from previously separated cell walls of M. smegmatis by lysozyme treatment, according to Kotani's method (KOTANI *et al.*,

1963), mol. wt. 20.000. Its peptidoglycan derivatives were shown to be active (ELLOUZ *et al.*, 1974).

A polysaccharide-peptidoglycan was isolated by MIGLIORE-SAMOUR and JOLLES (1972) among watersoluble products of M. tuberculosis human Peuroi's cells after homogenization in water and ammonium sulfate treatment. The same authors have also prepared three other *"(Jolles) water soluble agents"* from M. tuberculosis (var. hominis), the polysaccharide moiety of this native polysaccharide-peptidoglycan, the acetylated peptidoglycan, and a tetrasaccharide-heptapeptide (WERNER *et al.*, 1976). All are adjuvants for delayed hypersensitivity when injected with Freund's incomplete adjuvant.

Another arabinogalactan-peptidoglycan, *neo-WSA* (ADAM *et al.*, 1973; JUY *et al.*, 1974) was extracted from the whole cells of M. smegmatis by the same method as for WSA, except that whole cells are used as the starting material.

The ability of mycobacteria to act as adjuvants appears to be related to various fractions mentioned above. Inclusion of each adjuvant fraction in the injected antigen preparation resulted in skin reactivity of delayed-type hypersensitivity to the test antigen. Most of the mycobacterial adjuvants also exhibit an adjuvant effect on the production of antibodies. However, most of them were described as active when administered in FIA.

More interesting, for practical use, is another watersoluble extract which is active not only when injected in FIA, but also when administered in saline. It is prepared from whole BCG cells and we call it *"Hiu II"* (HIU, 1972). It is a glycomucopeptide extracted from the cells by catalytic hydrogenolysis, mol. wt. 32,000. It is released after the action of gaseous hydrogen in ethanol containing acetic acid and palladium catalyst within 8 hrs. Its detailed study was compared with that of fresh BCG on p. 235.

We have submitted to our screening a few extracts of mycobacteria, including those which have just been mentioned (Table 7.XVIII) (MATHÉ *et al*, 1975a).

It can be seen in Table 7.XIX that MER was active under some conditions, and, interestingly, on the two tumours, when it was administered 3 days before neoplasia cell inoculation.

AE (HIU and AMIEL, 1971b) presents only an immunodepressive effect in the haemolytic plaque-forming cell (HPFC) assay.Hiu II (HIU, 1972), the action of which has been seen already on macrophage activation, is stimulating in the HPFC test. However,

Table 7.XVIII. Various extracts from mycobacteria submitted to ICIG-EORTC screening for immunity adjuvanticity

Agents	Obtained from	Composition	Dose per mouse
MER	ICIG Weiss's method	BCG without lipid fraction extracted in methanol and acetone	1 mg
AE	ICIG	Component of the BCG membrane (non-soluble)	1 mg
HIU II	ICIG	Component of the BCG membrane (soluble)	0.5 mg
WSA F3684	E. Lederer (CNRS Gif.)	Soluble extract from M. smegmatis	0.5 mg

it is not active in our screening on the growth of the two tumours. WSA gives the same results as Hiu II.

7.2.1.2.4. Corynebacteria Preparations

The interest in corynebacteria started with the demonstration by HALPERN *et al.* (1963) of its stimulatory effect on the "phagocytic capacity of the reticulo-endothelial" system. It was shown to be an adjuvant in our sense, in that it boosts antibody response to a variety of antigens (BIOZZI *et al.*, 1968; RAYNAUD *et al.*, 1972; HOWARD *et al.*, 1973a, 1973b; MATHÉ *et al.*, 1973b), and it was demonstrated by WOODRUFF and BOAK (1966) that it may possess an antitumour action although cancer immunity is considered to be predominantly cell-mediated and corynebacteria induces a T-cell depression (SCOTT, 1972; CASTRO, 1974; RUITENBERG and STEERENBERG, 1973). This will be considered in more detail below.

Because of the difficulty in obtaining preparations of C. parvum with constant efficacy when repeating our tests, we applied to the Pasteur Institute for another species of corynebacteria, the taxonomy of which is, incidentally, a matter of controversy (BOUSFIELD, 1972). The Pasteur Institute made three preparations of C. granulosum (Table 7.XX) (KOUZNETZOVA *et al.*, 1974), the activity of which is, as seen in Table 7.XXI, satisfactory and almost constant.

CURRIE and BAGSHAWE (1970) used strain 10390, obtained from the National Collection of Type Cultures, Colindale, at a dose of 200 µg, with a positive antitumour action.

Table 7.XIX. Summary of results obtained with BCG and M. smegmatis extracts (i.v., i.p., s.c., i.d.)[a] (MATHÉ *et al.*, 1976b)

Tests / Agents	Jerne test		Immunoprophylaxis of: L 1210		LLT	
	Route of injection	Results	Route of injection	Results	Route of injection	Results
MER (BCG extract)	i.d. day +1[b]	I^c = 2.4 S^e 2% ↗	s.c. day -3	I^d = 1.5 S^e 5% ↗	i.v. day -3	I^d = 1.2 S^f 4% ↗
AE (BCG extract)	i.v. day -3[b]	I = 0.5 S 5% ↘				
HIU II BCG extract)	i.v. day -3[b]	I = 1.6 S 5% ↗	NS		NS	
WSA F3684 (M.smegmatis extract)	i.v. day -3[b]	I = 9.0 S 1 ‰				

[a] Only routes and times of injection giving significant results are indicated.

[b] After or before antigen injection.

[c] $I = \frac{\text{Mean No. of PFC/spleen of treated mice}}{\text{Mean No. of PFC/spleen of control mice}}$

[d] $I = \frac{\text{Median of survival of treated mice}}{\text{Median of survival of control mice}}$

[e] Statistics: Student Fisher test; S, significant; NS, not significant.

[f] Statistics: Wilcoxon's non-parametric test.

Table 7.XX. Corynebacteria submitted to ICIG = EORTC screening for immunity adjuvanticity

Agents	Obtained from	Composition	Dose per mouse
C. Parvum lot 90591	Pasteur Institute	Suspension of dead bacteria	400 γ
C. Granulosum A lot 90717	Pasteur Institute	Killed bacilli	400 γ
C. Granulosum B lot 70009	Pasteur Institute	Membranes treated by phenol in the cold	100 γ
C. Granulosum C lot 90799	Pasteur Institute	Membranes prepared by mechanical action	100 γ

7.2.1.2.5. Polynucleotides

Among the agents recently revealed as adjuvants are the double-stranded synthesis polynucleotides (FIELD *et al.*, 1967) of which two have been the object of much research: poly I-poly C (Fig. 7.10) (BRAUN *et al.*, 1971a, 1971b; MATHÉ *et al.*, 1971c), and poly A-poly U (YOUN *et al.*, 1968; LEVY *et al.*, 1969; SARMA *et al.*, 1969; LACOUR *et al.*, 1975).

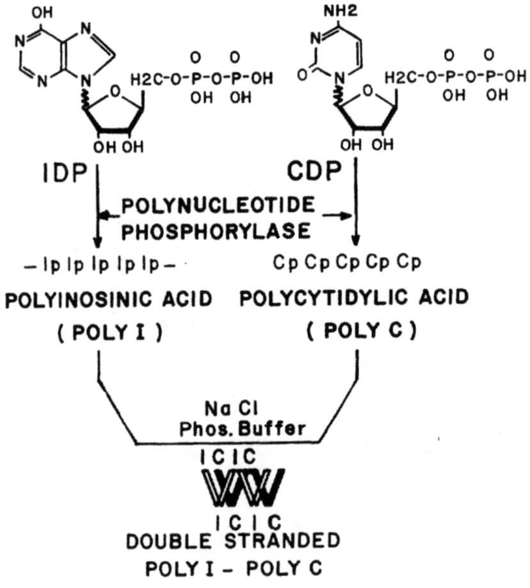

Fig. 7.10. Synthesis of polynucleotides

In our screening (Tables 7.XXII and 7.XXIII) *poly I-poly C* is immunostimulating when administered after antigen in the HPFC test, while it is immunodepressive when injected before. It possesses a prophylactic action on L 1210 leukaemia (MATHÉ *et al.*, 1973b).

Poly A-poly U is only active on the HPFC and only when administered after the antigen (MATHÉ *et al.*, 1973b).

7.2.1.2.6. Other Agents

We have tested a few more agents: two mitogens considered by some as adjuvants, *PHA* (NASPITZ *et al.*, 1968) and *LPS* (ALLISON and DAVIS, 1971), *lentinan*, a polysaccharide extracted from a mushroom, Lentinus edodes, and shown to be an adjuvant by CHIHARA *et al.* (1969, 1970), *levamisole* (see Table 7.V), also described as an immunopotentiator by RENOUX and RENOUX (1971a, 1971b): all act as immunosuppressors in the haemolytic plaque-forming cell test and do not affect the growth of the two tumours.

We also submitted to our screening *Vitamin A* (an adjuvant according to DRESSER 1968), *oestradiol*, and *caffeine* (see Table 7.V): they did not induce any effect in any test.

7.2.1.2.7. Experimental Pharmacology

These few results, their possible disparity from one test to another in our screening, and their possible discordance with those of other authors, underline the role played by several factors. A choice of parameters was obligatory for technical and financial reasons. The agents which are active in this screening on the two operational tests will be the object of preclinical and clinical studies. The others can be used in more sophisticated research, and they may be resubmitted to our screening with other parameters.

The role of these parameters and of various factors that may explain the discordance between the results of different groups will be illustrated.

(a) The *origin and modality of preparation of the agent* tested as adjuvant-candidate has been sufficiently discussed when studying the different strains of mycobacteria and the different species and preparations of corynebacteria.

(b) *Antigenicity of the adjuvant.* A correlation between an increase of the immunostimulating action of an adjuvant and a decrease of its antigenicity has been observed by TANAKA *et al.* (1968) and by HIU and AMIEL (1972). The phenomenon of antigenic competition may explain this fact.

(c) The *time factor* has also been illustrated in chapter 3. As seen in Table 7.VIII, 7.IX, 7.XIX, 7.XXI, and 7.XXIII and in Figure 7.6, the time gap between antigen and

Table 7.XXI. Significant effects obtained with corynebacteria administered by 4 routes (i.v., i.d., i.p., s.c.) of administration and at different intervals before or after the antigen[a]

Agents	Jerne test				L 1210				LLT			
	Injection (days before or after antigen)	Route of administration	I_1	Statistics p_1	Injection (days before antigen)	Route of administration	I_2	Statistics p_2	Injection (days before antigen)	Route of administration	I_2	Statistics p_2
C. Parvum lot no 90591	-14	i.v.	1.5	0.01	-3	i.v.	1.15	0.01	-3	i.v.	1.2	0.02
	- 5	i.p.	1.2	0.05								
	- 3	i.p.	1.3	0.01								
	0	i.v.	1.6	0.02								
	+ 1	s.c.	1.8	0.03								
	+ 2	i.v.	2	0.01								
	+ 3	i.v.	1.8	0.02								
C. Granulosum A lot no 90717	-14	s.c.	1.19	0.05	- 3	i.v.	1.2	0.01	-3	i.v.	1.2	0.02
	-14	i.v.	1.7	0.01	- 3	i.p.	1.4	0.05				
	- 5	i.p.	1.2	0.05	- 3	s.c.	1.115	0.02				
	- 3	i.p.	1.2	0.02								
	0	s.c.	1.32	0.05								
C. Granulosum B lot no 70009	-14	i.v.	1.75	0.05					-3	i.v.	1.2	0.01
	- 5	i.p.	1.5	0.05								
	- 3	i.p.	1.5	0.05								
	0	i.p.	1.7	0.01								
	0	s.c.	3	0.02								
	+ 2	i.v.	1.75	0.01								
	+ 2	s.c.	3	0.05								

	+ 3	i.p.	1.5	0.05								
	+ 3	s.c.	3.8	0.02								
C. Granulo-sum C lot 90799	- 5	i.v.	2.01	0.01	-3	i.v.	1.2	0.01	-3	s.c.	1.8	0.02
					-3	i.p.	1.2	0.05				

$I_1 = \frac{\text{Mean No. of PFC/spleen of the tested mice}}{\text{Mean No. of PFC/spleen of controls}}$

$I_2 = \frac{\text{Median of survival of experimental mice}}{\text{Median of survival of controls}}$

p_1 = Student Fisher test.

p_2 = Non-parametric test of Wilcoxon.

[a] Only intervals and routes of administration giving positive results are indicated in this table.

For complete results, see: MATHÉ *et al.*, (1973).

Table 7.XXII. Polynucleotids submitted to ICIG screening for immunity adjuvanticity

Agents	Obtained from	Composition	Dose per mouse
Poly I-poly C	Choay	Poly-inosinic acid polycytidylic acid	300 γ
Poly A-poly U	Choay	Poly-adenosinic acid Poly-uridylic acid	150 γ

adjuvant administrations is essential. One sees that the optimal time is variable, not only for different adjuvants, but also for various tests concerning the same agent. In the haemolytic plaque-forming cell test, for example, it is day -14 for living fresh BCG and day -3 for the hydrosoluble extract called Hiu II.

Still more important is the fact that if a given adjuvant potentiates the immune response when given at a certain time before or after the antigen, it may have no effect, or it may depress an immune response when given at a different time (MATHÉ *et al.*, 1973b). Similarly, SCOTT (1974a, 1974b) obtained an inhibitory effect of delayed hypersensitivity to sheep red blood cells by administering C. parvum before this antigen. This time factor may explain some immunodepression(s) induced in a test by an agent proven to be an adjuvant by others.

It may also explain negative results, in which the agent appears in a test to be neither stimulating nor depressing (MATHÉ *et al.*, 1973), as the kinetics of the immune modification induced by an adjuvant may comprise some inhibiting phases interspersed with some immunostimulating ones: DONNER and VAILLIER (1974) have shown that BCG and a polysaccharide extracted from Proteus vulgaris induce a three-phase effect on in vivo antitumour immunity: (a) the latter is strongly depressed in the early days after two injections of mycobacteria or P. vulgaris preparation; (b) in the second phase, immunity returns to a normal level; (c) the third period is characterized by a significant increase of the immune response of mice to tumour-associated antigens. This concerns anti-tumour immunity, and its decline is associated, according to the authors, with serum factors.

The reasons for the variability of the optimal gap between the injection of the antigen and that of the adjuvant are not known: in the case of B. abortus, for which the time is especially long, it can be reduced from 40 days to a few hours by the injection of a hyperimmune brucella antiserum (TOUJAS *et al.*, 1973, 1974).

Table 7.XXIII. Significant effects obtained with polynucleotides administered by 4 routes (i.v., i.p., s.c., i.d.) of administration and at different intervals before or after the antigen[a]

	Jerne test				L1210			
Agents	Injection (days before or after antigen)	Route of administration	I_1	Statistics P_1	Injection days before transplantation	Route of administration	I_2	Statistics P_2
Poly I-poly C	-14	i.p.	0.75	0.01 ↘	-3	i.v.	1.5	0.05 ↗
	- 3	i.v.	0.73	0.01 ↘				
	0	i.p.	1.8	0.05 ↗				
	+ 2	s.c.	1.8	0.01 ↗				
Poly A-poly U	+ 1	i.v.	2.2	0.01 ↗				
	+ 2	i.v.	2.1	0.02 ↗				
	+ 3	i.p.	1.7	0.05 ↗				

[a] See footnotes, Table 7.XXI.

Table 7.XXIV. Influence of the dose of fresh living BCG (injected i.v. 14 days before antigen) on the humoral response of mice sheep red blood cells (Jerne test) (MATHÉ, 1976a)

Dose (mg)	Index of stimulation		
10	I[a] = 0.06 ↘	S[b] >	1%
5	I = 1.3	NS	
1	I = 4.01 ↗	S >	2%
0.5	I = 4.01 ↗	S >	1%
0.1	I = 1.64	NS	
Control	I = 1		

[a] $\frac{\text{Mean No. of PFC/spleen of experimental mice}}{\text{Mean No. of PFC/spleen of controls}}$

[b] Student Fisher test. S, significant; NS, not significant.

(d) *Dose Factor*. The dose is also an essential factor for the action of an adjuvant. Table 7.XXIV shows that BCG is not active below 0.5 mg/mouse, is optimally efficient between 0.5 and 1 mg, is not any more active at the dose of 5 mg, and is immunodepressive at the dose of 10 mg (MATHÉ, 1975a); this immuno-inhibition has been shown, in our laboratory, to be due to T-suppressor lymphocyte stimulation (GEFFARD and ORBACH-ARBOUYS, 1976), as will be illustrated in Chapter 9.

The optimum dose has been found to be narrow for some adjuvants such as MER (WEISS *et al.*, 1966b).

In the case of poly I-poly C, we have shown, with Hayat (MATHÉ *et al.*, 1971c), that the variation of the effect with dose is not necessarily the same for the haemolytic plaque-forming cell test (only 1 mg/kg is efficient while 0.1 and 10 mg are not) (Fig. 7.11) and for the antitumour assay: 1 and 10 mg are active against L 1210 leukaemia (Fig. 5.4).

(e) *Multiple Versus Single Doses*. In our experiments on immunotherapy of L 1210 leukaemia, repeated injections of BCG were more effective than a single administration (MATHÉ *et al.*, 1969).

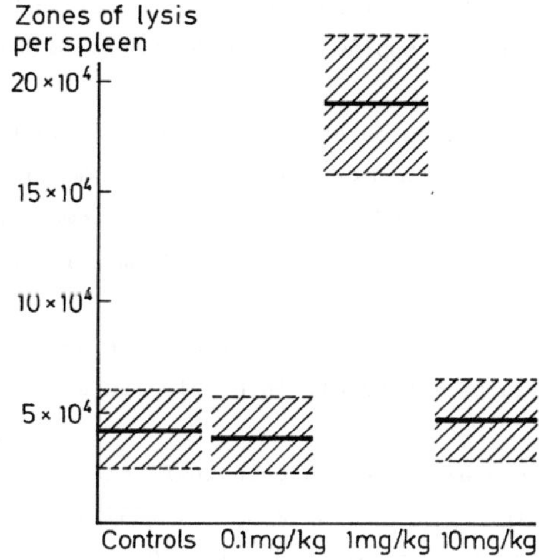

Fig. 7.11. Dose factor for the effect of poly I-poly C on the haemolytic plaque-forming cell test (MATHÉ et al., 1971c)

On the contrary, in immunotherapy of mice mastocytoma with C. parvum, SCOTT (1974b) observed that one single dose was as effective as multiple doses, if not more so.

(f) *Route of Adminstration.* The route of administration is so important that, like the time factor, we felt it necessary to include its systematic study in our screening.

We have seen in Figure 7.6 and Tables 7.VIII, 7.IX, 7.X, 7.XI, 7.XIX, 7.XXI, and 7.XXII that the efficient route of administration not only varies with the different agents but varies in different tests with a given adjuvant: when administered by the s.c. route, BCG is efficient in macrophage activity assay; it is not in the other tests, except on Lewis tumour, which it does enhance, while the same dose given i.v. protects against L 1210 leukaemia.

When using BCG in mice, there is no doubt that the i.v. route is the most efficient for all our tests (except macrophage activation), especially tumour assays.

We have felt it necessary to search for the possible effects of using routes other than i.v., i.p., s.d., and i.d. because, in man, the i.d. route does not permit administration of a great quantity, the i.p. route is not generally used, the s.d. route has been shown to be inefficient by BLUMING *et al.* (1972, 1974), and the i.v. route is not only poorly tolerated but immunodepressive, as will be described below (SCHWARZENBERG *et al.*, 1976).

We have been using scarifications for 13 years (MATHÉ *et al.*, 1968, 1976a); POWLES *et al.* (1973) have been using the Heaf gun injection method (Fig. 7.12), and

FALK *et al.* (1973) the oral route. We have, therefore, studied the effect of living fresh BCG applied by these various methods (Fig. 7.13) (MATHÉ, 1976a).

Application on scarifications has given significant results in immunotherapy after cyclophosphamide chemotherapy of murine L 1210 leukaemia. It should be mentioned here that PETERS *et al.* (1974) showed that BCG, applied to scarifications in guinea pigs, augments cell-mediated immunity and depresses humoral immunity.

Application with a multipuncture Heaf gun (40 punctures at 2 mm) also gave us a significant result in L 1210 leukaemia immunotherapy (MATHÉ, 1976a) (Fig. 7.13).

The oral route has been inefficient in this immunotherapy experiment (MATHÉ, 1976a) (Fig. 7.13). BALDWIN *et al.* (1975) were also unable to obtain significant results against sarcoma Mc 57 with BCG administered orally. This route has been studied by two other authors. ROSENTHAL (1974) compared the histological modifications in different organs, especially the liver, with those induced by i.d. inoculation: the oral route appeared to induce more changes. This does not mean that it is a good route of administration for immuno-adjuvanticity. BALDWIN *et al.* (1975) were unable to induce a general immunostimulation by orally administered BCG, but in situations where they had obtained marked tumour suppression by BCG and tumour cells, they observed that this effect could be lost by administering BCG orally, even in comparatively massive doses.

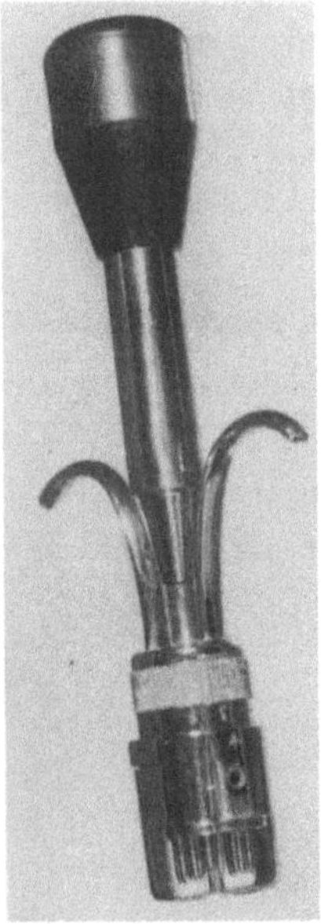

Fig. 7.12. The Heaf gun

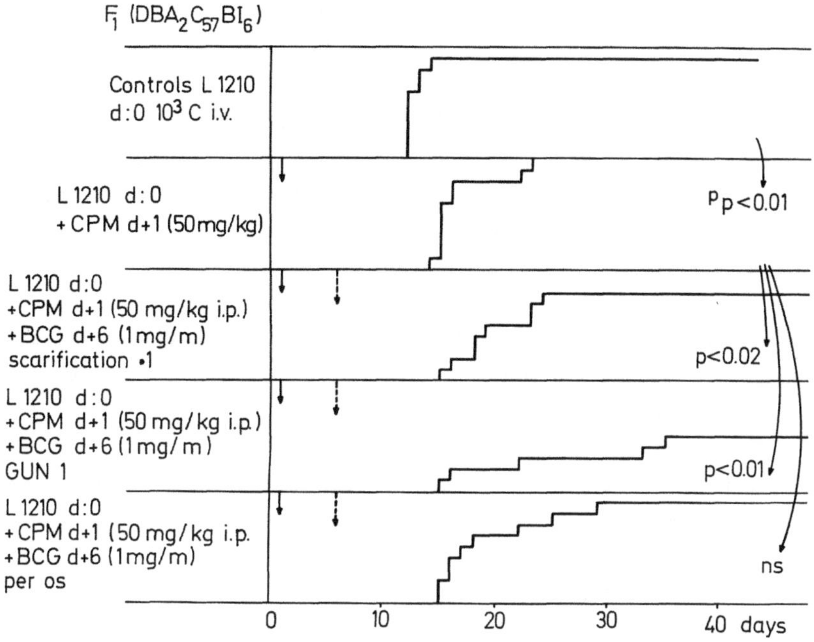

Fig. 7.13. Effect of BCG introduced by different routes of administration in post-chemotherapy active systemic immunotherapy of murine L1210 leukaemia (MATHE, 1976a) CPM = Cyclophosphamide; C = cells; d = day; m = mice, ns = not significant

Table 7.XXV. Jerne test: Administration of BCG by inhalation to (DBA_2 x $C_{57}Bl_6$)F1

Number of plaque forming cells (PCF)		
Controls: $31.10^3 \pm 14.10^3$		
BCG : $62.10^3 \pm 21.10^3$	$t^a = 2.48$	$p < 0.05$

[a] Student Fisher test.

Finally we have demonstrated in mice that the administration of BCG by inhalation is also active (Table 7.XXV).

The route of administration is not only essential in the case of BCG. As seen above (MATHÉ *et al.*, 1973b), it is important when using other adjuvants. It may, however, vary from model to model. When C. parvum is used, WOODRUFF (1974) showed that, as far as its antitumour effect is concerned, i.v. injection is a little more effective than i.p. which, in turn, is still more effective than s.c.

(g) *Combinations of Adjuvants.* The different adjuvants, as we shall see in Chapter 9, work via different mechanisms. Hence the idea arose of combining several.

However, there are no experimental data available to indicate that a combination of two active agents is more efficient than the application of the better of the two, applied at its optimal dose. This is what we have found, combining BCG and C. parvum in the treatment of L 1210 leukaemia; the result was not improved, compared to that obtained with BCG alone (not published).

Moreover, RENOUX and RENOUX (1971a, 1971b) have observed that resistance to Brucella infection by vaccination (which is, at the same time, a specific stimulus and an adjuvant) is reduced by previous administration of BCG.

However, it is suspected that what is true for the combination of two agents given at their optimal dose may also be true when both are administered at suboptimal doses.

(h) *Genetic Host Factors.* Working on immunoprophylaxis of methylcholanthrene (MCA) (1 injection of 0.1 mg s.c. into 5-6 week old mice) induced tumours with B. abortus as adjuvant (500 μg injected i.v., 8 days after MCA), PILET *et al.* (1974) observed significant tumour growth inhibitory effect on DBA/2 mice, slight favourable action on C3H and (C3H x DBA/2) F1, but enhancement of tumour growth in CBA and C57Bl/6 strains.

STIFFEL *et al.* (1974) have found, comparing the resistance of (high responders x AkR) F1 hybrids to that of (low responders x AkR) F1 hybrids to an AkR leukaemia, that the increased responsiveness induced by C. parvum is more important in the (low responders x AkR) hybrids. The authors suggest, as an explanation of this paradoxical phenomenon, a hypothesis according to which the "strong" could produce more "facilitating antibodies".

In conclusion, experimentalists are far from being in a position to give answers to all the questions which the clinician asks for his guidance in the use of adjuvants in man. Moreover, as will be seen for BCG, man is not a mouse.
Hence the necessity for developing clinical immunopharmacology.

7.2.1.3. Clinical Screening and Pharmacology

In man, as in animals, the only way to confirm that an agent which has been shown to be an adjuvant in other species, is also an adjuvant in ours, is to submit it to a battery of tests.

7.2.1.3.1. The ICIG-EORTC Clinical Screening for Systemic Immunity Adjuvants or Immunorestorators

We shall discuss in Chapter 11 the different tests which are available today for immunological investigation of human patients. It goes without saying that only a limited number can be used for screening.

Table 7.XXVI gives the battery proposed by BLUMING *et al.* (1974). Our battery will be described in detail in the appendix devoted to the monitoring of "general" immune functions.

Table 7.XXVI. Schedule of immunologic screening proposed by BLUMING *et al.* (1974).

	Admission	Prior to start	B C G		
			Day 1	During therapy	Day 32
Cellular reactivity					
1. Picryl chloride			+(Sens.)[a]	+(ch)[b]	+(ch)[b]
2. Recall antigens		+			+
3. DNCB	+(Sens.)[a]	+(ch)[b]			+(ch)[b]
4. Circulating lymphocyte counts			+	+[c]	+
Circulating antibody reactivity					
1. Tularin			+(Sens.)[a]	+[c]	+
2. Vi antigen	+(Sens.)[a]		+(Sens.)[a]	+[c]	+
3. Immunoglobulins			+	+	+

[a] Sensitizing dose applied.

[b] Challenge dose applied.

[c] Determinations performed on days 4, 7, 14 and 28.

The use of in vivo delayed hypersensitivity responses to recall antigens allowed us to classify patients into "responders" (allergic) and "non-responders" (anergic)

Screening for immunostimulator agents is conducted on the first group, that for immunorestorators on the second.

The other tests, such as skin tests to primary antigens are used to give complementary information: a dinitro-chlorobenzene (DNCB) negative patient becoming positive after the administration of a given agent, strongly suggests that this agent is efficient for immunorestoration. A subject positive to recall antigens, but negative to picrylchloride who, after application of an adjuvant-candidate, becomes positive to it, whilst the skin reaction to the recall antigens are stronger (the size of the induced nodules increases), presents a good indication that the candidate is really an adjuvant. In fact, almost 100% of the normal population responds to DNCB, while only a small proportion responds to picrylchloride (BLUMING *et al.*,1974).

The in vitro tests also give complementary information and we saw above that there may be dissociations between skin tests and mitogen transformation of lymphocytes in vitro on the one hand (SIMMLER *et al.*, 1976a), and between the skin tests and the leucocyte inhibition factor (LIF) on the other (BRULEY-ROSSET *et al.*, 1976b).

The count in the peripheral blood of T cells, B cells, and null cells, and the estimation of K cells, which are probably null cells, or part of them (as previously mentioned), are included in our battery.

When screening and studying an adjuvant-candidate, for other purposes, great care must be taken regarding the route of administration, dosage, and time of application.

These factors are particularly important, as the screening now under consideration is conducted on patients, and in view of the fact that an agent which is immunorestoring or immunostimulating in some conditions and/or some species may be immunodepressing in man and enhance tumour growth in cancer patients. As far as the duration of administration is concerned, it seems, from our previous studies, that at least 6 weeks are necessary.

7.2.1.3.2. Clinical Pharmacology

I shall describe the agents which have already been used in clinical immunotherapy or immunorestoration as described in Chapter 3, 4 and 6, with comments on their different pharmacological factors.

7.2.1.3.2.1. BCG Preparations

Since only living fresh or well preserved Pasteur BCG preparations are efficient in all the tests of our experimental screening in mice, our group has never used other preparations for systemic immunotherapy or immunorestoration.

The authors, who have used several preparations in man, consider the Pasteur one is the most efficient (GUTTERMAN, personal communication). The fresh Pasteur preparation used until now maintained the same concentration of viable units for 10 days at + 4° C. The new preparation, mentioned in Table 7.XIV and called "immuno-BCG Pasteur F", maintains its concentration for 3 months, and its efficiency at the same level for 3 months as controlled in a double blind study conducted by us.

BLUMING *et al.* (1972) showed the beneficial action of the fresh Pasteur preparation in melanoma patients and, in the same clinical study, the inefficacy of the Glaxo preparation, which is in accordance with our experimental results. The inability of the British Medical Research Council (1971), which used Glaxo BCG alone, to confirm the effect of "immunotherapy" in acute lymphoid leukaemia (ALL), is also in agreement with our experimental conclusions (MATHÉ *et al.*, 1973c). As a matter of fact, the EORTC Haemopathies Working Party (1975) confirmed the efficacy of the fresh Brussells Pasteur preparation (which is similar to the Paris preparation) as a means of ALL treatment. The positive effect obtained by POWLES *et al.* (1973) using the Glaxo preparation in acute myeloid leukaemia, may be due to the leukaemia cells he injected in combination at the same time.

A positive effect was obtained with the Tice strain by VOGLER and CHAN (1974), and by GUTTERMAN *et al.* (1973). This strain was originally the Pasteur strain.

The *route of administration* is probably a very important factor as was seen in Chapter 6 for immunorestoration of anergic patients (SIMMLER *et al.*, 1976b, 1976c), and as is shown in Table 7.XXVII for skin responders (SCHWARZENBERG *et al.*, 1976).

We have used, from the beginning of our work in 1963 (MATHÉ *et al.*, 1968), the method of BCG application on scarifications (Fig. 7.14).

Table 7.XXVII. The efficacy according to the different routes of BCG administration in normal skin responders (SCHWARZENBERG *et al.*, 1976)

Skin reactions[a]	1 Scarified area	4 Scarified areas	IV
Increased	1	3	0
Unchanged	0	4	1
Decreased	4	0	3

[a] To recall antigens measured by palpation in a double blind fashion.

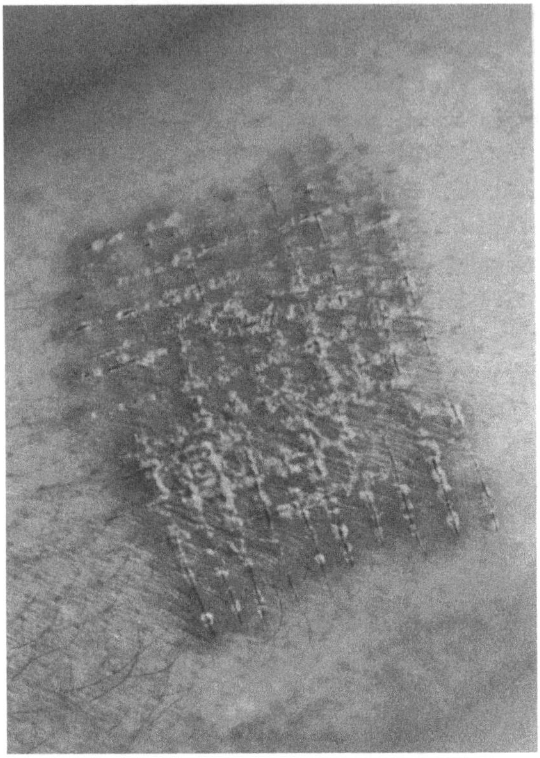

a

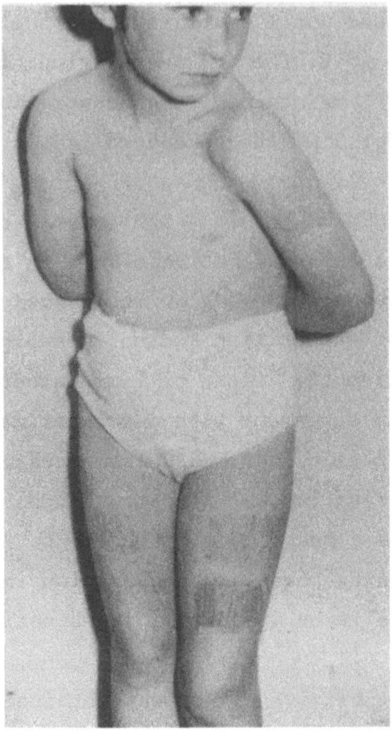

b

Fig. 7.14. (a) Aspect of a scarified area, 3 days after BCG application; (b) evolution of the macroscopical lesions induced by BCG application on scarified areas

We make 20 cutaneous scratches, each 5 cm long, arranged in a square, and 2 ml of a suspension containing 75 mg/ml of fresh Pasteur BCG are placed on the scarified area. The object is to facilitate penetration into the blood of the living microorganisms in order to obtain a bacteraemia, and penetration into the lymph.

In the beginning, we applied the BCG suspension to one area at a time (twice a week at first, once a week later) (MATHÉ, 1976b).

We then applied one dose of 75 mg on 4 scarified areas at the same time to the patients who are skin responsive to PPD while we still apply 75 mg on only one area to the patients who are not skin responders.

Fig. 7.14a shows the scarifications as they appeared the day after BCG application and reactions of different durations in order to visualize their evolution. The place where BCG is applied becomes red after 2 or 3 days.

The reaction becomes attenuated and disappears after a few weeks or months. The side effects observed will be described in the next chapter.

BCG has always been applied twice a week during the first month, then weekly in order to maintain an active BCGitis (MATHÉ, 1976b).

The British Medical Research Council (1971), using Glaxo BCG, injected it i.d. by Heaf gun. POWLES *et al.* (1973) administered the same preparation (10^6 living organisms, weekly, with the same apparatus). The reactions are morphologically and chronoligically the same as those observed after application of BCG on scarifications, although they are usually less intensive.

SPARKS *et al.* (1973) injected (Tice) BCG i.d. by a modified multiple-puncture tine (36 times) technique. The vaccinating disc is applied in different areas.

DESNER *et al.* (1975) have proposed the use of an original apparatus consisting of two parts: (a) a cutting piece with ten individual slots which hold the 11 scalpel blades, and (b) a standardizing frame which acts as a guide for the cutting piece and has depressions of 0.5, 0.75, and 1 mm for adjustment of the amount of scalpel blade exposed.

According to GUTTERMAN *et al.* (1974), the site of skin administration of BCG given on scarifications may be important. Stage III melanoma patients submitted to surgical removal of all accessible tumours and then treated with BCG applied by scarifications on the arms and thighs benefited significantly from this immunotherapy if they had had a trunk or extremity primary tumour, but no effect in the case of a head and neck primary tumour. The authors suggest the importance of putting BCG into tumour-involved lymphatics.

A new approach to the administration of BCG is the oral one, which FALK *et al.* (1973) claimed to have used with promising initial results in subjects with perceptible tumours. THOMAS *et al.* (1974) claimed that they have obtained regressions in two melanoma patients with metastasis of the small bowel with oral BCG. We have given patients living fresh BCG by this route, 150 mg, twice a week for several weeks, and have observed no regressions of any tumour. The doses used in our screening study are mentioned in Table 6.XVIII.

Aerosol administration has been used for vaccination against tuberculosis (ROSENTHAL *et al.*, 1968) and was tried in phase I trials by GARNER *et al.* (1974) and also by us (SCHWARZENBERG *et al.*, 1976).

Finally, as we have found in our study that the i.v. route is the most efficient in mice (MATHÉ *et al.*, 1973b, 1973c; KHALIL *et al.*, 1975), and as it was shown by us, in a toxicological study in monkeys (JURECZYK-PROCYK *et al.*, 1976), to be innocuous, we conducted a phase I and II trial in man, administering by the intravenous route 0.6 - 1.4 mg of fresh BCG either to anergic patients in an attempt at immunorestoration (SIMMLER *et al.*, 1976b, 1976c), or to skin responders for immunostimulation (SCHWARZENBERG *et al.*, 1976).

Route of administration effect: a dose effect relationship? The results are interesting, not only those observed via this i.v. route, but in their comparison to those obtained by scarifications. They are summarized in Tables 6.XX and 7.XXVII.

The i.v. injection does not induce restoration in immunodepressed patients, as seen above, and it may suppress the reactions of allergic subjects. Application on one scarified area often restores immunodepressed patients, while that on four areas rarely does (SIMMLER *et al.*, 1976b, 1976c). Conversely, application on four scarified areas may stimulate immune reactions in normal responders, while that on one area very rarely does so and may decrease these reactions (SCHWARZENBERG *et al.*, 1976).

Our hypothesis (MATHÉ, 1976a) is that we may be facing a dose-effect relationship, which we have described in mice, according to which smaller and higher doses than the optimal one are not efficient, and higher doses are either ineffective or immunodepressive.

It seems that the application on one scarified area in immunodepressed patients induces a septicaemia corresponding to the optimal intensity, hence an immunorestoration, but does not do so in normally responsive subjects probably because of immune killing of the bacilli, hence the absence of immunostimulation in such individuals. Application on 4 areas may induce the optimal septicaemia in normal responders, hence the immunostimulation it can produce, whereas in immunodepressed patients it induces too intense an infection to restore them. Finally, in both immunodepressed and normally responsive subjects, the i.v. injection induces a septicaemia which is so intense that it does not restore the former and it may suppress immune responses in the latter, such as very high doses of BCG in mice.

Thus our study on the route of administration factor probably gives important information on the dose factor. The latter has also been shown by GUTTERMAN *et al* . (1975b) to be important as far as the minimal effective dose is concerned. Following treatment with BCG of the residual imperceptible disease remaining after surgical removal of all detectable lesions in stage III melanoma patients, these authors obtained significant prolongation of remission and survival when they applied 6 x 10^8 viable units (VU) on scarifications for each application, and no benefit when they applied 6×10^7 VU. The same authors (GUTTERMAN *et al.*, 1974) came to the same conclusion based on percentage of regressions on melanoma patients with perceptible tumours.

7.2.1.3.2.2. Mycobycteria Extracts

The only one which has been administered to patients is MER, which has been given by WEISS (1972) once a month by i.d. injections in 5-10 sites, the total dose given in all sites being 1-2 mg. It is too early to judge the clinical effect.

CUTTNER *et al.* (1975) injected i.d. 1 mg MER, divided between 5 to 10 sites on five patients with acute myeloid leukaemia in remission. The remission had persisted at time of publication for 92-193 days, which seemed, to the authors, an indication of efficacy. The patients increased their positive response to 5 recall antigens from 1-3 pre-MER to 3-5 post-MER.

ROBINSON *et al.* (1975) submitted 50 inoperable patients with solid tumours to a phase I trial of MER. The total dose did not exceed 1 mg. The immune status improved more frequently under the first method of administration.

7.2.1.3.2.3. Corynebacteria

We have administered C. parvum, and then C. granulosum, i.m., at doses of 1-2 mg/m^2 each week. As indicated above, the group of patients with acute lympoid leukaemia which received it as a complement of BCG has not yet differed significantly from the group receiving only BCG (MATHÉ *et al.*, 1975b).

ISRAEL (1972, 1974), who combined C. parvum with chemotherapy, administered this adjuvant by "hypodermic" injection at the dose of 2-4 mg twice a week for 2-20 weeks. It is difficult to draw any conclusions from his results, in as much as this agent is given at the same time as the oncostatics.

Our experience with i.v. injection is limited, the tolerance is good.

An intensive pharmacological study has been conducted by HERSH *et al.* (1974) on 44 patients with melanoma, who received i.v. C. parvum in doses ranging from 1-10 mg/m^2 given either once or twice per month, and 41 who received s.c. doses ranging from 2-15 mg/m^2 weekly or monthly. In vitro lymphocyte blastogenic responses to lymphocyte mitogens appeared to be increased only after i.v. injections of C. parvum. No definite change in delayed hypersensitivity responses to a battery of recall antigens was seen. Blood monocyte and lymphocyte counts decreased 4-24 hours following an i.v. dose.

It must be underlined, as will be explained later, that corynebacteria depress cell-mediated immunity (SCOTT, 1972); therefore, they should not be used in man without a careful control of immune reactions (see appendix) and, still better, before more scientific data on their effect on our species are available.

7.2.1.3.2.4. Polynucleotides

Poly I-poly C has been administered in our service at the dose of 1 mg/m^2/day i.v., for 30 days. It has induced remissions in patients presenting only a few leukaemic cells in their marrow (Table 3.VI, Fig. 3.44) (MATHÉ *et al.*, 1970). Given with BCG for imperceptible residual disease, it has slightly improved the result, but not significantly (not published). We have recently increased the dose up to 5 mg/day.

Poly A-poly U has also been injected i.v. to patients by our group at doses of 1-5 mg/m^2/day. It has been easily accepted (not published). J. and F. LACOUR (personal communication) administer it i.v., at the dose of 30 mg/week, six times.

7.2.1.3.2.5. Other Agents

Among other adjuvants, we have studied clinically the effects of levamisole. It was well tolerated but inactive. As mentioned above, positive results have been obtained by other authors (LEVO *et al.*, 1975; HIRSHAUT *et al.*, 1973; TRIPODI *et al.*, 1973; VERHAEGEN *et al.*, 1973).

7.2.1.3.2.6. Combinations of Several Adjuvants

ISRAEL (personal communication) claims that the combination of BCG and C. parvum is more efficient than each one given singly. We have not yet seen convincing data to support this opinion.

In acute lymphoid leukaemia patients, we combined, during the active immunotherapy phase of their treatment, BCG and C. parvum, and the apparent benefit over BCG alone is not significant, as already mentioned (MATHÉ *et al.*, 1975b).

Similarly, poly I-poly C, combined with BCG or with BCG and C. parvum, administered to patients in the same condition, did not significantly improve the percentage of patients maintained in remission (not published).

7.2.2. Hormones

7.2.2.1. Experimental

Several hormones modulate, more or less, the immune responses. We shall not review the effects of all agents conventionally called "hormones": this has been done by many authors (DMOCHOWSKI and HORNING, 1947; MATHÉ and NETTER, 1949; CRIEP and MAYER, 1951; NICOL and HELMY, 1951; GERMUTH and OTTINGER, 1951; DOUGHERTY, 1952; GERMUTH *et al.*, 1952; BERGLUND, 1956; BIOZZI *et al.*, 1957; BLECHER and WHITE, 1959; DOUGHERTY, 1959; MEYER *et al.*, 1959; ROSE, 1959; CHARLES and NICOL, 1961; DOUGHERTY *et al.*, 1964; BATCHELOR and CHAPMAN, 1965; BATCHELOR, 1971). Only oestrogens and thymosin will be considered.

7.2.2.1.1. Oestrogens

The sex difference in immune reactivity (PEARSON, 1959; BRENT and MEDAWAR, 1966; MEYER *et al.*, 1959; BATCHELOR, 1971) and the immunostimulating effect of orchidectomy (CASTRO *et al.*, 1973) and several experimental studies of the oestrogen effect on different immune reactions (DMOCHOWOSKI and HORNING, 1947; BIOZZI *et al.*, 1957; CHARLES and NICOL, 1961) suggest that oestrogens may be considered as immunopotentiators. However, submitting oestriadiol to our screening, we did not find any effect in any test.

7.2.2.1.2. Thymosin

Literature already exists on the endocrine role of the thymus, which started with GREGOIRE (1935) and COMSA (1940). The earliest reference to a substance endowed with lymphocytosis-stimulating properties is that of BOMSKOV and SLADOVIC (1940). The recent extensive literature has been reviewed by TRAININ (1974), who analysed the following effects of thymic extracts in thymus-deprived mice: (a) prevention of wasting and stimulating of the lymphoid system; (b) stimulation of cell-mediated immune responses; (c) stimulation of circulating antibody responses. (d) As far as the effect on tumour induction and growth is concerned, there may be two "fractions", one able to inhibit tumour development, to prolong survival of the hosts, and to reduce tumour incidence and mortality, the other presenting exactly the opposite properties (MAISIN, 1963, 1964a, 1964b; POTOP, 1966; POTOP *et al.*, 1967).

This hormone has also been shown experimentally to restore immune responses in thymus-deprived animals (TRENTIN, pers. comm.): it has been shown that spleen cells from neonatally thymectomized mice could attain competence to induce a graft versus host reaction (GVH) after in vitro exposure to isogeneic thymus extracts (TRAININ *et al.*, 1969).

The lymphocytopoietic factor is a protein of mol. wt. of about 10,000 according to GOLDSTEIN and WHITE (1970) ("thymosin"), of 6,000 according to TRAININ (1973) ("thymus humoral factor"). A circulating thymic hormone has been demonstrated by BACH *et al.* (1975) and characterized as a peptide with an M.W. of about 1,000. It is considered to be secreted by the non-lymphoid cells of the thymus, probably those stained with PAS (TRAININ, 1974), precisely those shown by KHALIL *et al.* (1975) (Fig. 9.5) to be hyperplastic after BCG administration as will be described below.

7.2.2.2. Clinical

7.2.2.2.1. Oestrogens

It would be interesting to take the effect(s) of oestrogens into account where they have been used in the treatment of prostate or breast cancer.

7.2.2.2.2. Thymosin

SCHAFER *et al.* (1975) administered, in vitro, fraction V of thymus extract to normal humans and cancer patients. They observed that, in cultures of lymphocytes stimulated by allogeneic irradiated cells, thymosin increased the reaction if its baseline was low, but did not affect or decrease it if the base line was high.

The same authors (SCHAFER *et al.*, 1975) have already administered thymosin to cancer and non-cancer immunodepressed patients: after an i.d. test dose, they injected subcutaneously and intramuscularly 1 or 10 or 30 mg/m^2/day for 7 days.

7.3. Local Immunotherapy

Local immunotherapy may be achieved by adjuvants, by antigens, or by haptens, and the approach to its pharmacological problems has been experimental as well as clinical.

7.3.1. Experimental Pharmacology

7.3.1.1. BCG

ZBAR *et al.* (1971, 1973) injected 10^7 living bacilli into guinea-pig tumours. The minimum number seemed to be 10^6 for curing a guinea-pig 95 mg hepatoma: above that dose, the faster the number, the higher the percentage of regressions (ZBAR and RAPP, 1974).

If fresh living bacteria are more efficient in systemic immunotherapy than lyophilized bacteria because they induce a more intense septicaemia due to their higher proliferation, one wonders if it is reasonable, in local immunotherapy, to use living bacteria and to take the risk of a severe septicaemia in patients whose resistance is low because they are carrying perceptible (voluminous) tumours.

BALDWIN *et al.* (1974) have proposed irradiated BCG (with γ rays at a dose of 1 to 10^6r). The result obtained with their material has been satisfactory. Nevertheless, MOORE *et al.* (1974), mixing living sarcomata cells and (Glaxo) BCG showed that, to inhibit the cell grafting, living mycobacteria were efficient, and heat-inactivated preparations were only able to partially inhibit the growth, while sonicated BCG or PPD were not active.

Most authors including ourselves use lyophilized preparations, and the accidents have been rare: nevertheless, they might be suppressed by adapting the dose to the general and the immune conditions of the patients; the poorer the condition, the smaller the dose. Moreover, it should be remembered that local immunotherapy with a given material only works if the patient is allergic to it: therefore, the first recommendation is not to use intratumoural injection(s) of BCG in PPD non-responsive patients and to try to restore immunity in such subjects before attempting local immunotherapy.

7.3.1.2. Extracts from Mycobacteria

WEPSIC *et al.* (1970) injected BCG cell walls into transferable Morris hepatoma with very variable effects. Inhibition of growth or enhancement was observed.

ZBAR and RAPP (1974), in an effort to standardize conditions of BCG for local immunotherapy, have investigated some properties of non-living BCG preparations. They found, with Ribi (Rocky Mountain Laboratory), that mycobacterial cell walls attached to mineral oil droplets are effective in causing regressions of established i.d. tumours and preventing lymph node metastasis (ZBAR *et al.*, 1973; ZBAR *et al.*, 1974). Preparations containing mixtures of cell walls and oil droplets, but not attached, were ineffective.

The nature of the cell walls and of oil are critical. Cell walls of M.tuberculosis, M. kansasii and M.bovis are effective. Those of M.smegmatis and N.asteroides are not.

Curiously enough, multiple injections of BCG cell walls suspended in tween-saline into mice, produced a specific unresponsive state for subsequent treatment of BCG cell walls-oil droplet mixture. Among the possible reasons for this phenomenon, the authors consider competition between T-helper cells, which would be stimulated by the first injection, and T-killer lymphocytes, which are involved in the second.

Another explanation could be the possibility that antigen-antibody complexes suppress delayed hypersensitivity (DH): AXELRAD and ROWLEY (1968) showed that sheep red blood cells given in saline resulted in partial DH unresponsiveness to a subsequent immunization with SRBC in complete Freund adjuvant; complete unresponsiveness was accomplished by administering high titre of anti-SRBC antibody plus SRBC.

Cell walls have been treated with proteolytic enzymes and then extracted with organic solvent. "Deproteinized", "delipidated" extracts on oil droplets are less effective than intact walls in tumour regression (ZBAR *et al.*, 1974). A tetralose mycolic acid ester added to the "deproteinized" "delipidated" walls restored tumour regression property (AZUMA *et al.*, 1974, MEYER *et al.*, 1974).

In conclusion, it seems that tetralose mycolic acid ester(s) are essential components for this action. Protein antigen(s) are probably, according to preliminary studies (ZBAR and RAPP, 1974), also necessary.

HOPPER *et al.* (1975) obtained remarkable regressions of rat sarcoma and hepatoma after intra-tumoral injections of *MER*. CHASSOUX and SALOMON (1975) injected this extract at the same dose as BCG, with the same results.

SALOMON and LYNCH (1976) also obtained remarkable results with *interphase material* (IMP) (300 µg) and *neo WSA* (300 µg).

7.3.1.3. Corynebacteria

SCOTT (1974a, 1974b), LIKHITE (1974), CHASSOUX and SALOMON (1975) and ourselves (MARTIN *et al.*, 1976) have injected rat or mice tumours successfully with C. parvum, the latter at the dose of 10^9 dead bacilli. ZBAR and RAPP (1974) failed to obtain a favourable effect with this organism in guinea pigs.

7.3.1.4. Bordetella Pertussis

Bordetella pertussis intra-tumour injections at the dose of 4.5 10^9 dead bacteria also induced some regressions in rats (SALOMON and LYNCH, 1976).

7.3.1.5. Other Materials

In ZBAR and RAPP's (1974) work, vitamin A, in oil form, which is an adjuvant, was not effective.

From TANAKA and SASAKI's (1974) experiments, in which the inoculation of living tumour cells mixed with heterospecific or allogeneic cells was not followed by growth, local immunotherapy with non-histocompatible cells, after immunization of the recipient, could be used.

A special form of local immunotherapy has been developed, realized by SPARKS and BREEDING (1974), who injected 50 µ neuraminidase into mice tumours and obtained regressions.

7.3.1.6. Mixtures

SALOMON and LYNCH (1976) observed an enhanced efficiency when using a mixture of BCG and C. parvum in some mice models.

SPARKS and BREEDING (1974), combining BCG and neuraminidase in their mouse model, obtained a better result than with neuraminidase alone.

7.3.2. Clinical Pharmacology

7.3.2.1. BCG

The materials injected into human tumours have been variable: Glaxo preparation[1] (MORTON *et al.*, 1970a, 1970b, 1972; LEVY *et al.*, 1972; SIEGLER *et al.*, 1972; PINSKY *et al.*, 1973; SPARKS *et al.*, 1973), Tice preparation[2] (NATHANSON *et al.*, 1972; BORNSTEIN *et al.*, 1973; PINSKY *et al.*, 1973; SMITH *et al.*, 1973), lyophilized Connaugh material[3] (KLEIN and HOLTERMAN, 1972), lyophilized Pasteur Institute preparation[4] (personal study, unpublished).

The doses varied from 0.05-0.2 ml of Glaxo preparation[5] injected once (MORTON *et al.*, 1970a, 1970b, 1972), from 1-5 ampoules of Glaxo or Tice[6] material diluted 1/20, repeated according to the results (SPARKS *et al.*, 1973) to 3 mg living Pasteur BCG-injected into hepatomas (TON THAT TUNG *et al.*, 1975).

No differences have been observed among the preparations. This is in accordance with the mechanisms of action of local immunotherapy as will be described in Chapter 9.

---

[1] Supplied by Lilly Co. Indianapolis, Ind.

[2] Supplied by the Research Institute, Chicago, Ill.

[3] Supplied by Connaugh Medical Research Laboratory, Toronto, Ont.

[4] Supplied by Paris Pasteur Institute.

[5] 4.9×10^6 viable units per ml, lyophilized.

[6] 303×10^6 viable units per ml, lyophilized.

7.3.2.2. Other Materials

Other adjuvants could be used, but other antigens and haptens can also be utilized, as the treatment consists in creating a secondary type immune reaction in the tumour by injecting into it an antigenic material to which the subject should be systematically reactive.

The injection of vaccine virus[1] (0.1 ml sheep lymph containing 2×10^8 pock-forming units per ml) (HUNTER-CRAIG *et al.*, 1970), or of streptokinase-streptodornase (STEWART and TOLNAI, 1969), and the topical application of dinitrochlorobenzene (0.1% in an aqueous base) (HELM *et al.*, 1965; KLEIN, 1968; WILLIAMS and KLEIN, 1970) are discussed in Chapter 4 (see Fig. 4.I).

References

ADAM, A., CIORBARU, R., PETIT, J.F., LEDERER, E.: Proc. Nat. Acad. Sci. USA., 69, 851 (1972).

ADAM, A., CIORBARU, R., PETIT, J.F., LEDERER, E.: p. 179 in "Investigation and Stimulation of Immunity in Cancer Patients" (G. Mathé and R. Weiner, eds.). Heidelberg: Springer-Verlag 1974.

ADAM, A., CIORBARU, R., PETIT, J.F., LEDERER, E., CHEDID, L., LAMENSANS, A., PARANT, A., PARANT, M., ROSSELET, J.P., BERGER, F.M.: Infect. Immunity 7, 855 (1973).

ALEXANDER, P., EVANS, R.,: Nature New Biol., 232, 76 (1971).

ALLISON, A.C., DAVIES, A.J.S.: Nature (Lond)., 223, 330 (1971).

ALLISON, A.C., GREGORIADIS, G.: Nature 252, 252 (1974).

APFFEL, C.A., ARNASON, B.G., PETERS, J.H.: Nature 209, 694 (1966).

APFFEL, C.A., PETERS, J.H.: J. Theor. Biol. 26, 47 (1970).

ASSELINEAU, J.: C.R. Acad. Sci., Paris 229, 791 (1949).

ASSELINEAU, J., LEDERER, E.: Fortschr. Chem. Org. Naturst. 10, 170 (1953).

ASSELINEAU, J., PORTELANCE, V.: p. 214 in "Investigation and Stimulation of Immunity in Cancer Patients" (G. Mathé and R. Weiner, eds.) Heidelberg: Springer-Verlag 1974.

AXELRAD, M., ROWLEY, D.A.: Science 160, 1465 (1968).

AZUMA, I., RIBI, E.E., MEYER, F.J., ZBAR, B.: J. Nat. Cancer Inst., 52, 95 (1974).

BACH, J.F., DARDENNE, M., PLEAN, J.M., BACH, M.A.: p. 186 in "Thymus factors in immunity" (H. Friedman, ed.). Vol. 249, New-York, Annals New-York Academy of Siences 1975.

BAGSHAWE, K.D., CURRIE, G.A.: Nature 218, 1254 (1968).

BAIRD, L.G., KAPLAN, A.M.: in preparation (1976).

BALDWIN, R.W., BOWEN, J.G., EMBLETON, M.J., PRICE, M.R., ROBINS, R.A.: p.393 in "The Role of Immunological Factors in Viral and Oncogenic Processes" (R.F. Beers, R.C. Tilghman, E.G. Bassett, eds.). Baltimore, John Hopkins University press 1974.

[1] Supplied by Lister Institute.

BALDWIN, R.W., COOK, A.J., HOPPER, D.G., PIMM, M.V.: Int. J. Cancer 13, 743 (1974).
BALDWIN, R.W., HOPPER, D.G., PIMM, M.V.: Brit. J. Cancer 31, 124 (1975).
BALDWIN, R.W., PIMM, M.V.: Europ. J. Clin. Biol. Res. 16, 875 (1971).
BALDWIN, R.W., PIMM, M.V.: Nat. Cancer Inst. Monogr. 39, 11 (1973).
BARTLETT, G.L., ZBAR, B.: J. Nat. Cancer Inst., 48, 1709 (1972).
BARTLETT, G.L., ZBAR, B., RAPP, H.J.: J. Nat. Cancer Inst. 48, 245 (1972).
BASHAM, T.Y., WAXDAL, M.J.: J. Immunol. 114, 715 (1975).
BATCHELOR, J.R.: p. 276 in "Regulation of the Antibody Response (B. Cinader, ed.). 2nd ed., Springfield: C.C. Thomas 1971.
BATCHELOR, J.R., CHAPMAN, B.A.: Immunology 9, 553 (1965).
BEKESI, J.G., HOLLAND, J.F.: p. 357 in "Investigation and Stimulation of Immunity in Cancer Patients" (G. Mathé, R. Weiner, eds.). Vol. 47 R.R.C.R., Heidelberg: Springer-Verlag 1974.
BEKESI, J.G., HOLLAND , J.F., YATES, J.W., HENDERSON, E., FLEMINGER, R.: Proc. Amer. Assoc. Cancer Res., 16, 121 (abstract No. 481) (1975).
BEKESI, J.G., ARNEAULT, St.G., WALTER, L., HOLLAND, J.F.: J. Nat. Cancer Inst. 49, 107 (1972).
BEKIERKUNST, A.: p. 259 in "Investigation and Stimulation of Immunity in Cancer Patients" (G. Mathé, R. Weiner, eds.). Vol. 47 R.R.C.R., Heidelberg: Springer-Verlag 1974.
BEKIERKUNST, A., LEVIJ, I.S., YARKONI, E., VILKASE, E., LEDERER, E.: Science 174, 1240 (1971).
BELPOMME, D., LE BORGNE DE KAOUEL, C., AJURIA, E., JASMIN, C., DORE, J.F.: Nature 222, 890 (1969).
BELPOMME, D., MINOWADA, J., MOORE, G.E.: Cancer 30, 282 (1972).
BEN-EFRAIM, S., CONSTANTINI-SOUROJON, M., WEISS, D.W.: Cell. Immunol. 7, 370 (1973).
BEN-EFRAIM, S., TEITELBAUM, R., OPHIR, R., KLEINMAN, R., WEISS, D.W.: Israel J. Med. Sci. 10, 972 (1974).
BENJAMINI, E., SCIBIENSKI, R.T.: p. 408 in "Investigation and Stimulation of Immunity in Cancer Patients" (G. Mathé, R. Weiner, eds.). Vol. 47 R.R.C.R., Heidelberg: Springer-Verlag 1974.
BEREBBI, M., BARSKI, G.: C.R. Acad. Sci., Paris 272, 351 (1971).
BERGLUND, K.: Acta Path. Microbiol. Scand. 38, 311 (1956).
BIOZZI, G., HALPERN, B.N., BILBEY, D., STIFFEL, C., BENACERRAF, G., MOUTON, D.: C.R. Soc. Biol. 151, 1326 (1975).
BIOZZI, G., STIFFEL, D., MOUTON, Y., BOUTHILLIER, A., DECREUSEFOND, C.: Immunology 14, 7 (1968).
BLECHER, M., WHITE, A.: p. 391 in "Recent Progress in Hormone Research" (G. Pincus, ed.). Vol. 15, New-York: Academic Press 1959.
BLOOM, H.J.G., PECKHAM, M.J., RICHARDSON, A.E., ALEXANDER, P.A., PAYNE, P.M.: Brit. J. Cancer 27, 267 (1973).
BLUMING, A., VOGEL, L., ZIEGLER, J.L.: p. 415 in "Investigation and Stimulation of

Immunity in Cancer Patients" (G. Mathé, R.Weiner, eds.). Vol. 47, R.R.C.R., Heidelberg: Springer-Verlag 1974.

BLUMING, A.Z., VOGEL, C.L., ZIEGLER, J.L., MODY, N., KAMYA, G.: Ann. Intern. Med. 76, 405 (1972).

BOMSKOV, G., SLADOVIC, L.: Deut. Med. Woschr. 66, 589 (1940).

BONA, C., HEUCLIN, C., CHEDID, L.: p. 196 in "Investigation and Stimulation of Immunity in Cancer Patients" (G. Mathé and R. Weiner, eds.). Heidelberg: Springer-Verlag 1974.

BOONE, C.W.: p. 384 in "Investigation and Stimulation of Immunity in Cancer Patients" (G. Mathé, R. Weiner, eds.). Vol. 47 R.R.C.R. Heidelberg: Springer-Verlag 1974.

BOONE, C., BLACKMAN, K., BRANDCHAFT, P.: Nature 231, 265 (1971).

BØRNSTEIN, R.S., MASTRANGELO, M.J., SULIT, H., CHEE, D., YARBO, J.W., PREHN, L., PREHN, R.T.: Nat. Cancer Inst. Monogr. 39, 213 (1973).

BOUSFIELD, I.J.: 71, 441 (1972).

BRAUN, W.: p. 17 in "Process in Antimicrobial and Anticancer Chemotherapy". Vol. 1, Tokyo: University of Tokyo 1970.

BRAUN, W., ISHIZUKA, M., YAJIMA, Y., WEBB, D., WINCHURCH, R.: p. 139 in "Biological Effects of Polynucleotides" (R.F. Beers, W. Braun, eds.). 1 Vol., Heidelberg: Springer-Verlag 1971a.

BRAUN, W., PLESCIA, O.J., RASKOWA, J., WEBB, D.: Israel J. Med. Sci., 7, 72 (1971b).

BRENT, L., MEDAWAR, P.B.: Proc. Royal Soc. London, 165, 281 (1966).

BRUGAROLAS, A., TAKITA, H., MOORE, G.E.: J. Surg. Oncol. 4, 123 (1972).

BRULEY-ROSSET, M., BOTTO, H.G., GAUTHIER, H., POUILLART, P., FLORENTIN, I., HUGUENIN, P.: Israel J. Med. Sci, in press (1976b).

BRULEY-ROSSET, M., FLORENTIN, I., KHALIL, A., MATHÉ, G.: Int. Arch. Allergy and Applied Immunol., in press (1976a).

BRUM, V.C.: Proc. Amer. Assoc. Cancer Res. (abstract No. 64) 16, 16 (1975).

BUNCH-CHRISTENSEN, K., LADEFOGED, A., GULD, J.: Bull. World Health Organ. 39, 821 (1968).

BUZZI, S., BUZZI, L.: Cancer Research, 34, 3481 (1974).

CAMPANILE, F., GASTON, M., BONMASSAR, E., GOLDIN, A.: Proc. Amer. Assoc. Cancer Res. 16, 36 (abstract No. 142) (1975).

CASTRO, J.E.: Europ. J. Cancer 10, 115 (1974).

CASTRO, J.E., MEDAWAR, P.B., HAMILTON, D.N.H.: p. 237 in "Immunopotentiation" (Ciba Found. Symp. No. 18). 1 Vol., Amsterdam: Elsevier Excerpta Medica 1973.

CHARLES, L.M., NICOL, T.: Nature 192, 565 (1961).

CHASSOUX, W., SALOMON, J.C.: Int. J. Cancer 16, 515 (1975).

CHEDID, L., LAMENSANS, A.: p. 166 in "Investigation and Stimulation of Immunity in Cancer Patients" (G. Mathé and R. Weiner, eds.). Heidelberg: Springer-Verlag 1974a.

CHEDID, L., LAMENSANS, A.: International Symp. on Bacterial Immunostimulants, Institut Pasteur, Paris 14-15th October1974b (abstract).

CHEDID, L., PARANT, M., PARANT, F., GUSTAFSON, R.H., BERGER, F.M.: Proc. Nat. Acad. Sci. 69, 855 (1972).

CHIHARA, G., HAMURO, J., MAEDA, Y., ARAI, Y., FUKUOKA, F.: Nature (Lond.) 225, 943 (1970).

CHIHARA, G., MAEDA, Y., HAMURO, J., SASAKA, T., FUKUOKA, F.: Nature, 222, 687 (1969).

CHOUCROUN, N.: Ann. Rev. Tuberc. 56, 203 (1947).

CHUNG, E.B., ZBAR, B., RAPP, H.J.: J. Nat. Cancer Inst., 51, 241 (1973).

CIKES, M.: Transplant. Proc. 3, 1161 (1971).

CLAMAN, H.N., CHAPERON, A.E., TRIPLETT, R.F.: Proc. Soc. Exp. Biol. Med. 122, 1167 (1966).

COMSA, J.: C.R. Soc. Biol. 133, 24 (1940).

COUDERT, A., AJURIA, E., DORE, J.F., HADJIANNAKIS, M.H., MARHOLEV, L.: C.R. Acad. Sci., Paris 274, 2833 (1972).

CRIEP, L.N., MAYER, L.D.: J. Allergy 22, 314 (1951).

CUNNINGHAM, T.J., OLSON, K.B., LAFFIN, R., HORTON, J., SULLIVAN, J.: Cancer 24, 932 (1969).

CURRIE, G.A.: Brit. J. Cancer 28, 25 (1973).

CURRIE, G.A., BAGSHAWE, K.D.: Brit. J. Cancer 22, 843 (1968).

CURRIE, G.A., BAGSHAWE, K.D.: Brit. J. Cancer 23, 141 (1969).

CURRIE, G.A., BAGSHAWE, K.D.: Brit. J. Cancer 1, 541 (1970).

CUTTNER, J., HOLLAND, J.F., BEKESI, J.G., RAMACHANDAR, K., DONOUAN, P.: Proc. Amer. Assoc. Cancer Res. 16, 264 (abstract No. 1175) (1975).

CZAJKOWSKI, N.P., ROSENBLATT, M., CUSHING, F.R., VAZQUEZ, J., WOLF, P.L.: Cancer 19, 739 (1966).

CZAJKOWSKI, N.P., ROSENBLATT, M., WOLF, P.L., VAZQUEZ, J.: Lancet 2, 905 (1967).

DESNER, M.R., LICHTENFELD, J.L., DENNISON, C.L., KALMBACH, K.O., MARDINEY, M.R. jr.: J. Nat. Cancer Inst. 54, 57 (1975).

DMOCHOWSKI, L., HORNING, E.S.: J. Path. Bact. 59, 307 (1947).

DONNER, M., VAILLIER, D.: p. 308 in "Investigation and Stimulation of Immunity in Cancer Patients" (G. Mathé , R. Weiner, eds.). Vol. 47 R.R.C.R., Heidelberg: Springer-Verlag 1974.

DORE, J.F., AJURIA, E., MATHÉ, G.: Europ. J. Clin. Biol. Res. 15, 81 (1970).

DORE, J.F., GUIBOUT, C.: 1st International Workshop on Human Leukaemia associated antigens. Duke University, June 1974, unpublished.

DORE, J.F., HADJIYANNAKIS, M.J., COUDERT, A., GUIBOUT, C., MARHOLEV, L., IMAI, Y.: p. 387 in "Investigation and Stimulation of Immunity in Cancer Patients" (G. Mathé, R. Weiner, eds.). Vol. 47 R.R.C.R., Heidelberg: Springer-Verlag 1974.

DOUGHERTY, T.F.: Physiol. Rev. 32, 379 (1952).

DOUGHERTY, T.F.: p. 264 in "The Study of Kinetics of Cellular Proliferation" (F. Stholman, ed.). New-York: Grune and Stratton 1959.

DOUGHERTY, T.F., BERLINER, M.L., SCHNEEBELL, G.L., BERLINER, D.L.: Ann. N.Y. Acad. Sci., 113, 825 (1964).

DRESSER, D.W.: Nature (London) 217, 527 (1968).
DUBOS, R.J., SCHAEDLER, R.W.: J. Exp. Med. 106, 703 (1957).
ELLOUZ, G., ADAM, A., CIORBARU, R., LEDERER, E.: Biochem. Biophys. Res. Commun. 59, 1317 (1974).
E.O.R.T.C.: Hemopathies Working Party. Third Meeting of International Society of Haematology European and African Division, London, 24-28 August 1975,abstract 21:06.
ESBER,H.J., HAGOPIAN, M., BOGDEN, A.E.: J. Nat. Cancer Inst. 53, 209 (1974).
ETEMADI, A.H., MIQUEL, A.M., LEDERER, E., BARBER, M.: Bull. Soc. Chim. Fr. 3274 (1964a).
ETEMADI, A.H., OKUDA, R., LEDERER, E.:Bull. Soc. Chim. Fr. 868 (1964b).
FALK, R.E., MANN, P., LANGER, B.: Arch.Surgery, 107, 261 (1973).
FAUVE, R.M., HEVIN, B., JACOB, H., GAILLARD, J.A., JACOB, F.: Proc. Nat. Acad. Sci. USA, 71, 4052 (1974).
FENYO, E.M., KLEIN, E.G., KLEIN, G., SWIECH, K.: J. Nat. Cancer Inst. 40, 69 (1968).
FIELD, A.K., TYTELL, A.A., LAMPSON, G.P., HILLEMAN, M.R.: Proc. Nat. Acad. Sci. 58, 1008 (1967).
FREUND, J.: Ad. Tuberculosis Res. 7, 130 (1956).
FREUND, J., McDERMOTT, K.: Proc. Soc. Exp. Biol. N.Y. 49, 548 (1942).
FRIDMAN, W.H., KOURILSKY, F.M.: Nature 224, 277 (1969).
FRIEND, C.: J. Exp. Med. 109, 217 (1959).
GALL, D.: Immunology, 11, 369 (1966).
GARNER, F.B., MEYER, C.A., WHITE, D.S., LIPTON, A.: Immunotherapy employing aerosol BCG. Proc. Amer. Assoc. Cancer Res. 15, 166 (abstract No. 729) (1974).
GEFFARD, M., ORBACH-ARBOUYS, S.: Cancer Immunol. Immunoth. 1, 41 (1976).
GERMUTH, F.G., OTTINGER, B.: J. Exp. Med. 94, 139 (1951).
GERMUTH, F.G., OTTINGER, B., OYAMA, J.: Proc. Soc. Exp. Biol. Med. 80, 188 (1952).
GHEORGHIU, M., CHAMBON, L.: Cancer Immunol. Immunoth. 1, 11 (1976).
GHEORGHIU, M., KOSLOFF, F., PARIS, C.: p. 221 in "International Symposium on BCG vaccine". Vol. 17, Munich: Karger 1971.
GOLDSTEIN, A.L., WHITE, A.: p. 465 in "Biochemical Actions of Hormones (G. Litwak, ed.). 1 Vol., New-York: Academic Press 1970.
GOWANS, J.L.: Brit. Med. J. 1, 557 (1974).
GREGOIRE, C.: Arch. Biol. Liège 46, 717 (1935).
GRESSER, I.: p. 327 in "Investigation and Stimulation of Immunity in Cancer Patients (G. Mathê and R. Weiner, eds.). Heidelberg: Springer-Verlag 1974.
GRESSER, I., BOURALI, C., LEVY, J-P., FONTAINE-BROUTY-BOYE, D., THOMAS, M.T.: Proc. Nat. Acad. Sci. (Wash.) 63 51 (1969).
GROSS, L.: Oncogenic viruses. 2nd ed. Oxford: Pergamon press 1970.
GROSS, L., DREYFUSS, Y.: Proc. Nat. Acad. Sci. 71, 3550 (1974).
GUERIN, C.: The history of BCG, BCG vaccination against tuberculosis (S. Rosenthal, ed.). Boston: Little Brown & Cy. 1957.

GUTTERMAN, J.U., HERSH, E.M., McCREDIE, K.B., RODRIGUEZ, V., BODEY, G.P., FREIREICH, E.J.: Proc. Amer. Assoc. Cancer Res. 16, 127 (abstract No. 505) (1975a).
GUTTERMAN, J.U., McBRIDE, C., FREIREICH, E.J., MAVLIGIT, G., FREI, E., HERSH, E.M.: Lancet 1, 1208 (1973).
GUTTERMAN, J.U., MAVLIGIT, G., KENNEDY, A., HERSH, E.: Proc. Amer. Assoc. Cancer Res. 16, 245 (abstract No. 1096) (1975b).
GUTTERMAN, J.U., MAVLIGIT, G., McBRIDE, C.M., FREI, E. III, HERSH, E.M.: p. 476 in "Investigation and Stimulation of immunity in Cancer patients" (G. Mathé, R. Weiner, eds.). Vol. 47 R.R.C.R., Heidelberg: Springer-Verlag 1974.
HALLE-PANNENKO, O., BOURUT, C., KAMEL, M.: Cancer Immunol. Immunoth. 1, 17 (1976).
HALPERN, B.: p. 262 in "Investigation and Stimulation of Immunity in Cancer Patients (G. Mathé and R. Weiner, eds.). Heidelberg: Springer-Verlag 1974.
HALPERN, B.N., PREVOT, A.R., BIOZZI, G.: J. Reticuloendothel, Soc. 1, 77 (1963).
HARRIS, H.: Cell fusion. 1 Vol., Oxford: Pergamon Press 1970.
HATTORI, T., NIIMOTO, M., YAMAGATA, S., TOHGE, T.: 9th International Congress of Chemotherapy, London (abstract C 81) July 1975.
HELM, F., KEIN, E., TRANENKLE, H.L., RIVERA, E.P.: Invest.Dermat. 45, 152 (1965).
HENLE, W., HENLE, G., BURTIN, P., CACHIN, Y., CLIFFORD, P., DE SCHRYVER,A., DE THE, G., DIEHL, V., HO, H.C., KLEIN, G.: J. Nat. Cancer Inst. 44, 225 (1970).
HENLE, G., HENLE, W., CLIFFORD, P., DIEHL, V., KAFUKO, C.W., KIRYA, B.G., KLEIN, G., MORROW, R.H., MUNUBE, G.M.R., PIKE, P., TUKEI, P.M., ZIEGLER, J.L.: J. Nat. Cancer Inst. 43, 1147 (1969).
HENLE, G., HENLE, W., KLEIN, G., GUNVEN, P., CLIFFORD, P., MORROW, R.H., ZIEGLER, J.L.: J. Nat. Cancer Inst. 46, 861 (1971).
HERSH, E.M., REED, R.C., GUTTERMAN, J.U., MAVLIGIT, G.M., BURGESS: in Symposium International on "Les immunostimulants bactériens, structures chimiques, mécanismes d'action, applications". Paris: Institut Pasteur, October 1974.
HIU, I.J.: Nature New Biol. 238, 241 (1972).
HIU, I.J.: p. 214 in "Investigation and Stimulation of Immunity in Cancer Patients" (G. Mathé and R. Weiner, eds.). Heidelberg: Springer-Verlag 1974.
HIU, I.J., AMIEL, J.L.: Rev. Europ. Et. Clin. Biol. 16, 55 (1971b).
HIU, I.J., AMIEL, J.L.: J. Gen. Microbiol. 66, 239 (1971a).
HIU, I.J., AMIEL, J.L.: Experientia, 28, 953 (1972).
HIU, I.J., AMIEL, J.L., JOLLES, P., BERARDET, M.: Rev. Fr. Et. Clin. Biol. 14, 508 (1969).
HIU, I.J., AMIEL, J.O., NIKDJOU, K.: Europ. J. Clin. Biol. Res. 15, 887 (1970).
HIRSHAUT, Y., PINSKY, C., MARGUARDT, H., OETTGEN, H.F.: Proc. Amer. Assoc. Cancer Res. 14, 109 (abstract No. 436) (1973).
HOPPER, D.G., PIMM, M.V., BALDWIN, R.W.: Brit. J. Cancer 31, 176 (1975).
HOWARD, J.G., CHRISTIE, G.H., SCOTT, M.T.: Cell Immunol. 7, 290 (1973).
HOWARD, J.G., SCOTT, M.T., CHRISTIE, G.H.: p. 101 in "Immunopotentiation" (Ciba Found. Symp. No. 18) Amsterdam: Elsevier Excerpta Medica 1973b.

HUMPHREY, L.J., MURRAY, D.R., BOEHM, O.R.: Surg. Gynec. Obstet., 132, 437 (1971).

HUNTER-CRAIG, I., NEWTON, K.A., WESTBURY, G., LACEY, B.W.: Brit. Med. J. 2, 512 (1970).

IKONOPISOV, R.L., LEWIS, M.G., HUNTER-CRAIG, I.D., BODENHAM, D.C., PHILLIPS, T.M., COOLING, C.I., PROCTOR, J., HAMILTON FAIRLEY, G., ALEXANDER, P.: Brit. Med. J. 2, 752 (1970).

ISRAEL, L.: p. 486 in "Investigation and Stimulation of Immunity in Cancer Patients" (G. Mathé and R. Weiner, eds.). Vol. 47 R.R.C.R., Heidelberg: Springer-Verlag 1974.

ISRAEL, L., HALPERN, B.: Nouv. Presse Med. 1, 19 1972.

JACOTOT, H.: C.R. Acad. Sci. Paris 224, 1390 (1947).

JANOSSY, G., SHOHAT, M., GREAVES, M.F., DOURMASHKIN, R.R.: Immunology 24, 211 (1973).

JASMIN, C., PITON, C., ROSENFELD, C.: Int. J. Cancer 3, 254 (1968).

JERNE, N.K., NORDIN, A.A., HENRY, C.C.: p. 109 in "Cell-bound Antibodies. (B. Amos and H. Koprowski, eds.). Vol. 1, Philadelphia, Wistar Institute 1963.

JOLLES, P., PARAF, A.: Chemical and biological basis of adjuvants. Vol. 13. Molecular biology biochemistry and biophysics. Heidelberg: Springer-Verlag 1973.

JURCZYK-PROCYK, S., MARTIN, M., DUBOUCH, P., GHEORGHIU, M., ECONOMIDES, F., KHALIL, A., RAPPAPORT, H.: Cancer Immunol. Immunoth. 1, 55 (1976).

JUY, D., BONA, C., CHEDID, L.: C.R. Acad. Sci., Paris 278, 2859 (1974).

KAPLAN, A.M., MORAHAN, P.S., REGELSON, W.: J. Nat. Cancer Inst. 52, 1919 (1974).

KEMP, R.B.: Nature 218, 1255 (1968).

KHALIL, A., BOURUT, C., HALLE-PANNENKO, O., MATHÉ, G., RAPPAPORT, H.: Biomedicine 22, 112 (1975).

KILLION, J.J., LEFEVER, A., CANTRELL, J.L., BIERIG, K., KOLLMORGEN, G.M.: Proc. Amer. Assoc. Cancer Res. 16, 22 (abstract No. 86) (1975).

KIND. L.S.: J. Immunol. 79, 238 (1957).

KLEIN, E.: New-York State J. Med. 68, 900 (1968).

KLEIN, E., HOLTERMANN, O.A.: Nat. Cancer Inst. Monogr. 35, 379 (1972).

KOBAYASHI, H., HOSOKAWA, M., IMAMURA, M., NAKAYAMA, M., OIKAWA, T.: Proc. Amer. Assoc. Cancer Res., 16, 96 (abstract No. 383) (1975).

KOBAYASHI, H., SENDO, F., SHIRAI, T., KAJI, H., KODAMA, T., SAITO, H.: J. Nat. Cancer Inst. 42, 327 (1969).

KOKOSCHKA, E.M., CERNI, C., MICKSCHE, M.: European Immunology Meeting, Amsterdam, 17-19th September 1975.

KOTANI, S., HASHIMOTO, S., MATSUBARA, T., KATO, K., HARADA, K., KOGAMI, J., KITAURA, T., TANAKA, A.: Biken's J. 6, 181 (1963).

KOUZNETZOVA, B., BIZZINI, B., CHERMANN, J.C., DEGRAND, F., PREVOT, A.R., RAYNAUD, M.: p. 275 in "Investigation and Stimulation of Immunity in Cancer Patients" (G. Mathé and R. Weiner, eds.). Vol. 47 R.R.C.R., Heidelberg: Springer-Verlag 1974.

KRONMAN, B.S., WEPSIC, H.T., CHURCHILL, W.H. jr., ZBAR, B., BORSOS, T., RAPP, H.J.: Science 168, 257 (1970).
LACOUR, F., DELAGE, G., CHIANALE, C.: Science 187, 256 (1975).
LADEFOGED, A., BUNCH-CHRISTENSEN, K., GULD, J.: Bull. World Health Organ. 43, 71(1970).
LARSON, V.M., CLARK, W.R., HILLEMAN, M.R.: Proc. Soc. Exp. Biol. (NY) 131, 1002 (1969).
LAWRENCE, W., SIMONSEN, M.: Transplantation, 5, 1304 (1967).
LE BORGNE DE KAOUEL, C., BILLARDON, C., DORE, J.F., SYLVESTRE, D., MARHOLEV, L., NGUYEN VAN CONG, WEIL, D., PICARD, J.Y., REBOURCET, R., FREZAL, J.: Biomedicine 18, 59 (1973).
LE BORGNE DE KAOUEL, C., BILLARDON, C., NGUYEN VAN CONG, MARHOLEV, L., DORE, J.F., FREZAL, J.: Bull. Europ. Soc. Human Genetics 62 (1972).
LEE, A.: Proc. Soc. Exp. Biol., N.Y. 128, 891 (1968).
LE GARREC, Y., SABOLOVIC, D., TOUJAS, L., DAZORD, L., GUELFI, J., PILET, C.: Biomedicine, 21, 40 (1974).
LEVENTHAL, R.G., HALTERMAN, R.H., HERBERMAN, R.B.: Proc. Am. Assoc. Cancer Res., 12, 51 (abstract No. 203) (1971).
LEVO, Y., ROTTER, V., RAMOT, B.: Biomedicine 23, 198 (1975).
LEVY, N.L., AMOS, D.B., SOLOVIEFF, G.V., DOS REIS, A.P.: Nat. Cancer Inst. Monogr. 35, 5 (1972).
LEVY, H.G., LAW, L.W., RABSON, A.S.: Proc. Nat. Acad. Sci. 62, 357 (1969).
LIACOPOULOS, P., BIRRIEN, J.L., BLEUX, C., COUDERC: p. 201 in "Investigation and Stimulation of immunity in Cancer Patients (G. Mathé and R. Weiner, eds.). Heidelberg: Springer-Verlag 1974
LIKHITE, V.V.: Int. J. Cancer. 14, 684 (1974).
LIKHITE, V.V., HALPERN, B.N.: Int. J. Cancer 12, 699 (1973).
LIN, J.S.L., HUBER, N., MURPHY, W.H.: Cancer Res. 29, 2157 (1969).
LIN, J.S., MURPHY, W.H.: Cancer Res. 29, 2163 (1969).
LINDENMANN, J., KLEIN, P.A.: Immunologic aspects of viral oncolysis. Vol. 9 R.R.C.R., Heidelberg: Springer-Verlag 1967.
LINDENMANN, J., KLEIN, P.A.: J. Exp. Med. 126, 93 (1967).
LITTMAN, B.H., MELTZER, M.S., CLEVELAND, R.P., ZBAR, B., RAPP, H.J.: J. Nat. Cancer Inst. 51, 1627 (1973).
MacKHANN, C.F., GUNNARSSON, A.: Cancer 34, 1521 (1974).
MADYASTHA, K.R., BARTH, R.F.: Proc. Amer. Assoc. Cancer Res. 16, 137 (abstract No. 547) (1975).
MAISIN, J.H.F.: C.R. Soc. Biol. 157, 1519 (1963).
MAISIN, J.H.F.: Nature 202, 202 (1964a).
MAISIN, J.H.F.: Bull. Acad. Med. Belg., 4, 197 (1964a).
MARAL, R., FLOC'H, F., JOUANNE, M., WERNER, G.H.: in "European Society of Toxicology", Amsterdam: Excerpta Medica in press (1976).
MARKOVITS, J., PINTE, P., ETEMADI, A.H.: C.R. Acad. Sci., Paris 263, 960 (1966).

MARKOVITS, J., VILKAS, E., LEDERER, E.: Europ. J. Biochem. 18, 287 (1971).

MARTIN, M., BOURUT, C., HALLE-PANNENKO, O., MATHÉ, G.: Biomedicine, 23, 337 (1975).

MARTIN, M., KHALIL, A., HALLE-PANNENKO, O., BOURUT, C., MATHÉ, G.: Biomedicine, in press (1976).

MARTYRE, M.C.: Biomedicine, in press (1976).

MARTYRE, M.C., HALLE-PANNENKO, O., JOLLES, P.: p. 401 in "Investigation and Stimulation of Immunity in Cancer Patients" (G. Mathé and R. Weiner, eds.). Vol. 47 R.R.C.R., Heidelberg: Springer-Verlag 1974b.

MARTYRE, M.C., WEINER, R., HALLE-PANNENKO, O.: p. 405 in "Investigation and Stimulation of Immunity in Cancer Patients" (G. Mathé and R. Weiner, eds.). Vol. 47 R.R.C.R., Heidelberg: Springer-Verlag 1974a.

MATHÉ, G.: Rev. Fr. Et. Clin. Biol. 13, 881 (1968).

MATHÉ, G.: in "European Society of Toxicology", Amsterdam: Excerpta Medica, in press (1976a).

MATHÉ, G.: Cancer Immunol. Immunoth., 1, 3 (1976b).

MATHÉ, G., AMIEL, J.L.: La greffe. Aspects biologiques et cliniques. 1 Vol., Paris: Masson & Cie. 1962.

MATHÉ, G., AMIEL, J.L., SCHWARZENBERG, L.: Bone marrow transplantation and leucocyte transfusions. 1 Vol., Springfield: C.C. Thomas 1971a.

MATHÉ, G., AMIEL, J.L., SCHWARZENBERG, L., SCHNEIDER M., CATTAN, A., SCHLUMBERGER, J.R., HAYAT, M., VASSAL, F. DE: Rev. Fr. Et. Clin. Biol. 13, 454 (1968).

MATHÉ, G., AMIEL, J.L., SCHWARZENBERG, L., SCHNEIDER, M., HAYAT, M., VASSAL, F. DE, JASMIN, C., ROSENFELD, C., SAKOUHI, M., CHOAY, J.: Rev. Europ. Et. Clin. Biol. 15, 671 (1970).

MATHÉ, G., BELPOMME, D., DANTCHEV, D., POUILLART, P., JASMIN, C., MISSET, J.L., MUSSET, M., AMIEL, J.L., SCHLUMBERGER, J.R., HAYAT, M., DE VASSAL, F., LAFLEUR, M.: Biomedicine 20, 333 (1974).

MATHÉ, G., HALLE-PANNENKO, O., AMIEL, J.L.: p. 48 in "Corynebacterium parvum: application in experimental and clinical oncology" (B. Halpern, ed.). 1 Vol., New-York: Plenum press 1975b.

MATHÉ, G., HALLE-PANNENKO, O., BOURUT, C.: Exp. Hematol. 1, 110 (1973a).

MATHÉ, G., HALLE-PANNENKO, O., BOURUT, C.: Nat. Cancer Inst. Monogr. 39, 107 (1973c).

MATHÉ, G., HALLE-PANNENKO, O., FLORENTIN, I., BRULEY-ROSSET, M., KAMEL, M., HIU, I.J., BOURUT, C.:Europ. J. Cancer, 11, 801 (1975a).

MATHÉ, G., HAYAT, M., SAKOUHI, M., CHOAY, J.: C.R. Acad. Sc. Paris, 272, 170 (1971c).

MATHÉ, G., HIU, I.J., HALLE-PANNENKO, O., BOURUT, C.: Israel. J. Med. Sci. in press (1976b).

MATHÉ, G., KAMEL, M., DEZFULIAN, M., HALLE-PANNENKO, O., BOURUT, C.: Cancer Res. 33, 1987 (1973b).

MATHÉ, G., NETTER, A.: Le Sang 7, 442 (1949).

MATHÉ, G., POUILLART, P., LAPEYRAQUE, F.: Brit. J. Cancer 23, 814 (1969).

MATHÉ, G., POUILLART, P., STERESCU, M., AMIEL, J.L., SCHWARZENBERG, L., SCHNEIDER, M., HAYAT, M., DE VASSAL, F., JASMIN, C., LAFLEUR, M.: Europ. J. Clin. Biol. Res. 16, 554 (1971b).

MATHÉ, G., VASSAL F. DE, DELGADO, M., POUILLART, P., BELPOMME, D., HAYAT, M., SCHWARZENBERG, L., AMIEL, J.L., MISSET, J.L., MUSSET, M., JASMIN, C.: Cancer Immunol. Immunoth. 1, 77 (1976a).

MAYES, R.B., GRAFF, R.J., PAREIRA, M.D.: Transplantation 5, 80 (1967).

MEDICAL RESEARCH COUNCIL. Brit. Med. J. 4, 189 (1971).

MERRITT, K., JOHNSON, A.G., J. Immunol. 94, 416 (1965).

MEYER, T.J., ANACKER, R.L., RIBI, E.: Cell. Immunol. 14, 52 (1974).

MEYER, R.K., RAO, M.A., ASPINALL, R.L.: Endocrinology 64, 890 (1959).

MIGLIORE-SAMOUR, D., JOLLES, P.: Febs Letters 25, 301 (1972).

MIGLIORE-SAMOUR, D., JOLLES, P.: p. 207 in "Investigation and Stimulation of Cancer Immunity in Cancer Patients" (G. Mathé and R. Weiner, eds.). Heidelberg: Springer-Verlag 1974.

MINOWADA, J., OHNUMA, T., MOORE, G.W.: J. Nat. Cancer Inst., 49, 891 (1972).

MIZUNO, D., YOSHIOKA, L., AKAMATU, M., KATAOKA, T.: Cancer Res. 28, 1531 (1968).

MOORE, G.E., GERNER, R.E., BRUGAROLAS, A.: Surg. Gynec. Obstet. 136, 359 (1973).

MOORE, M., LAWRENCE, N., WITHEROW, P.J.: Europ. J. Cancer 10, 673 (1974).

MOORE, G.E., McLIMANS, W.F.: J. Theoretical Biol. 20, 217 (1968).

MORTON, D.L.: Nat. Cancer Inst. Monogr. 35, 375 (1972).

MORTON, D.L., EILBER, F.R., JOSEPH, W.L., WOOD, W.C., TRAHAN, E., KETCHAM, A.S.: Ann. Surg. 172, 740 (1970a).

MORTON, D.L., EILBER, F.R., MALMGREN, R.A., WOOD, W.C.: Surgery 68, 158 (1970b).

MORTON, D.L., JOSEPH, W.L., KETCHAM, A.S., GEELHOED, G.W., ADKINS, P.C.: Ann. Surg. 178, 360 (1973).

MOTTA, R.: Europ. J. Clin. Biol. Res. 15, 161 (1970a).

MOTTA, R.: Europ. J. Clin. Biol. Res. 15, 510 (1970b).

NASPITZ, C.K., BLENNERHASSETT, J.B., SINGHAL, S.K., RICHTER, M.: Int. Arch. Allergy 33, 389 (1968).

NATHANSON, L.: Cancer Chemother. Rep. 56, 659 (1972).

NICOL, T., HELMY, I.D.: Nature 167, 199 (1951).

NILSSON, K., SUNDSTROM, C.: Int. J. Cancer 13, 808 (1974).

NOLL, H., BLOCH, ASSELINEAU, J., LEDERER, E.: Biochem. Biophys. Acta 20, 299 (1956).

NOTKINS, A.L., MERGENHAGEN, S.E., RIZZO, A.A., SCHEELE, C., WALDMAN, T.A.: J. Exp. 123, 347 (1966).

OLDHAM, R.K., WEINER, R.S., MATHÉ, G., BREARD, J., SIMMLER, M.C., CARDE, P., HERBERMAN, R.: Int. J. Cancer, 17, 326 (1976)

ORMAI, S., DE CLERCQ, E.: Science 163, 471 (1969).

OTH, D., SABOLOVIC, D., ROBERT, F.: Beitr. Path. 149, 60 (1973).

PARFENTJEV, I.A., GOODLINE, M.A.: J. Pharmacol. Exptl. Therapy 92, 411 (1948).

PARISH, C.R.: J. Exp. Med., 134, 21 (1971).

PEARSON, C.M.: p. 64 in "Mechanisms of Hypersensitivity" (J.H. Shaffer, G.A. Logrippo and M.W. Chase, eds.). London: Churchill Ltd. 1959.

PELLIS, N.R., KAHAN, B.D.: Proc. Amer. Assoc. Cancer Res. 16, 47 (abstract No. 185) (1975).

PERNIS, B., PARONETTO, F.: Proc. Soc. Exp. Biol. (NY) 110, 390 (1962).

PETERS, L.C., HANNA, M.G. jr., GUTTERMAN, J.U., MAVLIGIT, G.M., HERSH, E.M.: Proc. Soc. Exp. Biol. Med. 147, 344 (1974).

PILET, Ch., LE GARREC, Y., TOUJAS, L., SABOLOVIC, D., MONTEIL, J.C., ROTHIER, F., MISHRA, U., GHEBREHIWET, B., GUELFI, J.: p. 294 in "Investigation and Stimulation of Immunity in Cancer Patients"(G. Mathé and R. Weiner, eds.). Vol. 47 R.R.C.R., Heidelberg: Springer-Verlag 1974.

PINSKY, C., HIRSHAUT, Y., OETTGEN, H.: Nat. Cancer Inst. Monogr. 39, 225 (1973).

PORT, C.D., SPORN, M.B., KAUFMAN, D.G.: Proc. Amer. Assoc. Cancer Res. 16, 21 (abstract No. 83) (1975).

POTOP, I.: Rev. Roum. Endocrinol. 3, 185 (1966).

POTOP, I., JUVINA, G., MREANA, G., BOERU, B.: Arch. Gesch. 30, 190 (1967).

POWLES, R.: Cancer 34, 1558 (1974).

POWLES, F.L., CROWTHER, D., BATEMAN, C.J.T., BEARD, M.E.J., MacELWAIN, T.J., RUSSEL, J., LISTER, T.A., WHITEHOUSE, J.M.A., WRIGLEY, P.F.M., PIKE, M.: Brit. J. Cancer 28, 365 (1973).

POWLES, R.L., LISTER, T.A., OLIVER, R.T.D., RUSSELL, J., SMITH, C., KAY, H.E.M., McELWAIN, T.J., HAMILTON FAIRLEY, G.: Brit. Med. J. 4, 375 (1974a).

POWLES, R., McELWAIN, T.J., ALEXANDER, P., CROWTHER, D., FAIRLEY, G., PIKE, M.: p. 449 in "Investigation and Stimulation of Immunity in Cancer Patients" (G. Mathé and R. Weiner, eds.). Vol. 47 R.R.C.R., Heidelberg : Springer-Verlag 1974b.

PRAGER, M.D.: Bibl. Haemat. 39, 706 (1973).

PRAGER, M.D., BAECHTEL, S., RIBBLE, R.J., LUDDEN, C.M., MEHTA, J.M.: Cancer Res. 34, 3203 (1974b).

PRAGER, M.D., DERR, I., SWANN, A., COTROPIA, J.: Cancer Res. 31, 1488 (1971).

PRAGER, M.D., RIBBLE, R.J., MEHTA, J.M.: p. 379 in "Investigation and Stimulation of Immunity in Cancer Patients" (G. Mathé and R. Weiner, eds.). Vol. 47 R.R.C.R., Heidelberg: Springer-Verlag 1974a.

RAFFEL, S.: J. Infect. Dis. 82, 267 (1948).

RAI, K.R., HOLLAND, J.F., GLIDEWELL, O.: Proc. Amer. Assoc. Cancer Res. 16, 265, (abstract No. 1176) (1975).

RAMON, G., RICHOU, R., THIERY, J.P., GERBEAUX, C.: C.R. Acad. Sci. (Paris) 229, 278 (1949).

RAYNAUD, M., KOUZNETZOVA, B., BIZZINI, B., CHERMANN, J.C.: Ann. Inst. Pasteur, 122, 695 (1972).

RENOUX, G., RENOUX, M.: C.R. Acad. Sci., Paris 272, 156 (1971a).

RENOUX, G., RENOUX, M.: C.R. Acad. Sci., Paris 272, 349 (1971b).

RENOUX, G., RENOUX, M.: Nature New Biology, 240, 217 (1972).

RIOS, A., SIMMONS, R.L.: Transplantation 13, 343 (1972).

ROBINSON, E., HAASZ, R., BARTAL, A., COHEN, Y.: 9th International Congress of Chemotherapy. London (abstract C 82) July 1975.

ROSE, B.: p. 599 in "Mechanisms of Hypersensitivity" (J.H. Shaffer, G.A. LoGrippo G.A. and M.W. Chase, eds.). London: Churchill Ltd. 1959.

ROSENFELD, C., MACIEIRA-COELHO, A., VENUAT, A.M., JASMIN, C., TUAN, T.Q.: J. Nat. Cancer Inst. 43, 581 (1969).

ROSENFELD, C., PAINTRAND, M., VENUAT, A.M., CHOQUET, C.: In Vitro 9, 388 (abstract) (1974).

ROSENFELD, C., VENUAT, A.M., GOUTNER, A., GUEGAND, J., CHOQUET, C., TRON, F., PICO, J.L.: Proc. Amer. Assoc. Cancer Res. 16, 29 (abstract No. 115) (1975).

ROSENFELD, C., VENUAT, A.M., PAINTRAND, M., CHOQUET, C.: XIVth International Congress of Hematology, Sao-Paulo, Brésil (abstract 189) July 1972.

ROSENTHAL, S.R., CRISPEN, R.G., THORNE, M.G., PIEKARSKI, N., RAISYS, N., RETTIG,P.: p. 421 in "Investigation and Stimulation of Immunity in Cancer Patients" (G. Ma-Mathé and R. Weiner, eds.). Vol. 47 R.R.C.R., Heidelberg: Springer-Verlag 1974.

ROSENTHAL, S.R., McENERY, J.T., RAISUS, N.: J. Asthma Res. 5, 309 (1968).

RUITENBERG, E.J., STEERENBERG, P.A.: Nature, New Biol. 242, 149 (1973).

SALOMON, J.C., LYNCH, N.: Cancer Immunol. Immunoth. in press (1976).

SANFORD, B.H.: Transplantation 5, 1273 (1967).

SARMA, P.S., SHIU, G., NEUBAUER, R.H., BARON, S., HUEBNER, R.J.: Proc. Nat. Acad. Sci. 62, 1046 (1969).

SAUTER, C., GLIFFORD, G.E.: Nature 212, 626 (1966).

SAUTER, C., LINDENMANN, J., GERBER, A., MARTZ, G.: p. 455 in "Investigation and Stimulation of Immunity in Cancer Patients" (G. Mathé and R. Weiner, eds.). Vol. 47 R.R.C.R., Heidelberg: Springer-Verlag 1974.

SCHAFER, L.A., WASHINGTON, M.L., GOLDSTEIN, A.L.: Proc. Amer. Assoc. Cancer Res. 16, 233 (abstract No. 1051) (1975).

SCHMIDTKE, J.R., JOHNSON, A.G.: J. Immunol. 106, 1191 (1971).

SCHNEIDER, M., MATHÉ, G., SCHWARZENBERG, L., POUILLART, P., WEINER, R., AMIEL, J.L., HAYAT, M., JASMIN, C., DE VASSAL, F.: p. 42 in "Investigation and Stimulation of Immunity in Cancer Patients" (G. Mathé and R. Weiner, eds.). Vol. 47 R.R.C.R., Heidelberg: Springer-Verlag 1974.

SCHWARZENBERG, L.: p. 179 in "Cryoconservation des Cellules Normales et Néoplasiques" (R.S. Weiner, R.K. Oldham, L. Schwarzenberg, eds.). 1 Vol. Paris: INSERM 1973.

SCHWARZENBERG, L., SIMMLER, M.C., PICO, J.L.: Cancer Immunoth. Immunol. 1, 69 (1976).

SCOTT, M.T.: Cell. Immunol., 5, 469 (1972).

SCOTT, M.T.: J. Nat. Cancer Inst. 53, 855 (1974a).

SCOTT, M.T.: Seminars in Oncology, 1, 367 (1974b).

SEDLACEK, H.H., MESSMANN, H., SEILER, F.R.: Proc. Amer. Assoc. Cancer Res., 16, 141 (abstract No. 564) (1975).

SIEGLER, H.F., SHINGLETON, W.W., METZGAR, R.S., BUCKLEY, C.E., BORGOC, P.M., MILLER, D.S., FETTER, B.G., PHAUP, M.B.: Surgery 72, 162 (1972).

SIMMLER, M.C., RAMEAU, G., CHOU, M.J., MATHÉ, G.: Israel J. Med. Sci. in press (1976a).

SIMMLER, M.C., SCHWARZENBERG, L., MATHÉ, G.: Nouv. Presse Med. in press (1976b).

SIMMLER, M.C., SCHWARZENBERG, L., MATHÉ, G.: Cancer Immunoth. Immunol. in press (1976c).

SIMMONS, R.L., RIOS, A.: Nat. Cancer Inst. Monogr. 39, 57 (1973).

SIMMONS, R.L., RIOS, A.: Israel J. Med. Sci. 10, 925 (1974a).

SIMMONS, R.L., RIOS, A.: p. 370 in "Investigation and Stimulation of Immunity in Cancer Patients" (G. Mathé and R. Weiner, eds.). Vol. 47 R.R.C.R., Heidelberg: Springer-Verlag 1974b.

SIMMONS, R.L., RIOS, A., LUNDGREN, G.: Fed. Proc. 30, 246 (1971a).

SIMMONS, R.L., RIOS, A., LUNDGREN, G., RAY, P.K., McKHANN, C.F., HAYWOOD, G.R.: Surgery 70, 38 (1971b).

SIMMONS, R.L., RIOS, A., RAY, P.K., LUNDGREN, G.: J. Nat. Cancer Inst. 47, 1087 (1971c).

SIMON-REUSS, I., COOK, C.M.W., SEAMAN, C.V.F., HEARD, D.: Cancer Res. 24, 2038 (1964).

SJOGREN, H.O., HELLSTROM. I.: Exp. Cell Res. 40, 208 (1965).

SLONECKER, C.E., RIEKE, W.O.: Nature 207, 729 (1965).

SMITH, G.V., MORSE, P.A. jr., DERAPS, G.D., RAJU, S., HARDY, J.D.: Surgery 74, 59 (1973).

SPARKS, F.C., BREEDING, J.H.: Cancer Res. 34, 3662 (1974).

SPARKS, F.C., SILVERSTEIN, M.J., HUNT, J.S., HASKELL, C.M., PILCH, Y.H., MORTON, D.L.: New England J. Med. 289, 827 (1973).

STEWART, T., TOLNAI, G.: Cancer 24, 117 (1969).

STIFFEL, C., MOUTON, D., BIOZZI, G.: p. 239 in "Investigation and Stimulation of Immunity in Cancer Patients" (G. Mathé and R. Weiner, eds.). Vol. 47 R.R.C.R., Heidelberg: Springer-Verlag 1974.

STODOLA, F.H., LESUK, A., ANDERSON, R.J.: J. Biol. Chem. 126, 505 (1938).

STUCK, B., OLD, L.J., BOYSE, E.A.: Nature 202, 1016 (1964).

SVET-MOLDAVSKY, G.J., HAMBURG, V.P.: p. 323 in "Specific tumour antigens, UICC Monograph Series", Vol. 2, Copenhagen: Munksgaard 1967.

TANAKA, T., SASAKI, N.: Gann 65, 395 (1974).

TANAKA, K., TANAKA, A., SUGUYAMAS, S.: Int. Arch. Allergy, 34, 495 (1968).

TANAKA, T., TOKUNAGA, T.: Gann 62, 433 (1971).

THOMAS, J.W., PLENDERLEITH, I.H., CLEMENTS, D., LANDI, S.: p. 315 in "XIth International Cancer Congress", Vol. 1, Florence: Conferences, Symposia, Workshops October 1974.

THOMPSON, R.B., ALBEROLA, B., MATHÉ, G.: Europ. J. Clin. Biol. Res. 17, 900 (1972).

TOKUNAGA, T., KATAOKA, T., NAKAMURA, R., YAMAMOTO, S., TANAKA, T.: Jap. J. Med. Sci, Biol. 26, 71 (1973).
TON THAT TUNG, NGUYEN NHU BANG, NGUYEN VAN VAN, TON THAT BACH, NGUYEN DANG TAM, SALOMON, J.C.: Lancet 1, 527 (1975).
TOUJAS, L., DAZORD. L., GUELFI, J.: C.R., Acad. Sci. Paris 276, 433 (1973).
TOUJAS, L., DAZORD, L., GUELFI, J.: p. 302 in "Investigation and Stimulation of Immunity in Cancer Patients" (G. Mathé and R. Weiner, eds.). Vol. 47 R.R.C.R., Heidelberg: Springer-Verlag 1974.
TOUJAS, L., SABOLOVIC, D., DAZORD, L., LEGARREC, Y., TOUJAS, J.P., GUELFI, J., PILET, C., Rev. Europ. Etud. Clin. Biol. 17, 267 (1972).
TRAININ, N.: Physiol. Rev. 54, 272 (1974).
TRAININ, N., SMALL, M., GLOBERSON, A.: J. Exp. Med. 130, 765 (1969).
TRAININ, N., SMALL, M., KIMCHI, Y.: p. 135 in "Thymic Hormones" (T.D. Luckey, ed.). Baltimore: University Park Press 1973.
TRIPODI, D., PARKS, L.C., BRUGMANS, J.: New England J. Med. 289, 354 (1973).
TURCOTTE, R., QUEVILLON, M.: J. Nat. Cancer Inst. 52, 215 (1974).
TURCOTTE, R., QUEVILLON, M.: Cancer Immunol. Immunoth., 1, 25 (1976).
UNANUE, E.R., ASKONAS, B.A., ALLISON, A.C.: J. Immunol., 103, 71 (1969).
University of California and National Cancer Institute. Special virus cancer pro-programme. Catalogue of available cell cultures, October 1970. Cell culture division naval supply Center, Oakland, California. University of California Naval Biochemical Research Laboratory.
VERHAEGEN, H., DECREE, J., DECOCK, W., VERBRUGGEN, F.: New England J. Med., 289, 1148 (1973).
VENUAT, A.M., ROSENFELD, C.: Proc. Amer. Assoc. Cancer Res., 14, 101 (abstract No. 403) (1973).
VENUAT, A.M., ROSENFELD, C.: In Vitro, 9, 378 (abstract) (1974).
VOGLER, W.R., CHAN, Y.K.: Lancet 2, 128 (1974).
WATKINS, J.F., CHEN, L.: Nature 223, 1018 (1969).
WEBB, D., BRAUN, W., PLESCIA, O.J.: Cancer Res., 32, 1814 (1972).
WEISS, L.: J. Cell Biol. 26, 735 (1965).
WEISS, D.W.: Nat. Cancer Inst. Monogr. 35, 157 (1972).
WEISS, D.W., BONHAG, R.S., LESLIE, P.: J. Exp. Med. 124, 1039 (1966b).
WEISS, L., MAYHEW, E., ULRICH, K.: Lab. Invest. 15, 1304 (1966a).
WEPSIC, H.T., HARRIS, S., SANDERS, J., ALAIMO, J.: Proc. Amer. Assoc. Cancer Res. 16, 183 (abstract No. 731) (1975).
WEPSIC, H.T., ZBAR, B., RAPP, H.J., BORSOS, T.: J. Nat. Cancer Inst. 44, 955 (1970).
WERNER, G.H., MARAL, R., FLOC'H, F., MIGLIORE-SAMOUR, D., JOLLES, P.: in preparation (1976).
WEST, W., HERBERMAN, R.B.: Cell. Immunol. 14, 139 (1974)
WESTPHAL, O., FISCHER, H., MUNDER, P.G.: p. 319 in 8th Internat. Congr. Interlaken, Switzerland: 3-9th Sept., 1970.

WHITE, R.G.: Brit. Med. Bull. 23, 39 (1967).
WHITE, R.G., BERNSTOCK, L., JOHNS, R.G.S., LEDERER, E.: Immunology 1, 54 (1958).
WHITE, R.G., JOLLES, P., SAMOUR, D., LEDERER, E.: Immunology 7, 158 (1964).
WILLIAMS, A.C., KLEIN, E.: Cancer 25, 450 (1970).
WISSLER, R.W., CRAFT, K., KESDEN, D., POLISKY, B., DZOKA, K.: p. 539 in "Advance in Transplantation" (J. Dausset and F.J. Rapaport, eds.). 1 Vol., Copenhagen: Munksgaard 1968.
WOODRUFF, M.F.A.: p. 272 in "Investigation and Stimulation od Immunity in Cancer Patients" (G. Mathé and R. Weiner, eds.). Vol 47, R.R.C.R., Heidelberg: Springer-Verlag 1974.
WOODRUFF, M.F.A., BOAK, J.L.: Brit. J. Cancer 20, 345 (1966).
WOODRUFF, J.J., GESNER, B.M.: J. Exp. Med. 129, 551 (1969).
WUNDERLICH, J.R., MARTIN, W.J., FLETCHER, F.: Fed. Proc. 30, 246 (1971).
YOUN, J.K., BARSKI, G., HUPPERT, J.: C.R. Acad. Sci. Paris 267, 816 (1968).
YOUN, J.K., BLANCHARD, M.G., BARSKI, G.: Europ. J. Cancer 10, 491 (1974).
ZBAR, B.: p. 350 in "Investigation and Stimulation of Immunity in Cancer Patients" (G. Mathé and R. Weiner, eds.). Vol. 47 R.R.C.R., Heidelberg: Springer-Verlag 1974.
YOUNG, C.W.: Med. Clin. N. Amer. 55, 721 (1971).
ZBAR, B., BERNSTEIN, I.D., BARTLETT, G.L., HANNA, M.G. jr., RAPP, H.J.: J. Nat. Cancer Inst. 49, 119 (1972).
ZBAR, B., BERNSTEIN, I.D., RAPP, H.J.: J. Nat. Cancer Inst. 46, 831 (1971).
ZBAR, B., BERNSTEIN, I., TANAKA, T., RAPP, H.J.: Science 170, 1217 (1970).
ZBAR, B., RAPP, H.J.: Cancer 34, 1532 (1974).
ZBAR, B., RIBI, E., MEYER, T., AZUMA, I., RAPP, H.J.: J. Nat. Cancer Inst. 52, 1571 (1974).
ZBAR, B., RIBI, E., RAPP, H.J.: Nat. Cancer Inst. Monogr. 39, 3 (1973).

Chapter 8
Toxicology and Adverse Effects

Compared to chemotherapy, immunotherapy is minimally toxic. It is not, however, a branch of homeopathy. We have seen that a contrary effect to the one sought can be obtained (tumour growth enhancement). Direct toxic effects, minor or severe and possibly lethal, have been observed. Thus the toxicology (MATHÉ, 1976; SCHWARZENBERG *et al.*, 1976) of this new weapon in cancer treatment must be known.

8.1. Systemic Active Immunotherapy

Clinical toxicology is not developed and it has been useful to obtain some information from animal studies.

8.1.1. Experimental

8.1.1.1. Specific

Literature makes little mention of complications resulting from the use of material for immunizing animals specifically against neoplasias.

The development of an active tumour may result from the injection of living cells, not irradiated and not treated with mitomycin C or another oncostatic in vitro, even when they are mixed with adjuvant preparations (MOORE *et al.*, 1974). Sensitivity to irradiation varies from tumour to tumour (personal observation). Also variable are, in the case of the use of autologous cells, tumour antigenicity and/or immunogenicity (BALDWIN and PIMM, 1973), and, in the case of the use of allogeneic material, the H2 specificity expression of neoplastic cells (MOTTA, 1974).

8.1.1.2. Systemic Immunity Adjuvants

8.1.1.2.1. BCG

PORTELANCE (1974) and PORTELANCE *et al.*(1976) have described the toxicity of high doses in the case of one of their sub-strains of BCG (derived from the original Calmette-Guerin strain).

He showed that it varies with route of administration. *In mice and guinea pigs*, the oral route is without risk; the reactions are more intensive with the s.c. route, still more with the i.p. route, and reach a maximum with the i.v. route. For each route, the mortality rates depend on dosage: in mice, LD_{50} between days 0-7 is 2000 mg/kg for the s.c. route, between 200 and 600 for the i.p. and about 60 for the i.v. Administered i.v., 2000 mg/kg does not kill any animals. In guinea pigs, lethality is not very different.

Death can be attributed to anaphylactic shock for the s.c. route, to intestinal blockage for the i.p., and to pulmonary thromboembolisms for the i.v.. Loss of weight depends on the route (per os $<$ s.c. $<$ i.p. $<$ i.v.), but not on the dose: this indicates that there is a maximal quantity of BCG which penetrates into the organism, beyond which the reactions are no more intense.

Acid alcohol resistant bacteria have been found rarely, and only after use of i.p. and s.c. routes. Microbacteriological examinations made on dead guinea pigs showed that huge doses can lead to intercurrent infections: this observation was made in animals injected i.v. with 200-480 mg/kg of BCG.

No less interesting is the lymphopaenia observed in mice treated with seven injections of considerable doses of BCG (4 mg s.c. to 40 mg i.v.).

In describing the dose-effect relationship in mice (Table 7.XXIV) (MATHÉ, 1976), I mentioned that high doses such as 10 mg of BCG i.v. are immunosuppressive and that GEFFARD and ORBACH-ARBOUYS (1976) showed, in our laboratory, that this effect is due, at least partially, to stimulation of suppressor T-lymphocytes, as will be discussed in Chapter 9.

BCG does not sensitize mice to histamine and does not produce passive anaphylaxis (SUTER and MUNOZ, 1963). However, mycobacteria cause mice to be very susceptible to endotoxins (SUTER *et al.*, 1958).

With KHALIL *et al.* (1975), we have compared the pathological modifications (Fig. 8.1) of mice treated i.v. with living fresh Pasteur BCG at the dose of 1 mg per animal i.v., which is active in our screening, with animals treated by the same dose s.c.,

Fig. 8.1. Microscopic aspect of systemic lesions observed in mice treated with BCG (1mg i.v.): (a) splenic granulomas of the red pulp of the spleen; (b) Lymph node granulomas with active phagocytosis; (c) Hepatic granulomas in the portal tract and in the sinusoids; (d) Intra-sinosoidal granuloma of the liver (KHALIL et al., 1975)

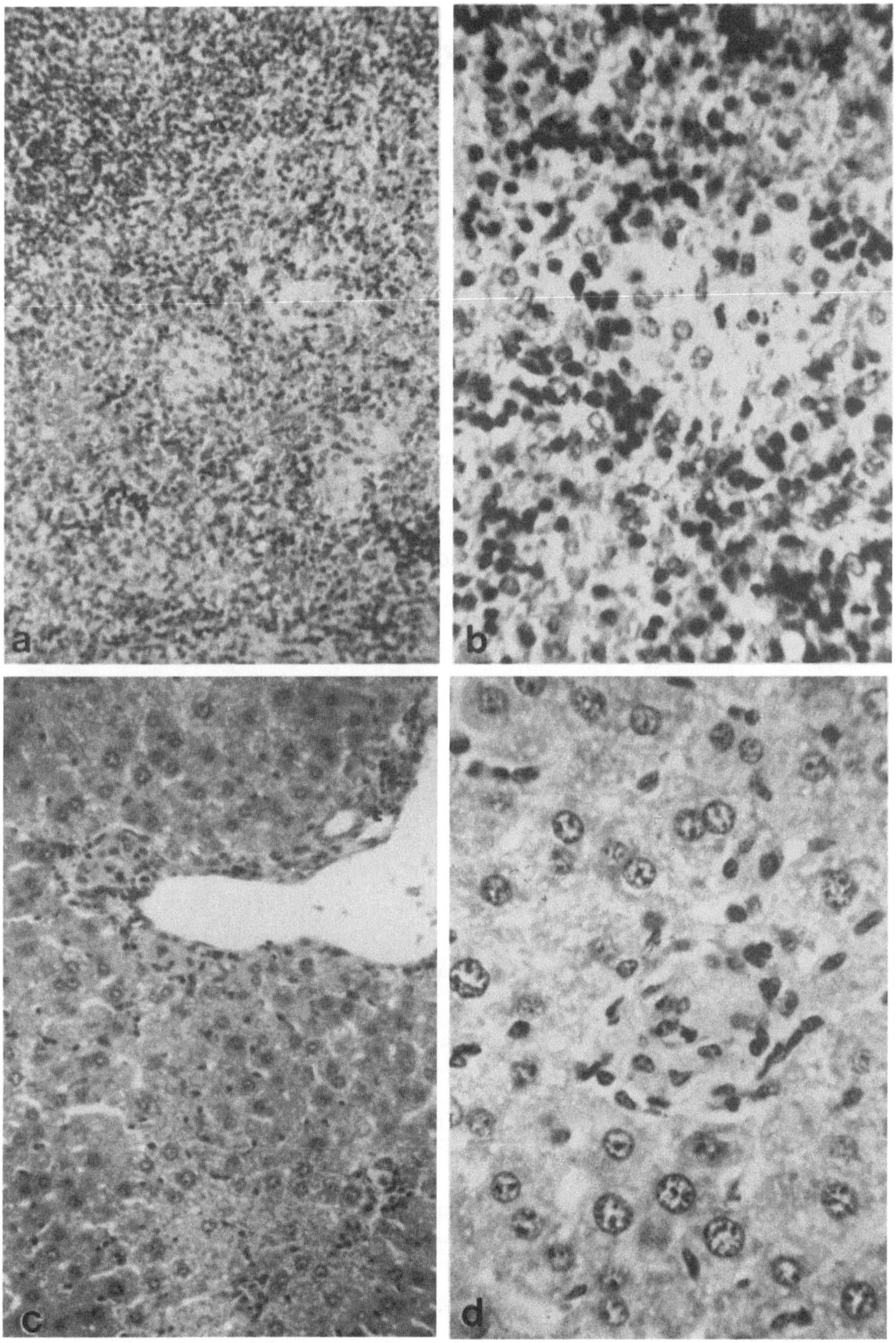

Fig. 8.1. (Legend on page 296)

which is not active (MATHÉ *et al.*, 1975a). On day 6 after i.v. injection, one can see a multifocal infiltration involving the red pulp and the parafollicular and periarteriolar zones of the spleen. It is more pronounced in i.v. than in s.c. injected animals. Beginning on day 14, a pronounced histiocytic granulomatous reaction is observed in the liver. On day 18, this reaction is also seen in the spleen, lymph nodes and lungs of the i.v. injected animals. It persists until day 45. Similar lesions have been described by MUGGLETON *et al.* (1975) in the lung and the liver of guinea pigs into which BCG was injected i.v.

In the s.c. injected mice, we saw histiocytic granulomas at the site of injection and in the regional lymph nodes.

Moreover, modifications of the thymus are observed: pyroninophilia of the cortex and hyperplasia of epithelial cells, which may not be toxic effects, but the expression of one of the mechanisms of action which will be discussed below.

In monkeys, repeated administration of 120 mg produce a blood lymphopaenia and an atrophy of thymodependent areas of the lymph nodes and spleen (PORTELANCE, 1974). This dose, administered to this species, corresponds to a dose of 3 g for a man.

We have devoted a study to the i.v. route of administration in baboons (JURCZYK-PROCYK *et al.*, 1976). We compared several doses, from 1-100 mg injected once or 5 times (every 8 days) to 1 g given once (Table 8.I).

The observations are summarized in Table 8.II. Only the animal which received 1 g presented a transitory weight loss, pulse and respiration rhythm acceleration, and fever (Fig. 8.2). Hepatic biological disorders were expressed by glutamic oxalo-acetic and glutamic pyruvic transaminases, as well as gamma-glutamyl transpeptidase activity increase (Fig. 8.3). Only the monkey which received 100 mg 5 times died spontaneously after the fifth injection on day + 30. The others were sacrificed for histopathological and bacteriological examinations.

In all, manifestations of septicaemia (granulomas, lymphocyte and macrophage proliferation and/or infiltration in the spleen, lymph nodes, and liver) were found (Fig. 8.4); bacilli were present, and the number of colonies given per gram of tissue was considerably higher in large doses than in small doses; for a given dose, it was much higher when it was repeated 5 times than when injected once (Table 8.11).

SCHAUF *et al.* (1975) observed a more important toxicity in marmosets into which he injected BCG s.c. or i.m. or into a tumour at the dose of 5 x 10^7 viable organisms (lyophilized). All animals died before day 60. Bacilli were found in liver, spleen and lungs, and pathological findings included nephritis, granulomatous hepatitis, and hepatic necrosis

Table 8.1. General description of i.v. BCG toxicology experiment (JURCZYK-PROCYK *et al.*, 1976)

No.	Monkey age	Dose of BCG	Laparatomy, biopsy[a]	Autopsy[a]	Monkey sacrificed[a]
1	DAI 11 yr	1 g	(a) day + 28 (b) day + 91	–	–
2	AIC 11 yr	100 mg x 5	–	died 1 day after 5th injection of BCG day + 30	–
3	MAN 7 yr	100 mg	–	–	day + 35
4	MIN 8 yr	10 mg x 5	day + 37	–	–
5	NOE 8 yr	10 mg	day + 37	–	–
6	RAQ 8 yr	1 mg x 5	day + 36	–	–
7	WAN 11 yr	1 mg	day + 37	–	–

[a] Day after first injection of BCG.

Table 8.II. Pathological changes and bacteriological examinations after high doses of BCG given i.v. in baboons (JURCZYK-PROCYK *et al.*, 1976)

Monkey	BCG	Histology		Bacteriology	
		1 month	3 months	1 month	3 months
DAI	1 g biopsy	Polymorphonuclear infiltration; early granulomas in the liver	Typical granulomas in liver	132,000 Conolies/g liver 6 130 Colonies/g lymph node	0/liver
AIC	100 mg x 5 dead	Acute diffuse infiltrations of macrophages, histiocytes, and lymphocytes with suggestion of formation of early granulomas		208,000 Colonies/g liver 36,300 Colonies/g spleen 9,260 Colonies/g thymus 31,300 Colonies/g adrenals	
MAN	100 mg	Numerous granulomas in the liver; spleen, lungs; diffuse histiocytic infiltration of lymph nodes		1,620 Colonies/g liver 5,850 Colonies/g spleen 0 Colony/g thymus 0 Colony/g adrenals	
MIN	10 mg x 5	Liver: typical small granulomas; Lymph nodes: proliferation and infiltration with histiocytes and macrophages		13,250 Colonies/g liver 2,300 Colonies/g lymph node	
NOE	10 mg	Liver: typical small granulomas; Lymph nodes: proliferation and infiltration with histiocytes and macrophages		130 Colonies/g liver 383 Colonies/g lymph node	
RAQ	1 mg x 5	Liver: infiltration of portal spaces with macrophages; Lymph nodes: proliferation of macrophages		1,900 Colonies/g liver 0 Colony/g lymph node	
WAN	1 mg	Liver: numerous typical granulomas; Lymph nodes: proliferation of macrophages in sinuses		125 Colonies/g liver 137 Colonies/g lymph node	

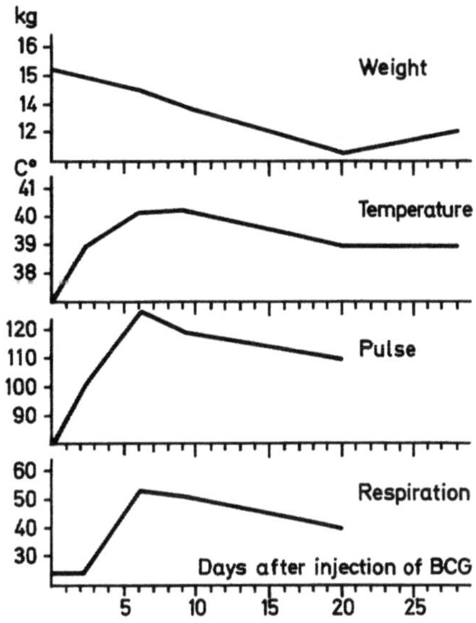

Fig. 8.2. Body weight, temperature, pulse, and respiratory rate of baboon during 4 weeks after receiving 1 g of BCG (JURCZYK-PROCYK et al., 1976)

In conclusion, comparing mice and monkey data, it appears that very high doses of BCG can be lethal, especially in certain species and via certain routes of administration. However, large doses are usually well tolerated, although they induce septecaemia the persistence of which is dependent on the repetition of the injections.

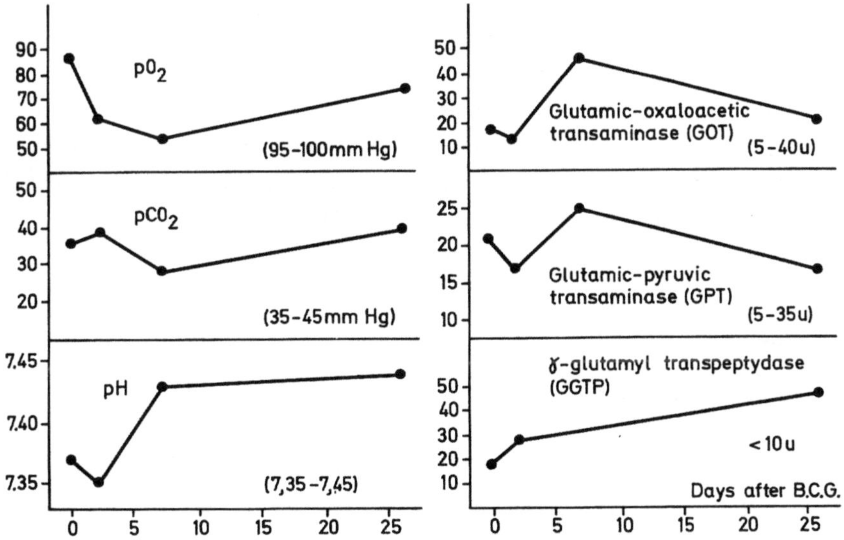

Fig. 8.3. Partial pressures of oxygen, carbon dioxide, pH, and liver enzymes in baboon during 6 weeks after receiving 1 g of BCG (JURCZYK-PROCYK et al., 1976)

A septicaemia of a given intensity seems necessary in mice to obtain the anti-tumour effect. Doses higher than the optimal one, even if they are not lethal, may be harmful and, specially, immunosuppressive.

8.1.1.2.2. Mycobacteria Preparations
The animal toxicity of the different mycobacteria preparations are summarized in Table 8.III.

8.1.1.2.3. Corynebacteria and Other Micro-Organisms
The side-effects of corynebacteria, Bordetella pertussis, and Brucella abortus are given in Table 8.IV.

8.1.1.2.4. Polynucleotides, Lentinan, Tetramisole, Vitamins
The side-effects of polynucleotides, lentinan, tetramisole, levamisol, and vitamins A and E are given in Table 8.V.

The three Tables summarize the data gathered from the extensive review devoted by MARAL *et al.* (1976) to "the toxicology of immunological drugs".

8.1.1.3. Miscellaneous

For toxic side-effects of other agents mentioned in Tables 8.III, 8.IV. and 8.V, as adjuvants "largo sensu", such as (a) endotoxins, (b) lectins, especially phytohaemagglutinin and pokeweed mitogen, (c) phospholipids extracted from different bacteria such as the "Extrait Bacterien Pasteur" (EBP), (d) polysaccharides other than lentinan, (e) mineral adjuvants, (f) polycations, such as diethyl-aminoethyl dextran, (g) interferon and interferon inducers such as bis-diethyl-aminoethyl-fluorenone, or tirolone and polyacrylic acids, and (h) aristolochic acid, the reader is referred to the review by MARAL *et al.* (1976).

8.1.2. Clinical

8.1.2.1. Specific

The i.d. and/or s.c. administration of non-modified 4 10^7 *allogeneic irradiated* in vitro (4000 rads), formalin treated (4 per 1000) and cryopreserved *cells* of acute lymphoid leukaemia (ALL), of acute myeloid leukaemia (AML), and of different solid tumours (ST), including melanoma, have not induced complications (MATHÉ, 1976; SCHWARZENBERG *et al.*,1976). A moderate skin reaction, of delayed hypersensitivity type, can be seen.

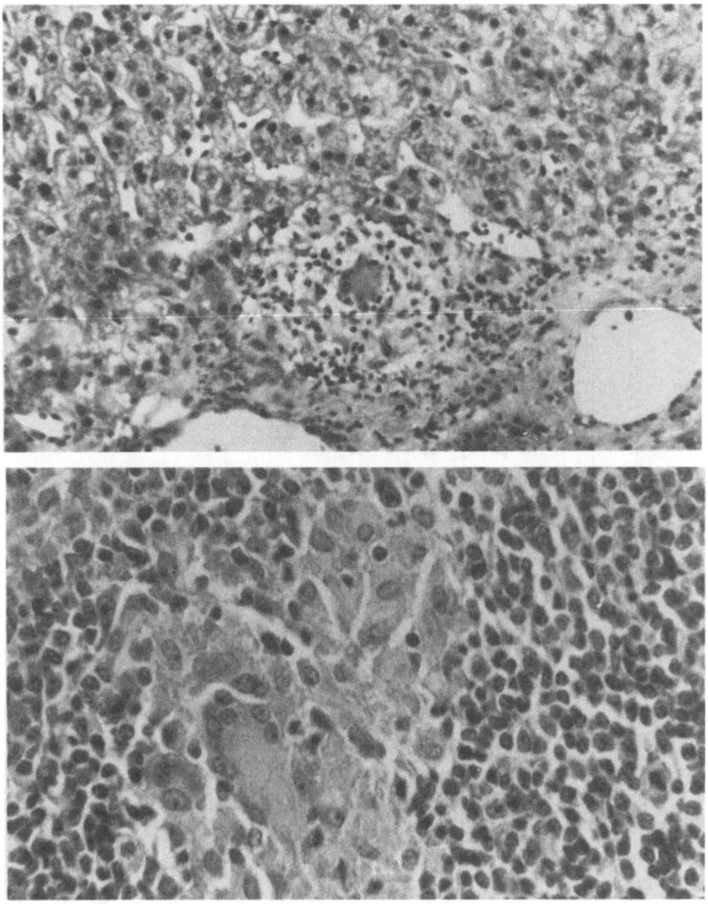

Fig. 8.4. Pathological aspects of liver and spleen in baboons which received i.v. BCG: (a) Liver biopsy portal tract granuloma 18 days after 1 g injection, (b) Splenic granuloma with phagocytosis (JURCZYK-PROCYK et al., 1976)

The injection of *non-irradiated allogeneic cells* from ALL can be followed by local growth if the leukaemia type of the donor is prolymphoblastic (Fig. 8.5). Since many of these cells are in the cycle, a possible explanation is their poverty in surface antigens and, since they are allogeneic, certain leukaemia cells may present a loss of certain histocompatibility antigens as shown by MOTTA (1974) with mouse leukaemia cells. Such growths have to be removed surgically and their site complementarily irradiated.

Table 8.III. Main side-effects in some animal species of mycobacterial preparations (From MARAL *et al.*, 1976)

Freund's Complete adjuvant	Freund's Incomplete adjuvant	Cord factor	Methanol extract residue (MER)	Insoluble waxes	Water-soluble extracts
Enhancement of anaphylactic reactions; production of auto-antibodies; induction of local granulomas (at site of injection); induction of arthritis		A single dose of 20 mg/kg is not toxic for mice; repeated injections of small doses: weight loss; peritoneal exudates and hemorrhagic lesions in lungs; i.v. injection induces pulmonary granulomata, liver disturbances, and enhances sensitivity to endotoxins	Not pyrogenic; can be injected by different routes and repeatedly; induction of local granulomata; BCG-like hyperthrophy of lymphoid tissue	Same side effects as Freund's complete adjuvant	Local inflammatory reactions; no arthritis; do not increase susceptibility to histamine and toxicity to endotoxins

Table 8.IV. Main side-effects in some animal species of corynebacteria and other microorganisms (From MARAL *et al.*, 1976)

Corynebacteria	Bordetella pertussis	Brucella abortus
Acute toxicity:	Acute toxicity:	
In mice (i.v. or i.p.): LD_{50} # 150 mg/kg In rabbit : weakly pyrogenic, local inflammatory reaction after 0.07 mg i.d..	10^{10} killed bacteria: i.v. = no mortality; febrile reactions, spleen hypertrophy, hyperlymphocytosis, hypoglycemia.	Local inflammatory reaction, hyperplasia of lymphoid organs.
Subacute and chronic toxicities:	Subacute and chronic toxicities:	
Sensitize mice to histamine and make them very susceptible to endotoxins and pentobarbital (impairment of liver microsomal functions).	Sensitize mice to histamine and make them very susceptible to endotoxins.	
In mice, loss of weight at 5 mg/kg i.p.. In rats and monkeys at 0.2 to 2 mg/kg: increase of polymorphs, anaemia; hypertrophy of liver and lymphoid organs (granulomata with epithelioid cells); local inflammatory reactions may result in abcesses; depression of T-lymphocytes.		

Table 8.V. Main side-effects in some animal species of polynucleotides, lentinan, tetramisole-levamisole, and vitamins (From MARAL *et al.*, 1976)

Poly I-poly C	Poly A-poly U	Lentinan	Tetramisole-levamisole	Vitamin A	Vitamin E
Toxicity variations with species Acute toxicity: LD_{50}: mice: 17 mg/kg rabbit: 0.6 mg/kg dog: 1 mg/kg; Febrile reaction, cataracts, embryotoxicity In Monkey: 1 mg/kg i.v. once: increase of alkaline phosphatases. In young chickens: at 1 mg/kg: cerebellar ataxia at 50 µg/kg; mitosis inhibition in regenerating liver	Acute toxicity lower than that of poly I-poly C	Non-toxic at active doses (0.1-5 mg/kg); But increases sensitivity of animals to serotonin, catecholamines, and histamine (with the latter risk of shock)	Therapeutic dose varies in animals, according to species, from 2-40 mg/kg per os or i.p. Action inhibited by theophyllin In rats: 140 mg/kg decreases food intake and growth rate. In dogs; 20 mg/kg per o.s. induces vomiting	Itching, loss of hair, fractures, teratogenic	In chickens, at enormous doses: waxy feathers, and slower weight gain

Subacute toxicity:

Dogs: 0.125 mg/kg i.v.
3 times a week:
Emesis, diarrhea,
convulsions; necrotic
lesions in liver and
bone marrow;
leucopenia; 6 times daily

Mice: 5 mg/kg
CFU [a] fell to 0.5%
of normal; marrow aplasia;
lymphoid aplasia
(including thymic)

---

[a] CFU = colony forming unit

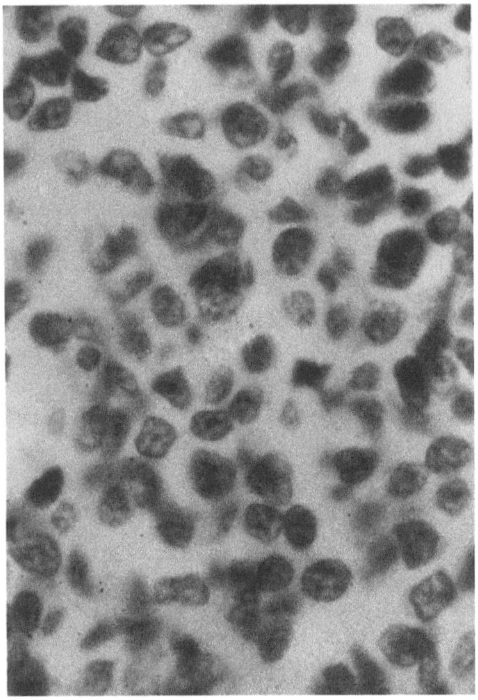

Fig. 8.5. Local growth at the site of injection of non-irradiated cells from prolymphoblastic leukaemia donor (MATHÉ 1976)

8.1.2.2. Systemic Immunity Adjuvants

8.1.2.2.1. BCG

In 277 patients, who were, in remission of acute lymphoid leukaemia treated with living fresh Pasteur BCG, applied on *scarifications* for 5 years (MATHÉ, 1976; SCHWARZENBERG *et al.*, 1976), tolerance at the local as well as at the general level was found to be excellent (Table 8.VI).

Locally, the reaction became positive after one or two applications if it was negative before. If the first reactions were positive or became positive, they increase in intensity during the first 2 or 3 months, at which time they are more raised, scabbed, and irritating than the first ones. Their intensity then becomes stabilized and remains the same for years. A reaction of a given scarified area persists for 2-3 weeks, after which it leaves very discreet scars, which are usually no longer visible after a few months.

It is exceptional that local reactions are cumbersome because of their exuberance. In such cases, no special treatment has been given and the course has been a little longer than usual, but benign.

Table 8.VI. Complications of BCG applied on one scarified area at a time in 277 cases of acute lymphoid leukaemia (MATHÉ, 1976; SCHWARZENBERG *et al.*, 1976)

	Complications	%	Necessity to stop treatment
Marked malaise	0	0	
Fever (temperature 39° C)	14	5	
Chills and/or sweating	24	8	
Pruritus	194	70	
Skin rash, localized	8	3	
Skin rash, generalized	1	0.4	
Ulcers	0	0	
Adenopathy	58	21	
Splenomegaly	46	16	
Hepatic dysfunction	20	7	
Hepatic granuloma	2/8	25	
Lung miliary lesions	1	0.4	
Keratoconjonctivitis	1	0.4	1

Regional reactions consisting of adenopathies are not rare (about 20%). The latter are palpable, causing some aching, and transitory. Biopsies have shown that they are typically BCG adenitis. They were never complicated, especially by fistula.

Systemic reactions consist mainly of chills and/or sweating (8%), fever (5%); 9 skin rashes were noted, one of which was generalized (Fig. 8.6).

Splenomegaly was noted in 16% of the patients at least once but, as the treatment was often applied at home, this percentage is probably below the actual number.

Hepatomegaly was not seen, except in patients who had hepatitis before BCG therapy, probably promoted by chemotherapy immunodepression. Moreover, in one patient who presented a subacute hepatitis during pre-immunotherapy chemotherapy, the manifestations of the disease disappeared soon after BCG was started. Biological hepatic dysfunctions (increased transaminase activity) were only noted in 7% of the patients. Out of eight voluntary subjects, whom we submitted to systematic hepatic biopsy, granulomas were found in two (Fig. 8.7).

No neuromeningeal manifestations were noted. We observed one case of choroiditis, which was considered as an allergic manifestation,and which disappeared after interrupting BCG treatment.

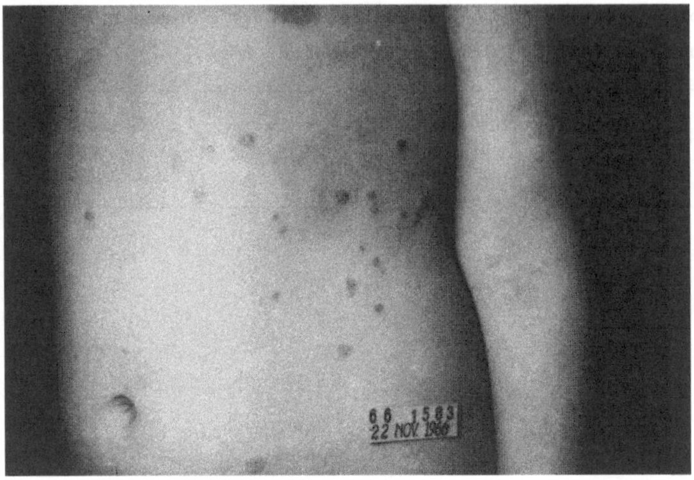

Fig. 8.6. Generalized skin rash in a BCG-treated patient

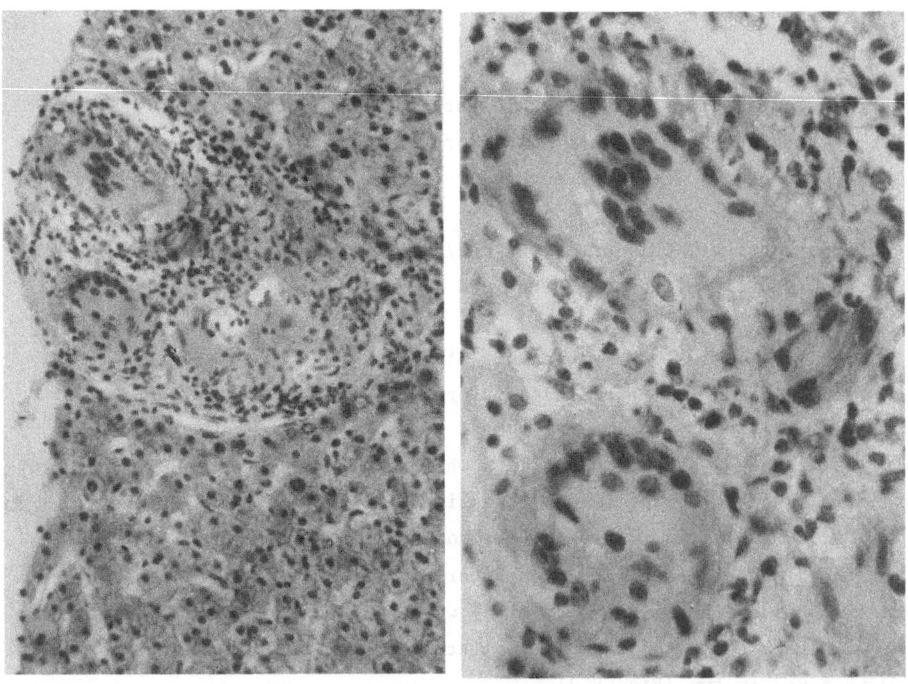

Fig. 8.7. Specific hepatic granuloma in a patient submitted to systemic immunotherapy with BCG

In another population of patients with solid tumours, who had been submitted to a "radical" operation, but who were considered, from the result of surgical examination, to have a poor prognosis and to be in need of complementary immunotherapy, tolerance of BCG on scarifications was also excellent, but local reactions were more intensive than in the leukaemic population (SCHWARZENBERG *et al.*, 1976).

In those first trials, we applied BCG on one scarified area at time. In a recent study (Tables 6.XVIII, 6.XIX and 7.XXVII) we administered fresh living Pasteur BCG on four scarified areas at a time, one on each member, twice a week for two weeks. Tolerance was not different from that observed after application to one area. However, in the anergic subjects, the conversion from negative to positive skin reactions was much less frequent than when BCG was applied on one scarified area. This observation, as discussed above, may be due to a more important penetration of the bacilli in the immunodepressed subjects and, therefore, to an absorbed dose superior to the optimal.

LOKICH (1975) mentioned a very peculiar side-effect of BCG applied on scarifications for melanoma which we have also seen: the development of a specific melanoma nodule within an area of scarification.

SPARKS *et al.* (1973) reported on 23 patients carrying different tumours submitted to systemic immunotherapy with Tice BCG given *i.d.* by the *multiple-puncture tine technique:* they found that it was well tolerated and they described the same manifestations as those we have observed with the scarification technique.

Thus these methods present considerably less side-effects than the simple *i.d. injections with one needle,* which we have tried: fever (40° C) was the rule; local severe inflammation and necrosis followed by ulceration of long duration, and eschars were observed, the healing of which could be delayed for up to 6-9 months SCHWARZENBERG *et al.*, 1976).

Because the *i.v. route* is the most efficient in mice, and after our observation that it was tolerated remarkably well in monkeys (JURCZYK-PROCYK *et al.*, 1976), we submitted 10 patients to i.v. injection of 0.6-14 mg BCG. There were no symptoms for 6-8 hours. The patients then developed chills, malaise, occasional vomiting, and high fever (40-41° C), which lasted for 24-48 hours in most subjects, and for 1 week in one. Only one patient presented no manifestations. Blood cultures for M. Bovis, made in the first hour, after injection, were all negative.

We have abandoned this route of administration, because not only did it not restore immunodepressed patients (see Table 6.XX) (SIMMLER *et al.*, 1976a, 1976b), but it induced, as seen in Table 7.XXVIII, immunodepression in normal skin responders (SCHWARZENBERG *et al.*, 1976), and probably growth enhancement in three patients (Fig. 3.50).

We administered Pasteur living fresh BCG *per os* at doses of 75 mg 3 times a week to 15 patients: in 13, tolerance was excellent; in 2, abdominal pain led to cessation of the treatment (SCHWARZENBERG *et al.*, 1976).

We have combined the administration of BCG on scarifications applied on four members and per os ingestion. There were no special side-effects (SCHWARZENBERG *et al.*, 1976).

THOMAS *et al.* (1974) have studied 7 patients at autopsy, who had been treated with BCG for malignant melanoma per os and on scarifications: they did not present any important pathological manifestations, which suggests that the dose introduced into the organism was small.

We also tried *aerosol* inhalation in three patients: chills were noted between the 4th and 8th hours, with fever lasting for several hours, as well as malaise, anorexia, and vomiting. The patients were symptom-free within 34-46 hours (SCHWARZENBERG *et al.*, 1976).

The manifestations of *BCG* applications may be more common when this agent is given *in association with chemotherapy:* SERROU *et al.* (1975) described a granulomatous hepatitis which was severe enough to demand an antituberculous treatment.

GORDON and EDYNAK (1975) studied lymphocyte stimulation with four mitogens before BCG application and suggest that low stimulation expresses a high risk of developing BCGitis, hence "contra-indicating the use of this adjuvant". I do not agree with these authors: (1) In our experience, the side-effects are no more important in immunodepressed patients than in subjects with normal reactions, and especially in PPD anergic patients than in tuberculin responders; (2) Table 6.XIX showed that BCG may restore immunodepressed subjects. However, we showed that relatively small doses may restore the skin reactions of the latter, while higher doses do not.

We have observed no more side-effects in patients receiving *BCG and tumour cells* than in those submitted only to mycobacteria. The experience of SPARKS *et al.* (1973) seems to be different: they have published the reactions observed in 34 patients who were submitted to both BCG and tumour cells i.d.: they complained of malaise more often (35%) than those receiving BCG alone (22%), of chills or sweating (21% versus 9%) and "change in performance" (18% versus 4%). The skin rash was, on the contrary, less frequent.

8.1.2.2.2. Mycobacterial Extracts

Complete Freund's adjuvant (CFA) which we attempted to use 15 years ago, before employing BCG, often induces severe local granulomata, complicated by ulcers with possible fistulas (SCHWARZENBERG *et al.*, 1976). Incomplete Freund's adjuvant, devoid of mycobacteria and used in humans on a large scale as a vehicle for influenza and cholera vaccines, presented the same side-effects (OGONUKI *et al.*, 1967, quoted by MARSAL *et al.*, 1975), which suggests that the severity of local granulomata induced by CFA are probably more related to mineral oil than to the bacilli.

MER injected i.d. is, according to WEISS (1972) and ROBINSON *et al.* (1975), well tolerated (Table 8.VII). The local reactions vary from mild and transient inflammation to limited granulomatas. RICHMAN (1975) mentions local pain in 19% of the patients and sterile abscesses in 55%. MOERTEL *et al.* (1975) have seen draining ulcerations of such lesions. Regional lymph nodes may be slightly enlarged for a short period. The general reactions are limited to mild fever (22% according to RICHMAN, 1975).

8.1.2.2.3. Corynebacteria

Applied *i.m. or s.c.* at a dose of 1-10 mg, C. parvum as well as C.granulosum (Table 8.VII), very often induce local pain from mild (ISRAEL and HALPERN, 1972) to severe in extent (MATHÉ, 1976; MATHÉ *et al.*, 1975b). The injections are followed by indurations in more than 50% of the patients in our experience, and HIRSHAUT *et al.* (1975) observed, after s.c. injection, chronic local reactions which could ulcerate and drain.

Fever is rare according to ISRAEL and HALPERN (1972); it is more frequent in our experience (25%), but of short duration (24 hrs). Nausea, vomiting, and headaches are not exceptional.

We find no difference between tolerance to C. parvum and that of C. granulosum (MATHÉ *et al.*, 1975b).

Applied simultaneously with BCG and/or poly I-poly C, they induce no more side-effects than when applied alone (personal experience).

The most extensive study of *i.v.* C. parvum toxicology is that conducted by HERSH *et al.* (1974) on patients with melanoma. Forty-four received 1-10 mg/m^2 C. parvum i.v. once or twice per month. Chills and fever $>$ 101° F developed in 87 and 75% of the patients. Transient mild hypotension was noted in 6% of the subjects. Among other side-effects were nausea and vomiting, and also headache in 55 and 24%. Toxic side-effects did not increase in frequency, intensity, and duration with increasing doses, although they could last for over 24 hrs. At 10 mg/m^2 i.v., 4 of 8 patients had "excessive" symptomatic toxicity.

Forty-one patients received 2-15 mg/m^2 weekly or monthly: soreness at the site of injections and mild fever $< 101^\circ$ F were observed in 100% and 90% of them. Other side-effects included local induration and/or nodule formation in 60% of the subjects.

There was, for both routes, no impairment of liver, kidney, or coagulation functions, except in one patient who presented a reversible thrombotic thrombocytopaenic purpura.

What must be underlined is that corynebacteria have been shown to inhibit T-lymphocytes (HOWARD *et al.*, 1973; SCOTT, 1974). Hence, enhancement of tumour growth must be feared when administering them, and a careful and repeated monitoring of the immune status of the patient is a necessary condition of their use. Figure 3.50 shows a case of enhancement observed in a melanoma patient after application of a combination of C. granulosum and BCG (not published).

8.1.2.2.4. Bordetella Pertussis and Brucella Abortus

These bacteria were not often used in man as immuno-potentiators, and their side-effects, at doses which would be necessary, are not known. The doses of the first organism which we used in a few attempts (personal observation) were not able to increase lymphocytosis significantly, as found in mice.

8.1.2.2.5. Polynucleotides

Poly I-poly C (Table 8.VII) has been fairly well tolerated in our trials involving 54 patients who were submitted to a dose of 1 mg/day i.v. for 1 month: 11 patients presented with fever $\geq 39^\circ$ C, dyspnea, and hypotension (MATHÉ *et al.*, 1970; and unpublished data). CORNELL *et al.* (1975), who administered i.v. 1 mg/kg, frequently noticed fever from 38-40° C lasting 24-72 hours. High doses administered in order to obtain interferon production, varying from 0.3-4 mg/kg, induced febrile reactions (CORNELL *et al.*, 1975), but no disturbances of liver, kidney, blood cells, and clotting functions were noted (FIELD *et al.*, 1971, quoted by MARAL *et al.*, 1976).

Poly A-poly U (Table 8.VII) was less well tolerated in our experience than poly I-poly C: fever, dysnea was observed in 6 patients out of 6. In a trial conducted by J. and F. Lacour (personal communication), it did not induce side-effects at doses of 3-13 mg once a week; it frequently produces fever at 38° C for 24 hours at a dose of 30 mg per injection.

8.1.2.2.6. Other Agents

Levamisole (Table 8.VII) induces some side-effects at doses of about 2 mg/kg: giddiness, fatigue, skin rashes (WEBSTER and HUGHES, 1975), nausea, loss of appetite

Table 8.VII. Main side-effects in man of adjuvants other than BCG and used in clinical active immunotherapy (MATHÉ, 1976)

Mycobacterial extracts	Corynebacteria		Polynucleotides		Levamisole
MER (i.d.)	(i.m.) 2-4 mg	(i.v.) 1-2 mg	Poly I-poly C	Poly A-poly U	
Local reaction varying from mild and transient inflammation to a limited granuloma	Local pain, local induration (> 50%)				Giddiness, Tiredness
Possible enlargement of regional lymph nodes	Nausea, vomiting headache, fever	Nausea: 55% vomiting: 24% fever: 75% hypotension:6%	Fever (11/54),	Fever,	Nausea, loss of appetite, Skin rashes
Possible transient febrile periods after repeated injection	(for all administrations if present after the first)	Reversible thrombotic thrombocytopenic purpura	Hypotension (11/54), Dyspnea (11/54)	Hypotension Dyspnea	Fever, convulsions, increase of libido
(WEISS, 1972)	(Personal experience, 1970-1975; ISRAEL and HALPERN, 1972)	(HERSH *et al.*, 1974)	(Personal experience) MATHE *et al.*, 1970)	(Personal experience)	(WEBSTER and HUGHES, 1975; AMERY, 1975; LIONEL *et al.*, 1969, quoted by MARAL *et al.*, 1976; BERTRAND *et al.*, 1974, quoted by MARAL *et al.*, 1976)

(AMERY, 1975), fever, convulsions (LIONEL *et al.*, 1969, quoted by MARAL *et al.*, 1976), temporary increase of libido (BERTRAND *et al.*, 1974, quoted by MARAL *et al.*, 1976).

Clinicians who may be tempted to prescribe *vitamin A*, largely as an immunopotentiator, must know that it is very toxic at doses slightly higher than the dietary normal dose (MAYGH, 1975, quoted by MARAL *et al.*, 1976). The hypervitaminosis A syndrome (BOLLAG, 1974, quoted by MARAL, 1976) comprises itching, loss of hair, headaches, and blurring of vision. One must also remember that it is teratogenic in animals (COCHLAN *et al.*, 1953, quoted by MARAL, 1976).

8.1.2.2.7 Combination of Several Adjuvants

We did not observe more side-effects when combining BCG on scarifications with C. parvum or C. granulosum i.m. or BCG± a corynebacterium and poly I-poly C. However, probable growth enhancement has been seen in a melanoma patient submitted to the combination of BCG on scarifications and C. granulosum administered i.m. (see Fig. 3.50).

8.1.2.3. Hormones

Thymosin. At doses between 1-30 $mg/m^2/day$ for 7 days, thymosin (fraction V extract of calf thymus) has been well tolerated in the trials which have so far been conducted: only local erythema was observed by SCHAFER *et al.* (1975). COSTANZI *et al.* (1975), injecting i.m. doses of 50-250 $mg/m^2/day$ for 7 days, noticed only a transient urticarial rash and detected no cardiopulmonary, hepatic, renal, or haemopoietic toxicity.

8.1.2.4. Other Means

Transfer factor has induced few side-effects: pain at the site of injection and erythema (FUDENBERG, 1974, quoted by MARAL *et al.*, 1976); transmission of viral hepatitis and transfer of adverse delayed hypersensitivity reactions are considered as possible by MARAL *et al.* (1976).

8.2. Local Active Immunotherapy

8.2.1 Experimental

A few authors have studied the pathological lesions in guinea pig (HANNA, 1974) and mouse (KHALIL *et al.*, 1975) tumours injected with BCG. They will be described in Chapter 9.

The general toxicity is about the same as that observed after i.v. injection of the same dose of BCG.

8.2.2. Clinical

The histological aspect induced by local intra-tumoral injections of *BCG* will also be described in Chapter 9.

Toxicology studies demonstrate that it is far from being a perfectly tolerated treatment (MATHÉ, 1976).

HUNT *et al.*(1973), SPARKS *et al.* (1973) and ROSENBERG *et al.*(1974) have observed many side-effects in patients treated with local injections of (Glaxo or Tice)BCG in the tumour: malaise, 100%; fever, 96%; chills or sweating, 80%; "change in performance", 64%; ulceration, 24%; they have described a "granulomatous hepatitis" with hepatic "dysfunction," increase of transaminases, and alkalin phosphatases. The diagnosis was made after percutaneous liver biopsy. They advise treating such manifestations, if they persist, with isoniazid.

Three deaths have been reported by McKHANN *et al.* (1975) and SCHWARZENBERG *et al.* (1976) in melanoma patients, following intralesional injection of BCG: in the three cases, death was preceded by high fever and hypotension.

Lyophilized BCG, which has been more generally used, is better tolerated than living preparations.

Most often the use of non-living agents other than BCG induce pain and intense erythematous reactions at the site of the tumour which, in the case of application of streptokinase and streptodornase, resemble erysipelas (STEWART and TOLNAI, 1969).

The intra-tumoural injection of *C. parvum* is usually well tolerated: HIRSHAUT *et al.* (1975) indicate that their maximal tolerated dose was of the order of 20 mg. Injected lesions usually became erythematous, and mildly painful.

I have mentioned earlier the possible *enhancement of non-injected visceral tumours* in patients, who had had a local skin nodule injection (LEVY *et al.*, 1972; BORNSTEIN *et al.*, 1973; LEVY, 1973).

References

AMERY,W.: Lancet 1, 574 (1975).
BALDWIN, R.W., PIMM, M.V.: Nat. Cancer Inst. Monogr. 39, 11 (1973).

BORNSTEIN, R.S., MASTRANGELO, M.J., SULIT, H., CHEE, D., YARBO, J.W., PREHN. L., PREHN, R.T.: Nat. Cancer Inst. Monogr. 39, 213 (1973).

CORNELL, C.J. jr., SMITH, K.A., CORNWELL, G.G. III, GRACE, W.R., McINTYRE, O.R.: Proc. Amer. Assoc. Cancer Res. 16, 265 (abstract No. 1178) (1975).

COSTANZI, J., GAGLIANO, R., LOUKAS, D., SAKAI, H., THURMAN, G., HARRIS, N., GOLDSTEIN, A.: Proc. Amer. Assoc. Cancer Res. 16, 135 (abstract No. 537) (1975).

GEFFARD, M., ORBACH-ARBOUYS, S.: Cancer Immunol. Immunoth. 1, 41 (1976).

GORDON, B.L., EDYNAK, E.M.: Proc. Amer. Assoc. Cancer Res. 16, 123 (abstract No. 489) (1975).

HANNA, M.G. jr.: Seminars in Oncology 1, 319 (1974).

HERSH, E.M., REED, R.C., GUTTERMAN, J.U., MAVLIGIT, G.M., BURGESS, A.A.: Symposium International sur "Les Immunostimulants Bactériens, Structures Chimiques, Mécanismes d'Action, Applications". Paris, Institut Pasteur, October 1974.

HIRSHAUT, Y., PINSKY, C., CUNNINGHAM-RUNDLES, W., RAO, B., FRIED, J., OETTGEN, H.: Proc. Amer. Assoc. Cancer Res. 16, 181 (abstract No. 721) (1975).

HOWARD, J.G., SCOTT, M.T., CHRISTIE, G.H.: p. 110 in "Immunopotentiation" (Ciba Foundation Symposium 18). 1 Vol., Amsterdam: Elsevier Excerpta Medica 1973.

HUNT, J.S., SPARKS, F.C., PILCH, Y.H., SILVERSTEIN, M.J., HASKELL, C.M., MORTON, D.L.: Lancet 2, 820 (1973).

ISRAEL, L., HALPERN, B.: Nouv. Presse Med. 1, 19 (1972).

JURCZYK-PROCYK, S., MARTIN, M., DUBOUCH, P., GHEORGHIU, M., ECONOMIDES, F., KHALIL, A., RAPPAPORT, H.: Cancer Immunol. Immunoth. 1, 55 (1976).

KHALIL, A., BOURUT, C., HALLE-PANNENKO, O., MATHÉ, G., RAPPAPORT, H.: Biomedicine 22, 112 (1975).

LEVY, N.L.: Nat. Cancer Inst. Monogr. 37, 85 (1973).

LEVY, N.L., MAHALEY, M.S. jr., DAY, E.D.: Int. J. Cancer 10, 244 (1972).

LOKICH, J.J.: Lancet 1, 331 (1975).

MacKHANN, C.F., HENDRICKSON, C.G., SPITLER, L.E., GUNNARSSON, A., BANERJEE, R., NELSON, W.R.: Cancer 35, 514 (1975).

MARAL, R., FLOC'H, F., JOUANNE, M., WERNER, G.H.: in "European Society of Toxicology". Amsterdam: Excerpta Medica, in press (1976).

MATHÉ, G.: in "European Society of Toxicology". Amsterdam: Exerpta Medica, in press (1976).

MATHÉ, G., AMIEL, J.L., SCHWARZENBERG, L., SCHNEIDER, M., HAYAT, M., VASSAL, F. DE, JASMIN, C., ROSENFELD, C., SAKOUHI, M., CHOAY, J.: Europ. J. Clin. Biol. Res. 15, 671 (1970).

MATHÉ, G., HALLE-PANNENKO, O., AMIEL, J.L.: p. 48 in "Corynebacterium parvum: application in Experimental and Clinical Oncology" New-York: Plenum press, 1975b.

MATHÉ, G., HALLE-PANNENKO, O., FLORENTIN, I., BRULEY-ROSSET, M., KAMEL, M., HIU, I.J., BOURUT, C.: Europ. J. Cancer 11, 801 (1975a).

MOERTEL, C.G., RITTS, R.E., SCHUTT, A.J., HAHN, R.G.: Proc. Amer. Assoc. Cancer Res. 16, 143 (abstract No. 569) (1975).

MOORE, M., LAWRENCE, N., WITHEROW, P.J.: Europ. J. Cancer 10, 673 (1974).
MOTTA, R.: Etude des antigènes de surface des cellules normales et leucémiques de souris et de leurs relations avec le système immunologique de l'hôte. Thesis, Université Paris VII, 1974.
MUGGLETON, P.W., PRINCE, G., HILTON, M.: Lancet 1, 1353 (1975).
PORTELANCE, V.: Colloque de Pédiatrie, Préventorium Calmette, Yerres, France, 1974.
QUEVILLON, M.: Cancer Immunol. Immunoth. in press (1976).
RICHMAN, S.P.: Proc. Amer. Assoc. Cancer Res. 16, 227 (abstract No. 1025) (1975).
ROBINSON, E., HAASZ, R., BARTAL, A., COHEN, Y.: 9th International Congress of Chemotherapy, London (abstract C 82) July 1975.
ROSENBERG, E.B., KANNER, S.P., SCHWARTZMAN, R.J., COLSKY, J.: Arch. Intern. Med. 134, 769 (1974).
SCHAFER, L.A., WASHINGTON, M.L., GOLDSTEIN, A.L.: Proc. Amer. Assoc. Cancer Res. 16, 233 (abstract No. 1051) (1975).
SCHAUF, V., LANDAU, W., FALK, L.: Proc. Amer. Assoc. Cancer Res. 16, 102 (abstract No. 406) (1975).
SCHWARZENBERG, L., SIMMLER, M.C., PICO, J.L.: Cancer Immul. Immunoth. 1, 69 (1976).
SCOTT, M.T., Cell. Immunol. 13, 251 (1974).
SERROU, B., MICHEL, H., DUBOIS, J.B., SERRE, A.: Biomedicine 23, 236 (1975).
SIMMLER, M.C., SCHWARZENBERG, L., MATHÉ, G.: Nouv. Presse Med. in press (1976a).
SIMMLER, M.C., SCHWARZENBERG, L., MATHÉ, G.: Cancer Immunol. Immunoth. in press (1976b).
SPARKS, F.C., SILVERSTEIN, M.J., HUNT, J.S., HASKELL, C.M., PILCH, Y.H., MORTON, D.L.: New England J. Med. 289, 827 (1973).
STEWART, T., TOLNAI, G.: Cancer 24, 117 (1969).
SUTER, E., MUNOZ, J.J.: Proc. Soc. Exp. Biol. Med. 114, 211 (1963).
SUTER, E., ULLMAN, G.E., HOFFMAN, R.G.: Proc. Soc. Exp. Biol. Med. 99, 167 (1958).
THOMAS, J.W., PLENDERLEITH, I.H., CLEMENTS, D., LANDI, S.: p. 315 in XIth International Cancer Congress, Florence, October 1974. Vol. 1, Conferences, Symposia workshops.
WEBSTER, D.J. T., HUGHES, L.E.: Lancet 1, 389 (1975).
WEISS, D.W.: Nat. Cancer Inst. Monogr. 35, 157 (1972).

Chapter 9
Mechanisms of Favourable and Adverse Effects and of Failures

9.1. Systemic Immuno-Intervention

9.1.1. Specific Stimulus

The mechanism of action of tumour-associated antigens administered on or in the cells or in their extract(s) have been discussed in Chapter 2.

We have also seen that cells or extracts may induce protection against tumour(s) as well as protection of the tumour(s), depending on the dose, at least of the cells (SEDLACEK *et al.*, 1975), or on the method of preparing the extract (MARTYRE *et al.*, 1974). Such data indicate that there are many factors which govern the direction of immunomodulation induced by a specific stimulus, and the problem is probably similar to that of histocompatibility antigens, the administration of which may induce rejection or acceptance of the graft, depending on different factors (HALLE-PANNENKO and MATHÉ, 1976). The mechanism of these factors is far from being explained.

9.1.2. Non-Specific

9.1.2.1. Adjuvants

Many workers have devoted their time to the mechanisms underlying adjuvant action. However, an overall view of the work done in this field from the pioneering stage (FREUND, 1951, 1953) until recently (Ciba Foundation Symposium on Immunopotentiation, 1973) leads us to conclude that there is little agreement about the mechanism of action. The great diversity of agents which possess adjuvant properties is mainly responsible for the confusion and the absence of a unified concept.

9.1.2.1.1. Historical Adjuvants

The first studies concerned agents incorporated with the antigenic material, the best known and the only one quoted in this monograph being Freund's complete adjuvant (FCA) (see Table 7.V) (FREUND, 1951, 1953), aluminium phosphate (RAMON, 1926), silica (PERNIS and PARONETTO, 1962), and bacterial endotoxins (JOHNSON *et al.*, 1956).

FREUND (1951, 1953) suggested that, in the case of his agent, such adjuvants may exert several actions: (a) protraction of the antigenic stimulus as a result of the slower release of the antigen from its depot in the water-in-oil emulsion, (b) stimulation of the macrophagic (reticulo-endothelial) system, the action of which has been demonstrated by RUPP *et al.* (1960), (c) stimulation of the immune response, which can, today, be translated as meaning stimulation of lymphocytes, as has been shown by UNANUE *et al.* (1969) and ALLISON and DAVIES (1971).

With regard to the mechanism of action of insoluble particles of aluminium phosphate or silica, the adsorption of the antigen by them was the mechanism usually quoted (RAMON, 1926), but later, it was shown that they act also on lymphocytes and macrophages: the spleen and lymph nodes are enlarged up to 10 times normal size after i.v. injections of silica into rabbits (PERNIS and PARONETTO, 1962); an enormous number of plasma cells can be seen at the site of injection of aluminium hydroxyde (MATHÉ *et al.*, 1963).

All of these studies, in fact, show that the chemical adjuvants act on all the elements involved in the elicitation of the immune response: antigen, lymphocytes, macrophages.

9.1.2.1.2. Systemic Action of Adjuvants

Effects of Adjuvants at the Antigen Level. The role of adjuvants at the antigen level has recently been analysed by JOLLES and PARAF (1973) as follows: Adjuvants may: (a) slow down the catabolism of the antigenic molecule, by retaining it in the focal granuloma, and/or by changing the antigen localization; (b) modify the antigenic part by changing the net electric charge or by altering the size of the specific site (conformational change); (c) transform a hapten (the antigenic site) into an immunogenic molecule, or make a poor immunogen more immunogenic by conferring adjuvanticity upon the antigen.

The last mechanism of action seems important according to DRESSER and PHILLIPS (1973) who wrote that "an antigen which is capable of eliciting an immune response without the need of additional adjuvants can be considered as possessing its own intrinsic adjuvanticity" (they call it an "immunogenic antigen"); in contrast, a non-immunogenic antigen will elicit a state of specific immunological paralysis ("tolerance"), unless the antigen is polymerized, or a substance with adjuvant-like properties (extrinsic adjuvanticity) is injected at the same time as the antigen (DRESSER, 1961, 1962; TAUB *et al.*, 1970). Classical adjuvants, therefore, can be considered as being substances with extrinsic adjuvanticity (GOLUB and WEIGLE, 1967; DRESSER 1968a, 1968b; FINGER *et al.*, 1968), and whatever other activities a substance with such adjuvanticity may have in an animal injected with a non-immunogenic antigen, the most important is to prevent the antigen from inducing immunological paralysis. It is possible that an antigen-sensitive cell requires two stimuli or signals to start dividing and differentiating into cells that produce antibodies (BRETSCHER, 1972). A cell which received only the specific signal resulting from the binding of monomeric or monovalent antigen becomes "paralysed" or is killed (DRESSER, 1962). The two signals necessary for the induction of immune differentiation might be provided either by an antigen-sensitive cell binding an antigen which possesses intrinsic adjuvanticity, or by such a cell binding antigen and, at the same time, receiving an "insult" (from an extrinsic adjuvant) resulting in a response which is an essential first step towards immune differentiation.

This interpretation concerns the action of an adjuvant producing an immunological response for an antigen which normally induces paralysis (or tolerance). It has been demonstrated in several experiments: deaggregated human gamma-globulin, which normally induces tolerance in mice, can induce humoral immunity if endotoxin (see Table 7.V) is injected during the 48 hours following the antigen injection (LOUIS *et al.*, 1973). An anti-TNP response to the hapten TNP-glycine can be obtained in vitro when endotoxin is present (WATSON *et al.*, 1973).

This action may be different from an adjuvant effect, increasing the level of the immune response. The difference may be at the level of adjuvant-like "insult" applied to the cell, or in the interval between the time of the adjuvant action and that of the antigen contact with the cell.

It is this last action, the amplification of the immune response, that we are looking for in cancer immunotherapy, or to be more precise, the amplification of the rejection branch of the immune balance.

Besides this action on the antigen, some adjuvants have intrinsic properties which can be expressed either independently of the action on the antigen, or superimposed onto it.

Effects on Cell Kinetics. DRESSER (1970) suggested that all adjuvant activity is mediated through the stimulation of cell division. Even if this mechanism is not the only one, it is surely an important one.

This mechanism is suggested by the increased volume and weight of lymphoid organs, which can be multiplied by ten (PERNIS and PARANETTO, 1962). It has been demonstrated by the monitoring of DNA synthesis (DUTTON and MISHELL, 1967). KOGA *et al.* (1969), using the JERNE *et al.* (1963) technique, showed a significant increase in the number of sheep red blood cell (SRBC) antibody-forming lymphocytes compared to controls identically immunized with SRBC. This test is of such significance that we use it in our screening of systemic immunity adjuvants (MATHÉ *et al.*, 1973, 1975). With the results obtained with this test, JOLLES and PARAF (1973) have classified adjuvants into two categories: (a) those which do not induce lymphocyte multiplication, e.g., phytohaemagglutinin, concanavalin A, or other lectins (see Table 7.V); in fact, the efficiency of lectins, like PHA, is not very well established; they either enhance or suppress humoral immunity and have no effect, or a suppressive effect, on cell-mediated immunity (LANDY and CHESSIN, 1969); (b) those which induce an intensive cell proliferation, such as silica, B. pertussis (see Table 7.V) and Freund's adjuvant.

At what stage do lymphocytes divide? According to DRESSER and PHILLIPS (1973), possible different effects of adjuvants may be due to the stimulation of division of cells, which are at different stages of differentiation. The increase of division may involve uncommitted precursor cells or committed cells that are already in the process of differentiation.

Studies made on B. pertussis showed that this adjuvant acts on the proliferation of antibody-forming cells. This could be due to a lengthening of the period during which the increase of PFC is taking place (DRESSER, 1970), or an increased rate of division of responding cells (ROWLEY *et al.*, 1968). It does not seem, according to ROWLEY *et al.* (1968) that an individual cell could synthetise more antibody when an adjuvant is administered.

Curiously enough, while it is suggested histologically (KHALIL *et al.*, 1975) that the number of macrophages increases in lymphoid organs after the administration of adjuvants, or at least most of them, a finding which is in good correlation with the intensification of phagocytic activity (BIOZZI *et al.*, 1953), there is no clear explanation of this phenomenon. There may be chemotaxis from other sites or resources of the body, and there is probably an acceleration of differentiation which will be considered later. However, no work has been devoted to the hypothesis of an increase in the division of the cells called "mononuclear macrophages" (VAN FURTH *et al.*, 1972). This accelerated multiplication would take place at the level of the promonocytes or of cells between the latter and stem cells.

Stem cell kinetics can also be influenced by adjuvants. Such an effect can be demonstrated at the mononuclear cell level in mice, and granulocytic cell level in man.

Mice given BCG i.v. 1 day after cyclophosphamide administration (250 µg/kg) showed a blood restoration of the mononuclear cell population (lymphocytes plus monocytes, which are so difficult to differentiate that it is more reasonable to consider them together), that was accelerated significantly compared to controls receiving no BCG. Furthermore, their number was higher than that reached by controls at the time of restoration and higher than the level of the mononuclear cell population in the non-treated animals (Fig. 9.1) (MATHÉ, 1974). A similar effect on granulocytic population is observed in man as was illustrated by Fig. 3.48.

This suggests that BCG acts on stem cells at a less differentiated level. According to POUILLART *et al.* (1975, 1976), BCG increases the number of stem cells in S phase (Table 9.I).

Similarly, it has been shown that complete Freund adjuvant can stimulate the multiplication and affect the distribution within the animal of cells which form granulocytic and monocytic colonies in vitro (McNEILL, 1970). Poly I-poly C, at a dose of 100 µg per mouse, produces an eight-fold increase in spleen colony-forming cells after 6 days. The mechanism of this action is not clear. Poly A-poly U gives only a negligible cellular response (McNEILL and KILLEN, 1971).

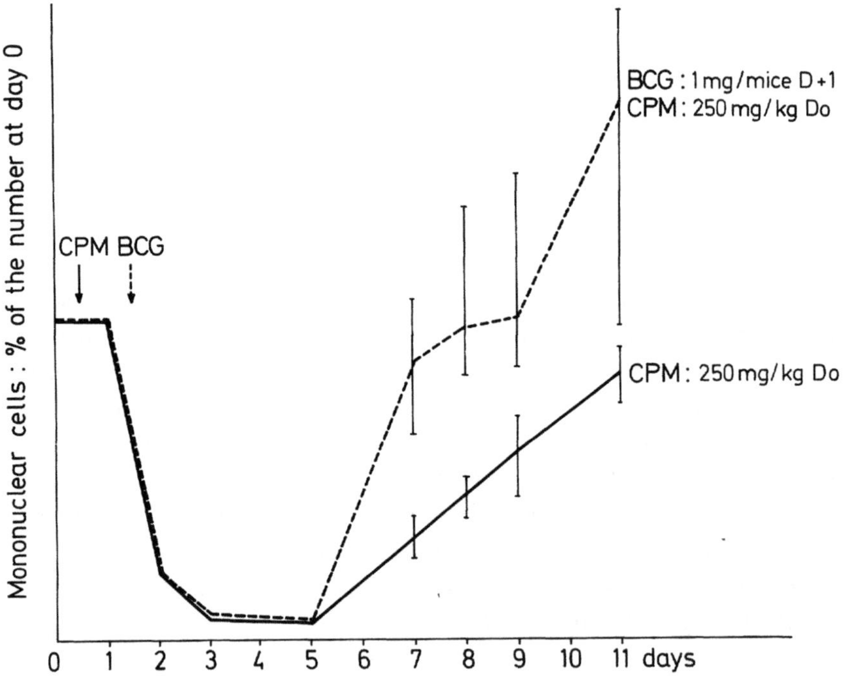

Fig. 9.1. Effect of BCG on lymphoid stem cells: the restoration of blood lymphocyte number after cyclophosphamide (CPM) aplasia is faster when BCG is administered 1 day after cytostatics (MATHE, 1974)

Table 9.I. Variations of the total number of CFU/femur after i.v. injections of 1 mg of fresh BCG to (DBA/2xC_{57}Bl/6)F1 mice (POUILLART *et al.*, 1975, 1976)

	Hours after i.v. injection 0	12	24	36	48	72
CFUs/femur	335	630	410		350	470
CFUs/femur after H_3TdR suicide technique	325	560	260		280	310
CFUa/femur	1090 ± 40	1524 ± 95	1410 ± 75	1540 ± 90	1620 ± 45	1910 ± 120
CFUa/femur after H_3TdR suicide technique	904 ± 54	1030 ± 36	1060 ± 60	1080 ± 45	1110 ± 60	1240 ± 60

CFUs : colony forming unit in the spleen

CFUa : colony forming unit in culture on agar

H_3TdR: tritiated thymidine

The CFUs are more sensitive to H_3TdR after 2 days, hence a higher fraction of them are in S phase.

The number of CFUa increases after 12 hours, and a third of them are, after this time, sensitive to H_3TdR, hence in S phase.

C. parvum was found to enhance the colony-forming capacity. This is attributed to its action on precursor cells capable of forming colonies in the bone marrow culture medium. This property is not particular to adjuvants: antigens, like red blood cells, and flagelline can enhance colony formation in vitro (McNEILL, 1970). Bovine serum albumin is not active in in vitro systems, but is very slightly active in vivo at high concentrations.

Latex (see Table 7.V) which is not antigenic, but which is known to undergo phagocytosis, is also able to stimulate the phagocyte colony-forming cell response (McNEILL, 1970).

This action of adjuvants on stem cells may result from two mechanisms which have to be elucidated by systematic studies: a direct stimulation of colony-forming cells (CFC); and an indirect stimulation of CFC by release of colony-stimulating factor(s).

Effect on Trapping. The trapping phenomenon has already been quoted to explain, at least partly, the increase of macrophages in lymphoid organs.

It has been shown that within the antigen-stimulated lymphoid organ, one of the earliest events in immunization may be the sequestration and recruitment of lymphocytes (HALL and MORRIS, 1963; FORD, 1969a, 1969b; ZATZ and LANCE, 1971; FROST and LANCE, 1973). This is true for particulate or high molecular weight substances, such as keyhole limpet haemocyanin, sheep red blood cells, Salmonella "H", pneumococcal polysaccharides, skin allografts, heterospecific grafts and a wide variety of tumours, whereas proteins, and also dinitrophenol (DNP) polylysine are ineffective (FROST and LANCE, 1973).

Adjuvants of very different kinds induce trapping. They differ from antigens in two ways: (a) the antigeniticy is not necessary. Substances like alum, vitamin A, alcohol (DRESSER, 1968; TAUB *et al.*, 1970), latex particles, carbon, silica, and carrageenan (see Table 7.V) (RIOS and SIMMONS, 1972), which have no antigenicity, induce trapping; (b) the duration of lymphocyte sequestration by adjuvants is much longer than that induced by antigens (FROST and LANCE, 1973).

Two questions arise: (a) Is the trapping an explanation for adjuvanticity? FROST and LANCE (1973) considered the possibility that adjuvants confer immunogenicity on "weak" antigens by initiating a potent sequestration of lymphocytes. They tested the immune response (antibody formation and anaphylaxis) of mice given different adjuvants (silica, carbon, complete Freund adjuvant) (see Table 7.V) and a weak antigen, the bovine gamma-globulin (BCG), which is an ineffective trapper; they also checked whether animals would trap lymphocytes after a second injection of

BCG. They found a positive correlation between the elicitation of an immune response and the trapping.

This study concerns local adjuvants, and we have seen that many adjuvants, if not all, can be effective when not physically mixed with the antigen, and even when the two are administered through different routes (DRESSER, 1961, 1968b; MATHÉ, 1968; MATHÉ *et al.*, 1969): we call them "systemic adjuvants" (MATHÉ, 1971). FROST and LANCE (1973) injected mice with complete Freund's adjuvant and bovine gammaglobulin (B γG); 1 day later there was no trapping. But after subsequent challenge with B γG, trapping occurred; the latter was the consequence of adjuvant action, since two injections of B γG without adjuvant do not induce trapping. This apparently paradoxical result is probably explained by the nature of B γG, which is rapidly diffused in the extra-cellular fluid compartment where it has a long half-life, so that the node draining the site of the s.c. administered adjuvant could serve as the organ of immunization with subsequent spread.

Trapping efficiency of BCG was studied in our laboratory by FLORENTIN *et al.* (1976). It was only observed with BCG given s.c., in the satellite lymph node of the BCG injection area. It began at day 8 after the BCG injection and increased progressively until day 35, when it reached 300% (Fig. 9.2). Surprisingly, BCG given i.v.

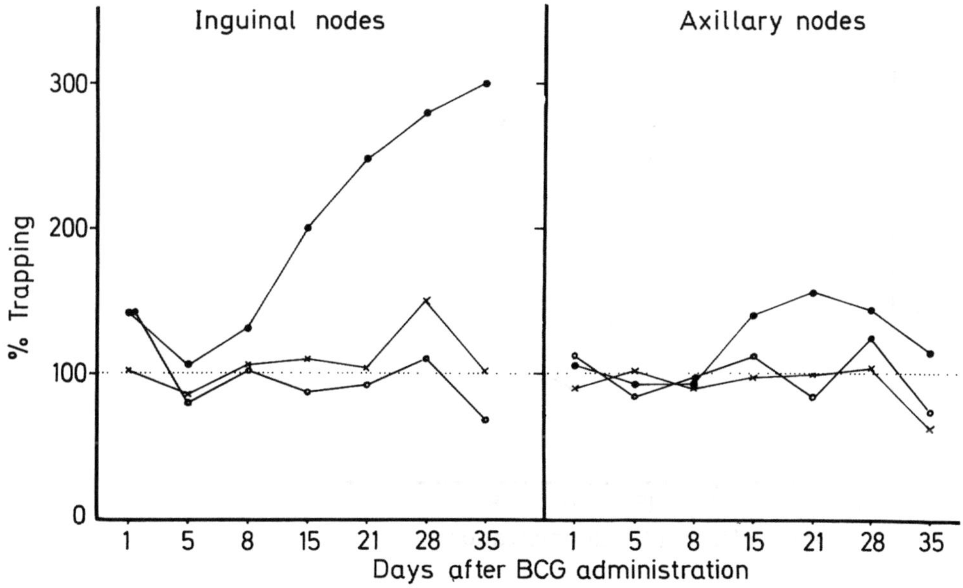

Fig. 9.2. Effect of time and route of BCG administration on trapping of ^{51}Cr *labelled isogeneic lymph node cells in the inguinal and axillary nodes.* x——x *BCG given intravenously;* •——• *ipsilateral node of subcutaneous BCG injection site;* o——o *contralateral node of subcutaneous BCG injection site (FLORENTIN et al., 1976)*

was not found to have any trapping efficiency in the spleen. Furthermore, it has a negative effect on the homing of lymphocytes in the bone marrow (Fig. 9.3). The significance of such a finding remains obscure.

The trapping power of other adjuvants according to FROST and LANCE (1973) is given in Table 9.II.

(b) What is the mechanism of lymphocyte trapping? FROST and LANCE (1973) have put forward data which strongly suggest that the cells responsible for the trapping phenomenon are not lymphocytes. They submitted the animals to several kinds of treatment (total body irradiation, corticosteroids, antithymocyte serum, cyclophosphamide, all of which have the ability to deplete lymphocyte populations, the first two depleting all types of lymphocytes, the third the T cells, and the fourth, predominantly the B-cells: the magnitude and kinetics of the trapping were not affected.

They wondered, therefore, if the responsible cells were not macrophages, as has already been suggested by FREI *et al.* (1965), DRESSER and MITCHISON (1969), and

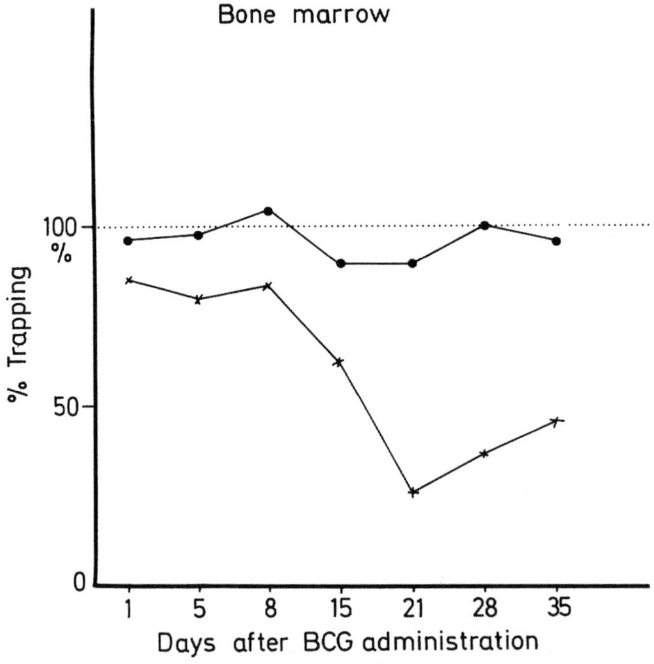

Fig. 9.3. Effect of time and route of BCG administration on trapping of ^{51}Cr labelled isogenic lymph node cells in the bone marrov.
x——x *BCG given intravenously;* •——• *BCG given subcutaneously (FLORENTIN et al., 1976)*

Table 9.II. Classification of adjuvants according to their potency to cause trapping (FROST and LANCE, 1973)

	Potency [a]	
C. parvum	4	FROST and LANCE, 1973
B. pertussis	3	DRESSER *et al.*, 1970
Vitamin A alcohol	2.5	FROST and LANCE, 1973
Complete Freund adjuvant	2.5	FROST and LANCE, 1973
Vitamin A palmitate	1.5	FROST and LANCE, 1973
Incomplete Freund adjuvant	1	FROST and LANCE, 1973
Alum particles	1	DRESSER *et al.*, 1970
Paraffin oil	0	FROST and LANCE, 1973
BCG	3	FLORENTIN *et al.*, 1976

[a] Defined as maximum trapping achieved.

UNANUE *et al.* (1969). They were impressed (a) by the fact that trapping is induced with particulate and high molecular weight antigens (such as sheep red blood cells, pneumococal polysaccharide, grafts or their extracted antigens), antigens which may need the macrophages for the reaction of immune response (UNANUE and CEROTTINI, 1970); (b) by the production of trapping by non-immunogenic particules which are intensively phagocytized (silica, carbon, carrageenan, latex, etc.); (c) by the absence of trapping after injection of bovine gammaglobulin (B γ G) (FROST and LANCE, 1973). Moreover, in an attempt to transfer trapping adoptively from mice immunized against B γ G to non-immunized ones in response to this antigen, they observed that peritoneal cells were more efficient than splenic cells, and that, among the latter, adherent cells were more effective than non-adherent ones.

SMITH and WOOD (1949) and FLOREY (1969) have suggested that adjuvants which can induce trapping could exert an effect on the endothelial cells of the efferent sinuses which swell and block lymph flow. According to FROST and LANCE (1973), the phagocytic properties of these cells, which are known to be poor (VAN FURTH *et al.*, 1972), make it improbable that they act as the initiating agent of trapping, and the data on adoptive transfer argue against their primary role.

The precise site at which lymphocytes are sequestered within lymphoid tissues has not yet been described. The phenomenon may be similar to the "plugging" described by DE SOUSA and PARROT (1969).

Cell Functions and Cell Interactions. (a) The effect of adjuvants on a function probably assumed by *macrophages,* e.g. the initiation of trapping has just been described.

Phagocytosis is one of the most important functions of this family of cells, the classification of which is still difficult (VAN FURTH *et al.*, 1972). This function is increased by adjuvants, as was initially shown by BIOZZI *et al.* (1953, 1955), and can be measured, for example, by carbon clearance (see JOLLES and PARAF., 1973).

When the adjuvant is particulate, it is easy to demonstrate its presence in macrophages by light or electron microscopy: this has been the case with B. pertussis, mycobacteria in Freund's complete adjuvant, and silica particules (see Table 7.V) (SPITZNAGEL and ALLISON, 1970).

Some soluble adjuvants form complexes with serum constituents which are taken up by macrophages: beryllium sulfate (see Table 7.V) combines with phosphates to form complexes which are taken up by phagocytes; those complexes can then exert toxic action on these cells (ALLISON, 1968) and/or increase antibody formation (UNANUE *et al.*, 1969). Other adjuvants such as retinol, (see Table 7.V), combine with serum proteins (ALLISON, 1973).

Is the uptake of adjuvant(s) by macrophages necessary? The response is positive for certain agents. UNANUE *et al.* (1969) took macrophages containing Maia squinado haemocyanin (MSH) in mice: they treated them in vitro with B. pertussis and injected them into other mice; they elicited, in the latter, a higher antibody titre than they did in the absence of adjuvant, and the increase of antibody was related to the dose of pertussis organisms. The conclusion of the author is that, in their system, the adjuvant exerted its stimulatory effect on antibody-producing cells after the uptake by macrophages (ALLISON, 1973). Macrophages treated with adjuvants and injected at the same time as other macrophages containing the antigen also increased the immune response; in other words, the adjuvant action did not require, at least in their model, the double presence of adjuvant and antigen in the same macrophages (same authors).

Hence macrophages do more than act as passive carriers of antigens. According to ALLISON (1973), adjuvants alter the balance between immunization and "tolerance" in favour of the former (perhaps by limiting the amount of free antigen that diffuses through the tissues); only within certain limits (as he demonstrated, by increasing the dose of free bovine serum albumin) is the amount of antibody formed dependent on the dose of antigen.

According to this author, another role of macrophages is to increase the immunogenicity by presenting antigen to lymphocytes in a way which favours immunity.

This concept is interesting but it does not explain our results on macrophage non-specific cytotoxicity, which have been shown above: peritoneal macrophages from mice given BCG i.v. or Hiu II i.v. became highly cytotoxic to tumour cells although the donor mice were never immunized against them (Fig. 7.6) (MATHÉ *et al.*, 1975).

KHALIL (see BRULEY-ROSSET *et al.*, 1976), in our laboratory, has shown that the macrophages of these animals contain a higher quantity of acid phosphatase than normal ones (Fig. 9.4). This is in addition to the observations of SAUTER and GIFFORD (1966) on the effect of endotoxins (which render macrophages cytotoxic) (ALEXANDER and EVANS, 1971) on lysosomal membranes, and those of SPITZNAGEL and ALLISON (1970b), who found that vitamins A and E, and E. coli lipolysaccharide (see Table 7.V) are able to labilize these membranes.

The effect we have seen probably corresponds to the transformation of the mononuclear cell designated in Figure 2.11 as a "histiocyte" into the so-called "activated histiocyte" (MACKANESS, 1970; NATHAN *et al.*, 1971; MATHÉ *et al.*, 1972).

JUY *et al.* (1974) have also shown that macrophages can be activated in vitro and in vivo by Freund's adjuvant, endotoxin from Salmonella enteridis (WESTHPHAL *et al.*, , 1952), and Neo-WAS (ADAM *et al.*, 1973) (see Table 7.V).

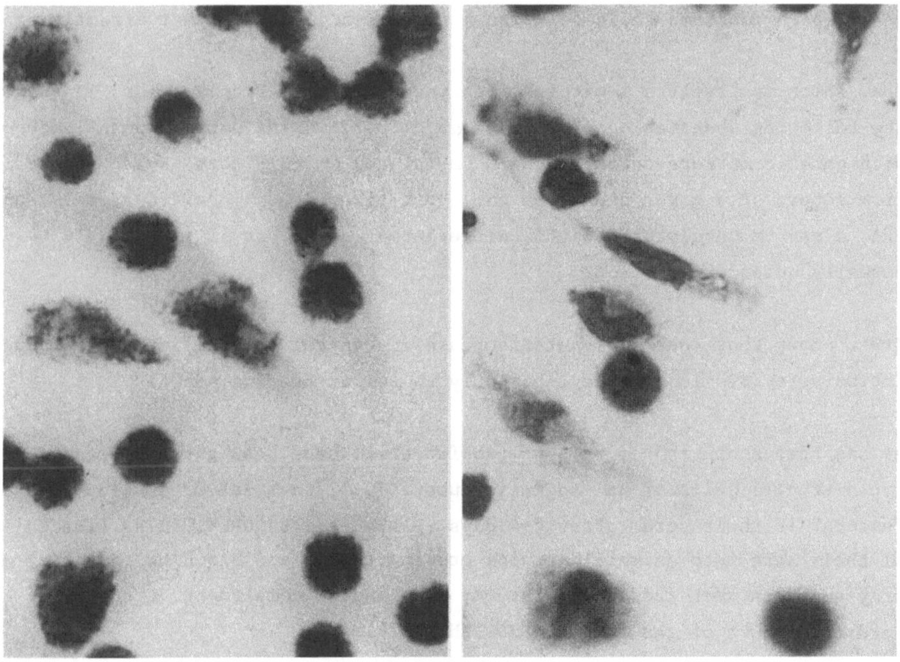

Fig. 9.4. BCG effect on phosphatase activity of macrophages: left: macrophages, right: controls from BCG-treated mice (Kindly provided by A. KHALIL)

It has been shown that macrophages can produce a lymphocyte activating factor (LAF) which potentiates the response of T cells to PHA. LAF production is greatly enhanced by endotoxins (GERY and WAKSMAN, 1972), and by MER (GERY and WAKSMAN, 1972).

A direct action of adjuvants on macrophages was recently illustrated by PIMM and BALDWIN (1975) who obtained an antitumour effect in athymic nude mice with systemic immunotherapy with BCG or C. parvum.

(b) Adjuvants not only work on macrophages; they (at least most of them) also act on *lymphocytes*.

There are some which undoubtedly stimulate *B-cells*. This is the case with lipopolysaccharide (LPS) from E. coli which is, at the same time, a thymus-independent antigen, a mitogen for B- and not for T-lymphocytes, and a B-adjuvant: MÖLLER *et al.* (1972) observed that LPS, added to sheep red blood cells, converts them from a thymus-dependent antigen to a thymus-independent one.

Stimulation of B-cells by C. parvum has been suggested by HOWARD *et al.* (1973a) who showed the increased immune response to pneumococcal polysaccharide, a B-dependent antigen, in the presence of this organism. According to DRESSER and PHILLIPS (1973), a high concentration of antigen plus B. pertussis can bypass a dependence on T-cell help, which confirms the conclusion that this adjuvant can act directly on B-cells.

PHA, which is itself a T-mitogen, can interfere with B cells. This is suggested by the following observation of LOBB *et al.* (1967): using normal human lymphocytes with genetic markers on the H-chain Gm (a) and Gm (b), they showed that cells taken from donors of a given phenotype can synthetize, after in vitro stimulation with PHA, a new immunoglobulin with a different allotype, as if it was able to derepress normally quiescent cistron.

It is known that most adjuvants increase in general humoral antibody production, including cytotoxic antibody, but only in tumour-bearing animals.

As has been shown above, rats or rabbits which have been given BCG and then mitomycin-treated cells of an isogeneic tumour 10-12 days later, develop "unblocking factors" in their serum after 2-4 days (BANSAL and SJÖGREN, 1972). The injection of their sera into animals carrying growing tumour and blocking factor(s) may not only suppress such factor(s), but may also induce complement cytotoxicity and retard the growth of the tumour (SJÖGREN, 1974).

Antibody Dependent Cell Mediated Cytotoxicity (PERLMANN and HOLM, 1969), the effector cells of which are B-lymphocytes (for the authors who described the phenomenon) and special (K) cells of unknown nature (ALLISON, 1973), can be potentiated by adjuvants. Freund's adjuvants increase effector cell activity in the spleens of treated mice (ALLISON, 1973).

These are examples of adjuvants which can stimulate B-cells directly. Most of them act on *T-lymphocytes*.

These cells have many functions, and many tests can be utilized to demonstrate that an adjuvant acts on them. TAUB *et al.* (1970) demonstrated that vitamin A induces a proliferation of cells in the thymus-dependent areas of the draining lymph node. KHALIL *et al.* (1975) described lymphoid hyperplasia of the T-dependent zones of the nodes and the spleen after i.v. injection of BCG (see Fig. 8.1).

Vitamin A, which is a non-immunogeneic adjuvant, induces a blastic transformation of lymphocytes in vitro (TAUB *et al.*, 1970). Poly A-poly U increases the incorporation of thymidine in lymphocytes (from sensitized subjects) exposed in vitro to tuberculin, and increases the mixed lymphocyte response. ALLISON (1973) has shown that this effect requires the presence of macrophages in the culture.

The action of adjuvants on T-cells is interesting because of their two main functions: the T-lymphocyte "helper" effect, and their role in cell-mediated immunity.

ALLISON and DAVIES (1971) have shown that more, or more active, T-lymphocytes give B-cells a greater, or more prolonged, "helper" effect. DRESSER and PHILLIPS (1973) used thymectomized reconstituted (B) mice and sham-operated controls with or without B. pertussis. They observed that the adjuvant effect of the latter was essentially mediated directly through the B-lymphocytes for high doses of sheep red blood cells (SRBC), and through T-cells for low doses of this antigen. The same experiment was conducted with lentinan (CHIHARA *et al.*, 1969, 1970) (see Table 7.V). DRESSER and PHILLIPS (1973) concluded that this adjuvant acts through T-lymphocytes, even at high doses of antigen.

This adjuvant has been shown to potentiate a second important function of T-cells, i.e. cell-mediated immunity (MAEDA and CHIHARA, 1971; MAEDA *et al.*, 1971). Similarly, Freund's complete adjuvant is well known for its role in enhancing the development of delayed hypersensitivity against serum proteins, while incomplete Freund's adjuvant, C. parvum, beryllium, and vitamin A (see Table 7.V) do not do so (ALLISON, 1973).

With KHALIL *et al.* (1975), we have shown that mice treated with fresh living BCG present a hyperplasia of the thymus cortex and medulla epithelial cells beginning at day 6 and that the hyperplasia is more pronounced after i.v. than after s.c. injection. Many of these cells have a PAS-positive cytoplasm and some are contiguous with the wall of small vessels and large cystic spaces containing PAS-positive secretions (Fig. 9.5). According to our hypothesis, this secretion may be the thymus lymphocyte stimulating factor (thymosin). Moreover, the thymic cortical lymphocytes show a pronounced pyroninophilia and increased mitotic activity without increase of thymic weight (Fig. 9.6).

Suppressive Effects of Adjuvants. Adjuvants can have suppressive effects, according to the type of protocol used for their study.

When B. pertussis is injected 1-3 days before the antigen (SRC) there is a slight suppression of the specific response (DRESSER *et al.*, 1970). Such suppression has been attributed to antigenic competition. BCG, administered at higher doses than

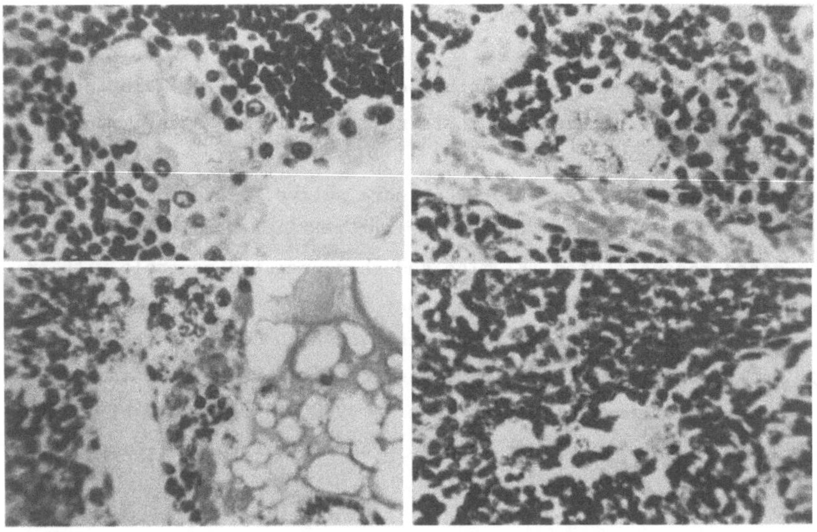

Fig. 9.5. (1) Thymus (cortico-medullary junction) of uninjected control animal (for comparison). Note small cluster of epithelial cells adjacent to blood vessel (H & E x 750). (2) Thymus (cortico-medullary junction) of animal killed on day 6 after i.v. injection of BCG. Epithelial hyperplasia adjacent to cystic space containing secretions (H & E x 750). (3) Thymus (cortico-medullary junction) of animal killed on day 6 after i.v. injection of BCG. PAS stain revealed numerous PAS-positive epithelial cells lining a cystic space filled with PAS-positive secretions (PAS stain x 750). (4) Thymic cortex of uninjected control animal. Only occasional PAS-positive cells are evident (PAS stain x 750)(KHALIL et al., 1975)

1 mg, produces suppression of anti-SRC response (Table 7.XXIV) (MATHÉ, 1976a) and an inhibition of the GVH reaction (GEFFARD and ORBACH-ARBOUYS, 1976). In this latter case, it was shown that the inhibition is due to suppressor T-cells. On the other hand, Mitchell has shown that BCG could elicit a suppressive function on macrophages (MITCHELL, 1976).

C. parvum injected i.v. leads to a depression of cell-mediated immunity due to the activation of macrophages (SCOTT, 1972a).

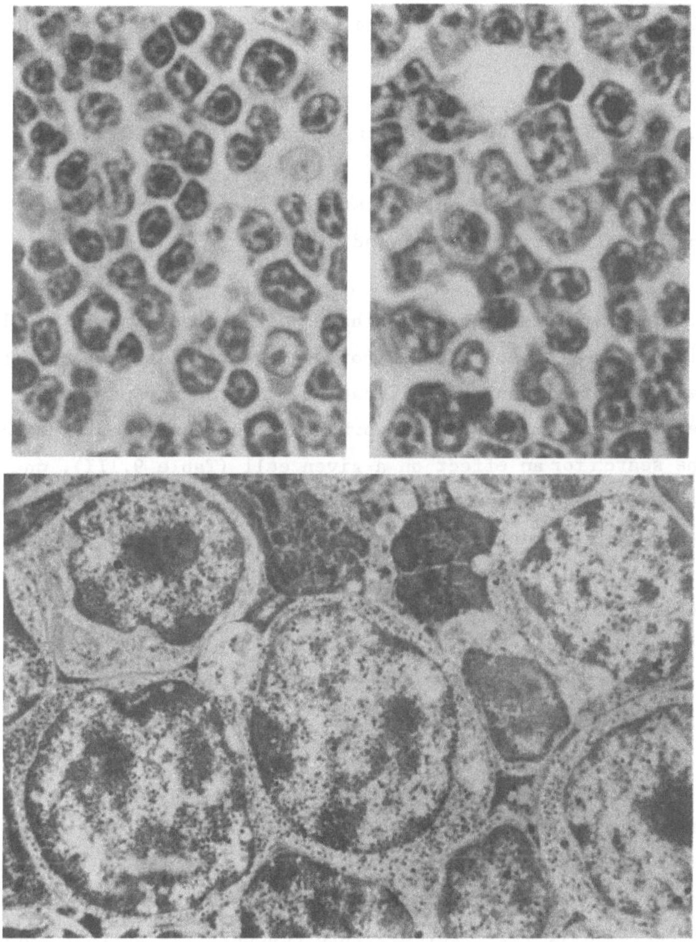

Fig. 9.6. Aspect of pyroninophilic cells seen in the thymus of i.v. BCG treated mice at light and electron microscopy. Upper: Light microscopy: on the left, control thymus; on the right, thymus of BCG treated mouse. Below: Electron microscopy: note numerous free ribosomes. (KHALIL et al., 1975)

The effect of B. pertussis on the immune response to Ø x 174 is a potentiation of B-cells and a suppression of the helper effect of T-cells (specially with low priming doses), which is either due to direct suppression of T-cell activity or overstimulation of B-cells, rendering them non-receptive to help by T-cells (DRESSER and TIEN-WEN TAO, 1975). Such an effect was not observed with SRC, since B. pertussis has been shown to potentiate the immune response of CBA mice to SRC by a stimulation of T-cells at low doses of antigen, and a direct effect on B-cells at high doses of antigen.

This observation shows the extreme complexity of the actions of adjuvants, which is illustrated by Figure 9.7, and which can vary from one antigen to another.

9.1.2.1.3. Classification of Adjuvants According to their Main Target Cells

In conclusion, it is only after having analysed all their possible effects, that one can attempt to classify adjuvants.

Several mechanisms of action have been quoted, and it is clear that not all mechanisms have yet been studied for all agents under all conditions. Thus one must accept very cautiously any classification based on any parameter, for example, the target cell(s) they are supposed to work on; if one can accept the positive results concerning the search for an effect on a given cell (Table 9.III), one cannot consider this cell as the exclusive target of the tested agent if the experiments were not systematically carried out on the other targets. We have set up a battery of tests to study the different actions of different adjuvants and the result obtained with BCG will illustrate this point (see p. 339) (FLORENTIN *et al.*, 1976).

Infra-Cellular Mechanisms. Most adjuvants work on several cells, macrophages, T- and B-lymphocytes, stem cells, and possibly others (Table 9.III).

A common mechanism of action may be found in the effects of most adjuvants on the cell (plasma and lysosome) membranes. Table 9.IV lists the agents studied for this action according to ALLISON (1973). The fact that the action of polynucleotides is potentiated by adenyl cyclase with increased level of 3'5' - cyclic phosphate (BRAUN *et al.*, 1971) favours this concept. The surface activity of adjuvants may increase contact between cells, macrophages and lymphocytes, T- and B-lymphocytes (ALLISON, 1973). It may also facilitate the release of factors, e.g., from T-lymphocytes which activate macrophages, or of some hormones, e.g., the thymic lymphocyte-stimulating factor (KHALIL *et al.*, 1975), as it may also increase the production of lymphocyte-activating factors by the macrophages (GERY and WAKSMAN, 1972).

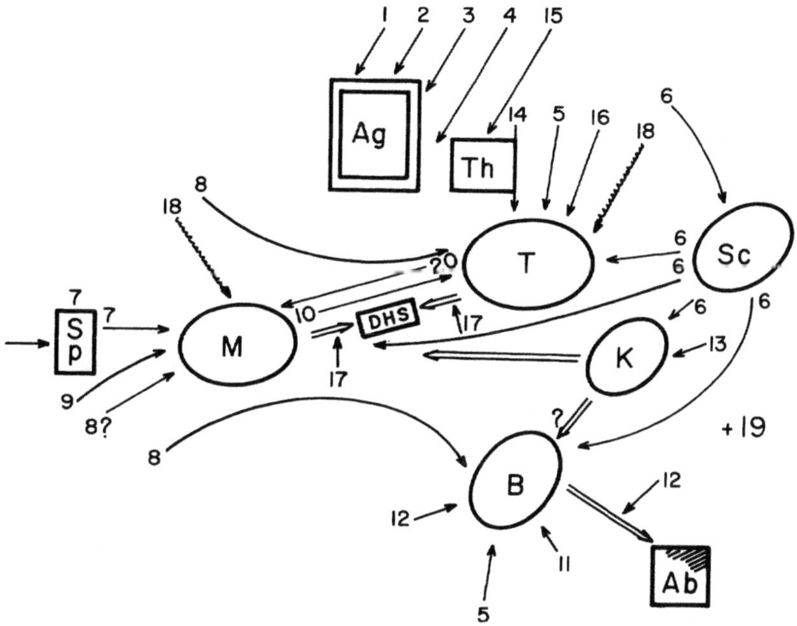

Fig. 9.7. Scheme of possible mechanisms of immunity adjuvant action:

(1) May slow down catabolism of Ag molecule

(2) May modify antigenic part by changing net electric charge or by altering specific antigenic sites

(3) May transform a hapten into an immunogenic molecule

(4) May make a poor antigen more immunogenic

(5) May "insult" antigen-sensitive cells which could represent the second signal possibly necessary, in the case of monomeric antigen, for these cells to divide. May derepress quiescent cistron(?)

(6) May have actions on stem cells, or colony stimulating factors (stimulation of cell division)

(7) May increase macrophage chemotaxis, activate and increase phagocytic activity (after combination with some serum protein(s)? (uptake by macrophages?)

(8) May increase trapping

(9) May increase macrophage cytotoxicity

(10) May increase lymphocyte-activating factor (LAF)

(11) May convert thymus-dependent antigens to thymus-independent antigens

(12) May increase proliferation of B-cells and of antibody production

(13) May increase effector (K) cells in antibody-dependent cell-mediated cytotoxicity

(14) May increase proliferation of thymus cells and of cells in the T-dependent areas of secondary lymphoid organs

(15) May increase thymus-lymphocyte stimulating factor

(16) May act on helper T-cells and on T-cell mediated immunity

(17) May enhance delayed hypersensitivity reactions

(18) May have suppressive effect via macrophages

(19) May increase contact between all those cells

(20) May increase release by T-lymphocytes of macrophage-activating factor(s)

Ag = antigen; Th = thymus; Sc = stem cells; T = T-lymphocyte; K = K-cell; B = B-cell; M = macrophage; Sp = serum protein; Ab = antibody

Table 9.III. Classification of adjuvants according to main target cells they have been shown to work on

Macrophages	
Important effect:	
Beryllium	ALLISON, 1968, 1973; UNANUE *et al.*, 1969
B. pertussis	UNANUE *et al.*, 1969
BCG	BRULEY-ROSSET *et al.*, 1976; MATHÉ *et al.*,1975
Hiu II	BRULEY-ROSSET *et al.*, 1976; MATHÉ *et al.*,1975
Endotoxin	ALEXANDER and EVANS, 1971
E. coli lipopolysaccharide	SPITZNAGEL and ALLISON, 1970a
Silica particles	SPITZNAGEL and ALLISON, 1970a
Freund's complete adjuvant	SPITZNAGEL and ALLISON, 1970a
MER	STEINKULLER *et al.*, 1969
Some effect:	
Polynucleotides	BRAUN *et al.*, 1971
Beryllium	UNANUE *et al.*, 1969
Retinol (vitamin A alcohol)	ALLISON, 1973
Wax D	JOLLES *et al.*, 1968
B-lymphocytes	
E. coli lipopolysaccharide	ANDERSON *et al.*, 1972a, 1972b; MÖLLER *et al.*, 1972
C. parvum	HOWARD *et al.*, 1973a
B. pertussis	DRESSER and PHILLIPS, 1973
PHA	LOBB *et al.*, 1967
BCG	FLORENTIN *et al.*, 1976
T-lymphocytes	
BCG	FLORENTIN *et al.*, 1976
Lentinan	MAEDA *et al.*, 1971; DRESSER and PHILLIPS, 1973
Retinol (Vitamin A alcohol)	TAUB *et al.*, 1970
Poly A-poly U	CONE and JOHNSON, 1971
Poly I-poly C	CONE and JOHNSON, 1971
Freund's complete adjuvant	see ALLISON, 1973
E. coli lipopolysaccharide	ANDERSON *et al.*, 1972a, 1972b
B. pertussis	DRESSER and PHILLIPS, 1973

Table 9.IV. Membrane active agents known as adjuvants (ALLISON, 1973)

Endotoxins	SPITZNAGEL and ALLISON, 1970b
Bordetella (endotoxin)	UNANUE *et al.*, 1969
Mycobacterium (wax D)	WHITE, 1967
Retinol	SPITZNAGEL and ALLISON, 1970b
Silica	SPITZNAGEL and ALLISON, 1970b
Beryllium	UNANUE *et al.*, 1969
Cationic detergents	GALL, 1966
Saponin	GALL, 1966

9.1.2.1.4. Particular Study of Adjuvants Already Used in Man

9.1.2.1.4.1. BCG

The discussion above, on the necessity of a BCG septicaemia (BCGitis) for antitumour action to be obtained from this organism and of a septicaemia of a given intensity (MATHÉ, 1976b), will not be repeated here.

Only the different immune mechanisms by which the factor(s) released by the living bacilli exert their action will be analysed.

BCG was found to increase resistance to bacterial infection in mice (DUBOS and SCHAEDLER, 1957), to cause hypertrophy of the spleen, and to increase the clearance of colloidal carbon.

Other studies have shown that BCG enhances the number of antibody plaque-forming cells (PFC) per spleen (STJERNSWARD, 1966). It was shown, furthermore, that this adjuvant could accelerate skin allograft rejection (BALNER *et al.*, 1962).

The recent concept of two T- and B-cell populations, which give better insight into the understanding of cell cooperation in the immune response, led us to investigate the role of BCG on the different cell components involved in different types of immunity (Table 9.V) (FLORENTIN *et al.*, 1976).

BCG is effective in humoral immunity only when given i.v., and this effect is dose-dependent. The antisheep red cell (SRC) response in mice given 1 mg of fresh living BCG (Pasteur Institute) 14 days before antigen injection is 3-4 times higher than in mice given BCG s.c. and in the controls. This augmentation affects both direct and indirect PFC (per spleen or per 10^6 nucleated cells). Similar effects are observed with the response to the thymus-independent antigen TNP-POL (polymer of flagellin).

Table 9.V. Effect of BCG given i.v. or s.c. on the immune status of (DBA/2x$C_{57}Bl_6$) F1 mice[a] (FLORENTIN *et al.*, 1976)

	BCG i.v.	BCG s.c.
Humoral immunity:		
Thymus-dependent antigen (SRC)	↗	0
Thymus-independent antigen TNP-POL (Polymer of flagellin)	↗	0
Cell-mediated immunity:		
Delayed skin hypersensitivity to picryl chloride	↗	0
Suppressor T cells[b]	↗	0
PHA responsiveness	↘	↘
LPS responsiveness	↗	0
Non-specific macrophage activation	↗	↗
Trapping	0	↗

[a] Fresh living BCG from the Pasteur Institute is given at day -16 (1mg) and the tests are performed at day 0. The i.v. route was compared to s.c. only because it is efficient in the immune prophylaxis of L 1210 leukaemia and of Lewis tumour whereas the latter is not active on L 1210 leukaemia growths and enhances those of Lewis tumour (MATHÉ *et al.*, 1973, 1975).

[b] BCG is given at day -7 (3 mg).

Delayed hypersensitivity skin reactions to picryl chloride were significantly increased in mice given BCG i.v.

However, PHA and PWM responsiveness of spleen cells from mice given BCG i.v. or s.c. were found to be lower than in the controls.

This led us to search for stimulation of T-suppressor cells. We have already discussed the dose-effect relationship showing that, while 0.5 mg-1 mg in mice were the optimal doses for the HPFC assay, 5 mg had no effect and 10 mg decreased the response, compared to controls (Table 7.24) (MATHÉ, 1976a).

GEFFARD and ORBACH-ARBOUYS (1976) used the graft-versus-host test. T-cells from parental mice ($C_{57}Bl_6$) treated with 3 mg BCG i.v. 1 week before, and injected into F1 hybrids (C_{57} x DBA_2) lethally irradiated, induce a graft versus host reaction, measured by IUDR incorporation, similar to that of the control. However, if the same F_1 recipients are given cycloheximide (a protein synthesis inhibitor) at the time of parental injection of spleen cells from BCG-treated mice, there is a 3-fold increase of IUDR incorporation (Table 9.VI). This is considered to reflect the inhibition of suppressor cell function(s). BCG, at a dose of 3 mg i.v., stimulates suppressor T-cells in mice.

Table 9.VI. Percent spleen ^{125}IUDR uptake in lethally irradiated ($C_{57}Bl_6xDBA_2$)F_1 hybrids 4 days after C_{57}Bl/6 T-cell injection (GEFFARD and ORBACH-ARBOUYS, 1976)

Treatment of donors	Treatment of recipients	
	None	Cycloheximide
No treatment	0.26	0.22
BCG: 3 mg i.v. two weeks before transfer	0.06	0.17

BCG given i.v. or s.c. activates peritoneal macrophages, which can induce a nonspecific cytolysis of tumour cells.

Trapping efficiency of BCG could be demonstrated with BCG given s.c. (Figs. 9.2 and 9.3) in the satellite lymph node of the BCG injection area. No trapping was found in the spleen when BCG was given i.v.. Furthermore, a decrease of lymphocyte trapping was found in the bone marrow of treated mice in comparison to controls.

According to MITCHELL (1973, 1975), BCG is capable of increasing the T cell-macrophage interactions. Enhancement by BCG of the production of the lymphocyte activating factor (LAF) by splenic macrophages was found to be greater when lymphocytes were present; conversely, macrophages are required for the stimulation of T-cells by BCG (MOHYR and MITCHELL, 1975).

A BCG-treated splenic lymphocyte population gives a better response to T and B-mitogens after reduction of the concentration of macrophages (MITCHELL *et al.*, 1973). Therefore, macrophages can act as regulator cells, as well as lymphocytes under the influence of BCG, and this mycobacterium can elicit a suppressor function of macrophages. BCG can act on the T-B interaction, and MacKANESS *et al.* (1974) also showed that systemic infection with BCG abolishes the function of T-cells blocking which is formed by the interaction of antigen and antibody.

MARDINEY *et al.* (1974), studying melanoma patients submitted to BCG, showed that this agent does not act on responses independent of previously sensitized lymphocytes (such as PHA transformation or mixed lymphocyte reaction), but enhances the responses to recall antigens and the production of immunoglobulins. This leads the authors to conclude that BCG mainly affects the secondary responses. This conclusion may be correct only for certain modalities of application.

Non-specific esterase activity of tissue extracts has been studied by YOUNG and BITTAR (1974) in animals treated by BCG: a 6-fold increase in diethyl-p-nitrophenyl phosphate-resistant esterases has been found in certain non-lymphoid tissues, such as the kidney. In mice injected with BCG, a diminution of catalase activity in the liver was observed simultaneously with an increase of activity of the reticulo-endothelial system and an increased resistance to leukaemia (LAMENSANS *et al.*, 1968).

I have already quoted the effect of BCG given i.v. on the number of colony-forming units (CFU) in the bone marrow studied by POUILLART *et al.* (1975, 1976) using the TILL and McCULLOCH (1961) (CFU s) and agar techniques (CFU a). The number of CFU s per femur was not increased in mice given BCG i.v.. However, a rapid increase of the percentage of CFU s in S phase was observed as early as the 4th hour, where 40% of the cells were in S phase; at the 72nd hour, 33% of the cells were still in S phase (versus 5% in the controls) (Table 9.I).

A different pattern was observed with CFUa; after a transitory peak occurring during the first hours, the number of CFUa per femur increased regularly until day 3. From this data, it was concluded that BCG has a central effect, acting directly on CFUs in the quiescent phase, pushing them into an active phase of the cell cycle. We have seen in Chapter 3 how this effect was used to shorten the intervals between intermittent chemotherapy cycles (Fig. 3.48) (MATHÉ, 1974).

Table 9.V gives a summary of the results we have found in our study of the effect of BCG on different immune functions (FLORENTIN *et al.*, 1976).

9.1.2.1.4.2. Corynebacteria

Since C. parvum and C. granulosum are now used in clinical practice (see Chapter 3), it is necessary to underline some of their characteristics.

As mentioned earlier, C. parvum stimulates the phagocytic activity of the reticulo-endothelial system (HALPERN *et al.*, 1963). It has been demonstrated that it activates macrophages (FAUVE and HEVIN, 1971; WILKINSON *et al.*, 1973; OLIVOTTO and BOMFORD, 1974).

Another interesting observation made by WOLMARK and FISHER (1974) is the great increase of bone marrow macrophage colony production after administration of C. parvum in tumour-bearing mice.

It boosts antibody responses to a variety of antigens (BIOZZI *et al.*, 1968; RAYNAUD *et al.*, 1972; HOWARD *et al.*, 1973b). But while many commonly used antigens require T-lymphocytes to effect antibody production (ALLISON and DAVIES, 1971), C. parvum is capable of boosting antibody levels to T cell-independent antigens (HOWARD *et al.*, 1973a, 1973b). For C. parvum, it is the activated macrophage that provides the added proliferative stimulus to the antigen-sensitive and antibody producing cells (HOWARD *et al.*, 1973a, 1973b).

Another singularity of C. parvum, which deserves consideration, is its effect on cell-mediated immunity, since antitumour immunity is recognized as being predominantly cell-mediated.

SCOTT (1974) showed that i.v. administration of C. parvum to mice inoculated with a mastocytoma did not increase cell-mediated immunity, and he considers that its systemic action is probably due to macrophages.

Systemic injection of C. parvum has been shown to depress various T-cell responses. Delayed hypersensitivity, blastic transformation by PHA, mixed lymphocyte reaction, graft-versus-host reaction (SCOTT, 1972b), and homograft reactivity (CASTRO, 1974) have been shown to be impaired. The mechanism has been elucidated: spleen cells from C. parvum-stimulated animals are refractory to stimulation by PHA, except if macrophages are removed; the conclusion is that T-cell activity is inhibited by C. parvum-activated spleen macrophages (SCOTT, 1972a, 1972b).

This role of C. parvum on the spleen was confirmed by splenectomy experiments (SCOTT, 1974). KIRCHNER *et al.* (1975) showed that the population of spleen macrophages activated by the administration of C. parvum could inhibit PHA responsiveness of normal lymphocytes. It should be noted that systemic C. parvum injection has to precede antigen sensitization to effect delayed hypersensitivity reaction against it.

In contrast, cells from lymph nodes that had been directly stimulated with a local injection of C. parvum are still capable of responding to PHA (SCOTT, 1972a, 1972b). Thus this inhibition of T cells does not seem to exist in lymph nodes.

These data may explain why delayed hypersensitivity to sheep red blood cells was only suppressed by i.v. injection of C. parvum and not by its s.c. administration (SCOTT, 1974).

9.1.2.1.4.3. Polynucleotides and Other Adjuvants

BRAUN and SHIOZAWA (1974) showed that the immunomodulating effects of poly I-poly C and poly A-poly U are associated with an alteration of the level of regulatory cyclic AMP in lymphocytes (ISHIZUKA *et al.*, 1971). Figure 9.8 recalls the situation of cAMP between ATP and AMP and its main functions; it shows that polynucleotides enhance adenyl-cyclase and reminds us that methylxanthines, such as theophylline and caffeine, inhibit phosphodiesterase which transforms cAMP into AMP. Figure 9.8 shows Braun's and Shiozawa's (1974) concept on the effect of cAMP on lymphocytes. According to these authors, an antigen signal alone would not suffice for B-cells to produce a full activity leading to detectable antibody secretion, whereas a combination of two signals, one on cAMP receptor sites and one on antigen receptor sites would do so. The activity of T lymphocytes is similarly enhanced by the presence of stimulators of cAMP formation or by stabilizers of its level, theophylline or caffeine, during exposure to antigen (BRAUN *et al.*, 1971; JOHNSON *et al.*, 1971; WEBB *et al.*, 1972). Macrophage activity is also stimulated by cAMP and by modifiers of its cell endogenous level (BRAUN *et al.*, 1971).

When analysing a result of tumour immunotherapy obtained by polynucleotides, one must remember that they are also interferon inducers and, in the case of virus-induced tumours (SARMA *et al.*, 1969; LACOUR *et al.*, 1972), the parts due respectively to antiviral interferon effect, antiviral immune effects, and antitumour cell effects should be considered. Moreover, a direct toxic effect on tumour cells has been demonstrated (WEBB *et al.*, 1972).

Poly I-poly C and poly A poly U do not act identically. A good example is their effect on stem cells. While the i.p. injection of the first results in an eight-fold increase in spleen colony-forming cells by the sixth day, the same dose of poly

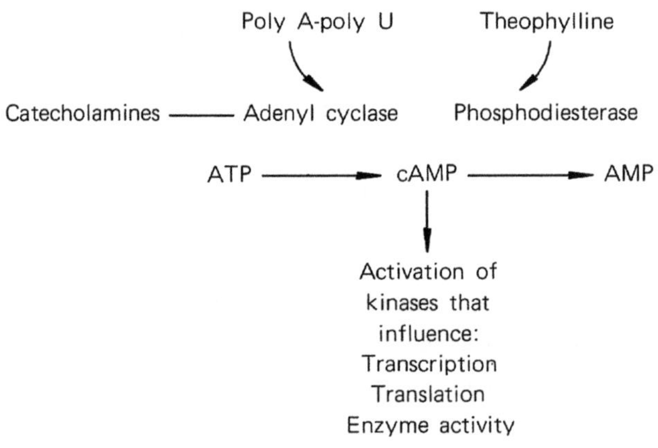

Fig. 9.8. Steps in cAMP formation and degradation, outlining the site of influence of polynucleotides and methylxanthines (BRAUN and SHIOZAURA, 1974)

A-poly U gives a negligible increase in the number of colonies (McNEILL and KILLEN, 1971).

Interferon and endotoxins work, according to BRAUN and SHIOZAWA (1974), similarly to polynucleotides on the cAMP regulatory systems.

It is probable that their antitumour effect results from the same three components discussed for polynucleotides (GRESSER, 1974).

Table 9.VII summarizes the main effects of the adjuvants just studied and that of others, B. pertussis, B. abortus, lentinan, levamisole and vitamin A.

9.1.2.2. Hormones

Here we shall only discuss *thymosin*. GOLDSTEIN *et al.*, quoted by MARX (1975), showed that administration of this polypeptide can be substituted, at least partially, for a functional thymus gland: its injection can maintain a T-cell count in a patient with thymic hypoplasia. When administered to newborn mice, it accelerated the development of their T-cell functions. It has been reported to give the capacity to bone marrow cells to form spontaneous sheep red blood cell rosettes (BACH *et al.*, 1971).

GOLDSTEIN *et al.* (quoted by MARX, 1975) has suggested that cyclic AMP may mimic some of the functions of thymosin. The latter could stimulate adenyl cyclase and increase the number of functional T-cells. If the cyclic AMP rises, thymosin functions could be inhibited. This could initiate a feedback mechanism according to TRAININ (quoted by MARX, 1975). This may explain why thymosin sometimes inhibits the response of lymphocytes.

Thymosin should not be confused with another thymic polypeptide which also induces T-cell differentiation, but which is not specific, because it also induces B-cell maturation and is found in tissues other than the thymus (GOLDSTEIN and BOYSE, quoted by MARX, 1975).

9.1.3. Failures and Adverse Effects

The failures and adverse effects of immuno-manipulation, like the failures and unfavourable results of any therapy, are much more common than the literature would indicate, especially in the clinical field. Only success attracts the attention of journals and the giving of grants. Similarly, in the experimental field, research workers do not retain models on which they obtain negative results, but concentrate, instead, on those which allow them to obtain positive results.

Table 9.VII. Effects of the main systemic immunity adjuvants on several components of the immune machinery

	Macrophages	T-cell stimulation	B-cell and humoral immunity	Suppressor lymphocytes
BCG	↗ FLORENTIN *et al.*, 1976	↗↙ FLORENTIN *et al.*, 1976	↗ FLORENTIN *et al.*, 1976	↗ [a] GEFFARD and ORBACH-ARBOUYS, 1976
MER	STEINKULLER *et al.*, 1969		↗ KUPPERMAN *et al.*, 1972	
Soluble mycobacteria extracts	↗ CHEDID *et al.* (1972, 1973) MODOLELL, 1974	CHEDID *et al.* (1972, 1973)	↗ CHEDID *et al.*, 1972, 1973	
Corynebacteria	↗↗ SCOTT, 1974	[b] HOWARD *et al.*, 1973 ↘ SCOTT, 1974	↗ WOODRUFF and DUNBAR, 1973	
Bordetella pertussis		↘ FLOERSHEIM, 1967 ↗ for SRC: DRESSER *et al.*, 1970 for ØX174: DRESSER and ↘ TIEN-WEN TAO, 1975	↗ KIND, 1957 DRESSER *et al.*, 1970 DRESSER and TIEN-WEN TAO, 1975	
Lentinan	0	↗↗ [c] MAEDA and CHIHARA, 1971		
Poly A-poly C		↗ JOHNSON *et al.*, 1971		
Tetramisole and levamisole		↗ TRIPODI *et al.*, 1973 HIRSAULT *et al.*, 1973	0	
Vitamin A			DRESSER, 1968	

[a] At doses which are higher than those which give augmentation of the immune responses.

[b] Due to macrophage activation.

[c] Only for antitumour activity. Otherwise it has no action on DHS and transplantation immunity.

As far as systemic immunotherapy is concerned, there are several causes of failure, and adverse effects which can be deduced from the above data.

9.1.3.1. Immune Inefficiency of Cancer Subjects

One of the most important may be the immune deficiency frequently observed in cancer patients. This syndrome and the attempts at immunorestoration have been described above. It goes without saying that immunotherapy is useless in an immunodepressed patient and that it should be preceded by an immunorestoration attempt. This was clearly shown experimentally by LEVY *et al.* (1974), and clinically in melanoma patients by GUTTERMAN *et al.* (1974) and by THOMAS *et al.* (1974).

9.1.3.2. Possible Immunodepressive Effect of Adjuvants

In the many tests of our screening (MATHÉ *et al.*, 1973, 1975) to which we have submitted several adjuvants, we sometimes found a decrease of the response or an adverse action in some conditions, while the agents were stimulating or without action in others (see Chapter 7).

When the Jerne test is used, a decrease of the response can only be interpreted as immunodepression.

It is possible that the spleen might be used for this test when there is a transitory immunodepression preceding or following another major phase characterized by immunostimulation. Studying the time factor, we have seen, in Chapter 7, that this can be the case (MATHÉ *et al.*, 1973, 1975).

DONNER and VAILLIER (1974) also showed that an immune response, after adjuvant administration, may follow kinetics with depressive phases interspersed with stimulating ones.

Among other hypothesis which could explain the immunodepressive effect of adjuvants, it should be recalled that a feedback control of the immune responses does exist: antibody formation is inhibited by IgG antibody directed against the same antigen (UHR and MÖLLER, 1968). Moreover, T-cell suppressors are known to be able to inhibit the production of antibodies against several antigens (ALLISON, 1973). I have already quoted GEFFARD and ORBACH (1976), who showed in our laboratory, that the stimulation of these cells explains, at least partially, the immunodepression induced by high doses of BCG (MATHÉ, 1976a) (Table 9.VI).

The possible immunodepressive effect of adjuvants has also been observed with C. parvum (SCOTT, 1974): administered before the antigen, it decreases delayed hypersensitivity to it. The mechanism of such an immunodepressive effect is unknown, but it has been shown that splenectomy, before administration of i.v. C. parvum, inhibits this effect.

The use of this corynebacterium as an adjuvant should be carried out with extreme care, as it has been demonstrated that: (a) it is able to increase B-cell proliferation without the cooperation of T-cells (HOWARD *et al.*, 1973a, 1973b); (b) it depresses T-cell activity, as suggested by in vitro experiments showing that PHA responsiveness of lymphocytes from mice treated with C. parvum is depressed (SCOTT, 1972a, 1972b); this T-cell inhibition is a necessary condition of impaired expression of delayed hypersensitivity to sheep red cells (SCOTT, 1974).

Another hypothesis for explaining a possible "immuno-deteriorating" effect of adjuvants, which is not exclusive, is lymphoid depletion induced in some areas by the trapping in the lymph node draining the site of s.c. injections, or in the spleen, in the case of i.v. administration. This could explain why i.v. injection of BCG in mice carrying Lewis solid tumours delays mortality, while s.c. injection accelerates it. In the case of such injections, there is an important trapping in the lymph node draining the site of administration, which seems to induce a lymphoid depletion in other parts of the organism, especially the bone marrow (Fig. 9.3) (FLORENTIN *et al.*, 1976).

9.1.3.3. Antigenic Competition

Antigenic competition has been suggested by BERNHEIM (1974) as being a possible mechanism of immune deterioration by adjuvants: as a matter of fact, most of them are antigenic, and the authors are under the impression that: (a) weak tumour antigens can be "crowded out" more easily than the potent ones that the adjuvants can carry (ABRAMOFF and WOLFE, 1956); (b) antigenic competition may or may not (ADLER, 1957; LAWRENCE and SIMONSEN, 1967) suppress a secondary response. Finally, antigen stimulated T-lymphocytes secrete a factor which activates other immunocompetent cells (GERSHON and KONDO, 1971a, 1971b). In favour of this possible immune deterioration by antigenic competition mechanism, BERNHEIM (1974) quotes METCALF (1961), who showed that more "reticular" neoplasias arise in mice subjected to repeated Salmonella adelaide antigen than to bovine serum albumine (BSA administration). Similarly, SIEGEL and MORTON (1966) showed that Rauscher virus leukemogenesis is increased by previous immunization with BSA in adjuvant.

This so-called antigenic competition is a complex phenomenon; with two antigens, the immune response to one can be enhanced, depressed, or not modified by the other.

9.1.3.4. Immunological Enhancement

Immune enhancement can be suggested as another explanation, either due: (a) to the potentiation of the production of so-called blocking antibodies (see Chapter 2); (b) to any antibodies able to form blocking complexes with the antigen(s); (c) to antigen(s) which may be liberated after the attack of the tumour by the immune machinery, and which could be able to block the recognition sites of the lymphocytes (see chapter 2).

9.1.3.5. Tumour Volume and Blocking Humoral Factors

Another factor of failure or adverse effect appeared, in many experiments and in clinical trials (see Chapters 3 and 4) to be the tumour volume. The reason for our constant failure to obtain a regression in tumours containing a high number of cells (MATHÉ, 1968) may be due to the presence of HELLSTRÖM *et al.*'s (1971) blocking activity, which has been said to inhibit the effect of the specific cytotoxicity lymphocytes. BANSAL and SJOGREN (1973a) did not increase the cytotoxicity effect of lymphocytes in animals bearing perceptible and growing tumours by the administration of BCG.

Thus immunotherapy of a perceptible tumour should ideally be applied either: (a) after suppression of this blocking activity; it is known that it increases with tumour growth and widespread metastasis, and that it rapidly disappears after complete removal of the neoplasia (it does not disappear if a part of the tumour is left behind) (HELLSTRÖM *et al.*, 1971; BALDWIN, 1973; HELLSTRÖM *et al.*, 1973); or (b) by administration of the so-called unblocking factor; we saw earlier that the latter can be obtained by injecting BCG into rats or rabbits, and 10-12 days later, mitomycin-treated tumour cells: their serum contains components which can inhibit the blocking activity in the same host-tumour relationship (BANSAL and SJÖGREN, 1972, 1973b, 1973c).

Not only may the blocking phenomenon render immunotherapy useless, but it may induce enhancement of tumour growth. Thus, this should be searched for systematically after systemic immunotherapy, as has been observed after local intra-melanoma tumour injections by LEVY *et al.* (1972).

The failure of BAGSHAWE and GOLDING (1971) to obtain good results in choriocarcinoma female patients with their husbands' skin grafts and BCG, may be due to such humoral factors, the three most likely hypotheses being tumour soluble antigen, antigen-antibody complexes and antibodies. In this latter tumour, we observed (MATHÉ *et al.*, 1964; AMIEL *et al.*, 1967) a correlation between unfavourable evolution, the prolongation of the husbands' skin grafts, and the presence of humoral antibodies against the husbands' leucocytes.

Blocking activity is not a phenomenon that is only associated with an enormous tumour volume; it becomes detectable simultaneously with the appearance of cytotoxic lymphocytes (BANSAL and SJÖGREN, 1974). This emphasizes that the most reasonable indication of active immunomanipulation is in the imperceptible residual or preperceptible disease (MATHÉ, 1968).

Can neoplastic cells or antigen(s) used for the specific branch of immunotherapy increase or induce blocking activity? BALDWIN *et al.* (1973) have suggested that the blocking phenomenon may be uniquely or partly due to tumour-associated antigen(s). The question is raised as to whether or not the injections of tumour cells, or of too many of them, may not sometimes increase this unfavourable humoral effect. This is suggested by SEDLACEK *et al.*'s (1975) experiments in which they treated dogs with spontaneous mammary tumours with mitomycin neuraminidase-treated cells: 2.10^7 cells were optimal; 1.10^8 cells induced acceleration of tumour growth.

Hence, when submitting a neoplasia to active immunotherapy, one must, before deciding to use tumour cells, consider their usefulness. When treating perceptible tumours, it is probably, in most conditions, useless, and it may be harmful to administer cells. On the contrary, when treating a residual imperceptible disease, it appears to be necessary to inject them. When the number of residual cells can be assumed to be very low (in the order of 10^5), as in acute lymphoid leukaemia, it seems logical to give a higher number.

The risk of giving too much tumour antigen may be especially dangerous when using solubilized material.

However, the time of giving such material in immunotherapy (MARTYRE, 1976) (Fig. 7.3) and the modality of preparation in immunoprophylaxis (MARTYRE *et al.*, 1974) also seem to be important, as is indicated above (Fig. 7.4).

9.1.3.6. Anatomic Sites of Tumours

Some sites may be inaccessible to immune tumour cell lysis, such as the brain, as suggested by GROOM and MORTON's (1975) observations.

Bone marrow may be a poorly immunogenic site, as many types of tumours of this tissue, i.e., leukaemia, are disseminated early (see MATHÉ and RAPPAPORT, 1976). The absence of T-cells in the marrow of certain strains of mice, e.g. CBA (RAFF, 1971), also favours this concept. We have seen that bone marrow trapping of lymphocytes may be decreased by BCG (FLORENTIN *et al.*, 1976).

9.1.3.7. Lack of Antigenicity and/or Immunogenicity of the Tumour

This lack may be another reason for failure of immunotherapy: this was demonstrated by PARR (1972), who studied four lymphosarcomas: the line with the greatest antigenicity was the only one of four to respond to BCG or BCG and irradiated tumour cells; it was, on the other hand, the one which had the least capacity for metastasis.

Lack of antigenicity may also be a crucial factor in enhancement: BALDWIN and PIMM (1973a, 1973b) observed that, after the same BCG treatment, the growth of neoplasms with strong antigenicity was inhibited, while that of weakly antigenic tumours was facilitated.

A change of antigenicity of tumour cells may take place during chemotherapy at the same time as resistance to the latter is induced; this has been shown for methylglyoxal bis (guanyl-hydrazone) and imidazole carboxamide (MIHICH, 1969; BONMASSAR *et al.*, 1972). This weak antigenicity may be much more frequent in man than might be expected from animal study data.

9.1.3.8. Secondary Resistance to Immunotherapy

Resistance to immunotherapy has been clearly demonstrated by the curves of Figure 3.2, which show that, after a plateau during which there is an immunologic control of the leukaemia growth, the curve of the latter rises again; this is the result of a secondary escape to this control (MATHÉ *et al.*, 1969).

A change and, particularly, a decrease or a loss of tumour-associated antigens may appear under immunotherapy. One of the most convincing facts in favour of this hypothesis is that of immunomodulation described by FENYÖ *et al.* (1968).

9.2. Local Immunotherapy

We saw that, in the clinical field, tumour regression after local injection of BCG has generally been restricted to subjects who were capable of eliciting an immune response to this mycobacterium or to PPD (MORTON *et al.*, 1970a, 1970b; NATHANSON, 1972; SMITH *et al.*, 1973).

Non-living extracts from mycobacteria, which do not develop sensitivity to tuberculin, do not inhibit experimental tumour growth after local injection (ZBAR, 1974).

This suggests that the regression is not due to the adjuvant molecule(s) of BCG, but to a delayed hypersensitivity reaction to one or several antigen(s) of BCG, especially PPD (MORTON *et al.*, 1970a, 1970b; ZBAR *et al.*, 1971, 1972). This interpretation is sustained by the positive effect of local immunotherapy using haptens or antigens as different as dinitrochlorobenzene (DNCB) (WILLIAMS and KLEIN, 1970), and PPD (KLEIN *et al.*, 1973), which are not adjuvants, or adjuvants other than BCG, such as C. parvum (CHASSOUX and SALOMON, 1976), streptokinase-streptodornase (STEWART and TOLNAI, 1969), and vaccinia virus (HUNTER-CRAIG *et al.*, 1970).

The pathological study made by HANNA (1974) on guinea pigs, and that which we conducted with KHALIL *et al.* (1976a, 1976b) on mice and humans confirm this concept, demonstrating delayed hypersensitivity-type lesions in the tumour.

HANNA (1974) observed, 24 hours after injection of BCG, an acute inflammatory reaction in the tumour with no alteration in the regional lymph nodes. Metastasis to the subcapsular marginal sinus of the latter was more extensive than in controls. Four days after BCG, the architecture of this node and especially the follicular structure were completely destroyed. Non-caseous granulomas were present throughout the cortex, bordering the tumour cells. Bacilli were occasionally seen in these granulomas. Their cells were macrophages and lymphocytes, which were also present in the medullary sinuses. On the 8th day after BCG injection, the lymph nodes consisted of coalesced granulomata and tumour cells showed signs of degeneration. On the 25th day, histiocytosis occupied most of the nodes; no tumour cells were seen, and the intradermal transplanted tumour was histiocytic and fibrotic.

LIKHITE (1974) examined histologically his isogeneic mammary adenocarcinoma 13762 injected with 1.5 mg (0.75 ml) of BCG; 6 days after this treatment, he saw, similarly, an intense infiltration by macrophages and some lymphocytes.

With KHALIL *et al.*(1976a), we observed the same microscopical modifications of mouse Lewis tumour injected with BCG.

More interesting still is the study we conducted with the same pathologists (KHALIL *et al.*, 1976) on the kinetics of histocytological manifestations in human skin melanomas treated by intra-tumoral injections of lyophilized Pasteur BCG (0.5-0.75 mg in each tumour). We injected several nodules on the same day and removed them after different time gaps to follow the kinetics of the reaction (Fig. 9.9)

Injection of 0.5 mg BCG into metastatic cutaneous melanoma resulted in massive necrosis containing degenerated cells with pycnotic nuclei. A great deal of extracellular melanin pigment was evident in the centre, and numerous pigment-laden macrophages were seen in the periphery of the necrosis. The surrounding inflammatory reaction consisted, at day 7, primarily of polymorphonuclear leucocytes with an admixture of large and medium-sized lymphoid cells with pyroninophilic cytoplasm ("immunoblasts"). There were also numerous macrophages. Higher doses (0.75 mg) of BCG caused a lesser proliferation of lymphoid cells and a higher proportion of macrophages. When the same dose was injected into a nodule scheduled for removal on day 10, the sections showed a diffuse cellular reaction composed almost entirely of mononuclear histiocytes. Sections from a nodule injection with 0.75 mg of BCG and excised on day 18 revealed a massive histiocytic reaction with formation of non-caseous tubercles which contain multinucleated giant cells of both the Langhans and foreign body types. Melanin-laden histiocytes were still abundant but less so than in the specimens obtained at day 7 and 10.

Hence the present concept is as follows: local active immunotherapy induces in situ chronic inflammation, histologically visible, resulting from, in the case of the use of BCG, the reaction against this mycobacterium, associated with an immune reaction against its antigen(s), one being essential, namely PPD. The cells involved in this reaction accomplish the specific work against BCG, but kill tumour cells by nonspecific mechanisms (specially the cytotoxicity of macrophages activated by BCG) (HANNA *et al.*, 1972a, 1972b, 1973).

When tumours other than the injected one regress, two mechanisms are possible: (a) the systemic effect of the adjuvant released from the injected site when an adjuvant is used; (b) the possible development of specific antitumour systemic immunity as demonstrated by ZBAR *et al.* (1970, 1971). SCOTT (1974) showed that tumour cell-mediated immunity is increased after successful intratumour injection of C. parvum into a mouse mastocytoma, while it is not increased after systemic injection.

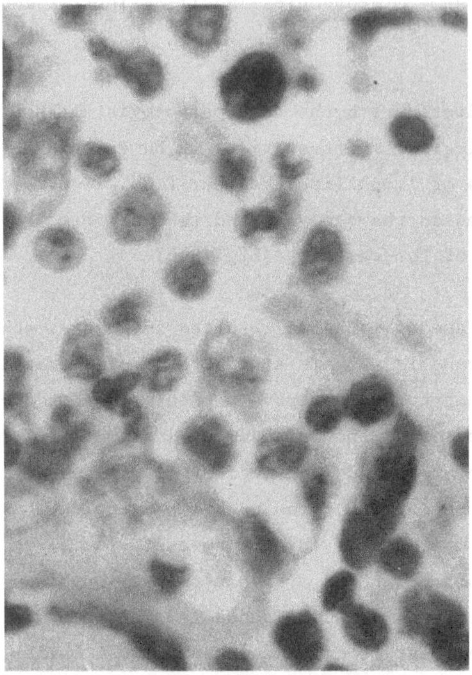

1

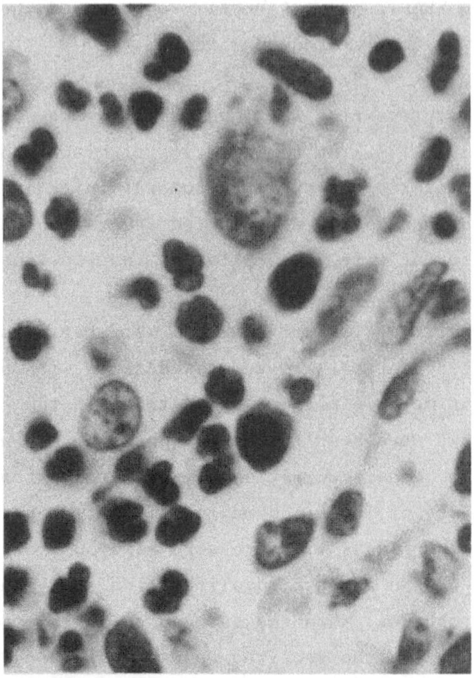

2

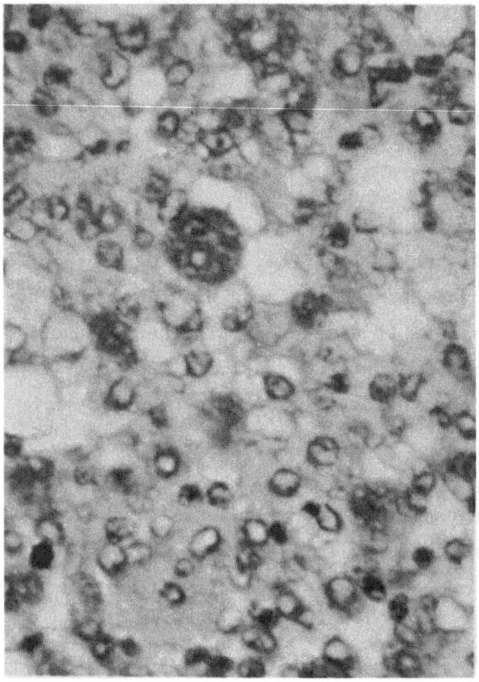

3

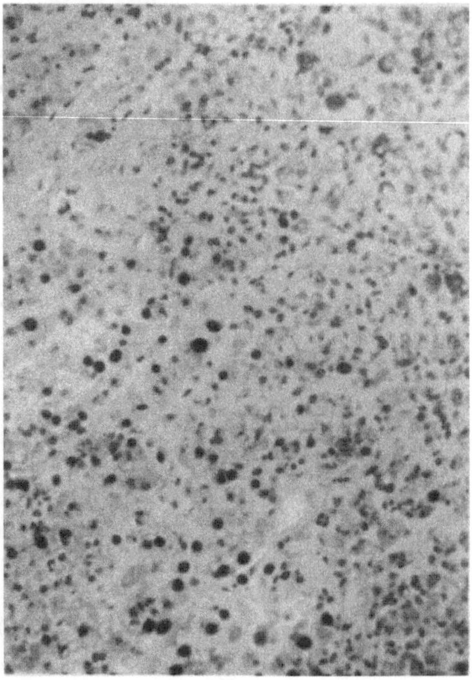

4

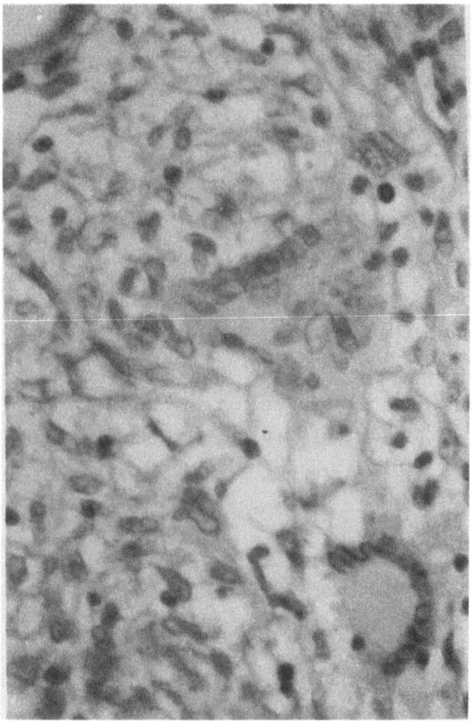

5

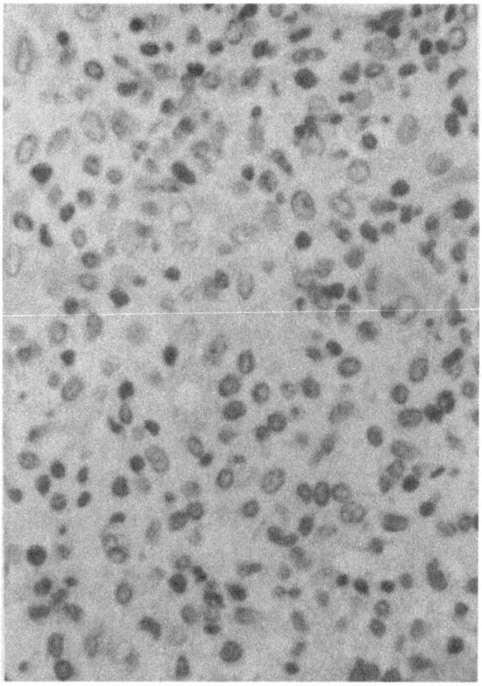

6

Fig. 9.9. Histological aspects of human skin melanomas after local injection of lyophilized (0.5–0.75 mg) BCG: (1) Human metastatic cutaneous melanoma excised at day 7 after local 0.5 mg BCG injection, showing abundant immunoblast proliferation at sites of injection; (2) Human metastatic cutaneous melanoma excised at day 7 after local 0.5 mg BCG injection showing a typical (tumour) cell encircled with polymorphonuclear leucocytes, histiocytes (one of them showed active phagocytosis), and lymphocytes; (3) Human cutaneous metastatic melanoma excised at day 7 after local injection of 0.75 mg of BCG, showing reaction of histiocytes, some of them laden with melanin; (4) Human metastatic cutaneous melanoma excised at day 7 after local 0.5 mg BCG injection, showing extensive central necrosis with the presence of many degenerated cells having pycnotic nuclei (tumour) and peripheral histiocytes laden with melanin; (5) Human cutaneous metastatic melanoma excised at day 18 after 0.75 mg BCG locally injected, showing granulomatous reaction formed of epithelioid cells, Langhan's and foreign body giant cells; (6) Human cutaneous metastatic melanomatous nodule excised at day 10 after local 0.75 mg BCG injection showing diffuse histocytic reaction (KHALIL et al., 1976b)

In a work devoted to the different structural elements of BCG, which suppress tumour development, RIBI *et al.* (1973) showed that, while tumour immunity may not have been necessary for regression of established tumours, a constant association was shown.

For this effect to be exerted, positive immune responses are also necessary, as in the case of systemic immunotherapy. Metastases progressed in patients who were, and remained, tuberculin-negative, while up to 90% of BCG inoculated lesions regressed when subjects were tuberculin-positive (MORTON *et al.*, 1970a, 1970b; NATHANSON, 1972).

Failures and Adverse Effects. In the case of local immunotherapy, the reasons for failure and deterioration are numerous, as in the case of the systemic type.

The failure to decrease the injected tumour mass is generally due to the seemingly negative delayed hypersensitivity to the agent used: the clinical statistics show evidence that regressions after BCG injections have always been restricted to patients who were tuberculin-positive (MORTON *et al.*, 1970a, 1970b; NATHANSON, 1972; SMITH *et al.*, 1973). It also applies to the use of PPD (KLEIN *et al.*, 1972) and DNCB (WILLIAMS and KLEIN, 1970).

Some failures may be of more complex, even unexplained, mechanisms: large i.v. doses of BCG may prevent the killing of the tumour induced by local injection (HANNA, quoted by ZBAR, 1974; ZBAR *et al.*, 1971) at the same time as it prevents the development of tuberculin sensitivity, as does thymectomy and total body irradiation (BARTLETT, 1972).

The development of tumour-specific immunity may result in a blocking phenomenon, perhaps caused by an excess of antigen release.

The usual insensitivity of distal metastases may be due to their size, which is too important for secondary systemic immunotherapy to be active.

When metastases start growing more rapidly than before after local immunotherapy (LEVY *et al.*, 1972; BORNSTEIN *et al.*, 1973; LEVY, 1973; SCHWARZENBERG *et al.*, 1976), it is tempting to evoke an "enhancement phenomenon", but it is difficult to be precise about the variety of phenomenon involved: it is also to be feared that the intense immune action in the injected tumour releases tumour antigen or antigen-antibody complexes, the role of which is suspected in blocking phenomena. A serum-mediated blocking factor of cell-mediated anti-tumour immunity has been demonstrated in a melanoma patient after intratumoral injection, while the latter had induced a clinical deterioration (LEVY *et al.*, 1972).

Another explanation may be the hypothesis we are working on, according to which a local intratumoral injection of an adjuvant may induce a trapping of the lymphocytes in the treated tumour and in the satellite lymph nodes, which may provoke a depletion of these cells in the other lymphoid areas and in the circulating blood, and then in the other tumours.

Local non-immunological factors may also cause enhancement, as it is known that topical application of inflammatory agents, such as turpentine, can accelerate tumour growth via unknown mechanisms (oedema?) (HANNA *et al.*, 1972a, 1972b).

References

ABRAMOFF, P., WOLFE, H.R.: J. Immunol., 77 (1956).

ADAM, A., CIORBARU, R., PETIT, J.F., LEDERER, E., CHEDID, L., LAMENSANS, A., PARANT, A., PARANT, M., ROSSELET, J.P., BERGER, F.M.: Infect. Immunity 7, 855 (1973).

ADLER, F.L.: J. Immunol. 78, 201 (1957).

ALEXANDER, P., EVANS, R.: Nature, New Biol. 232, 76, (1971).

ALLISON, A.C.: Sci. Basis Med. Annual Rev. 18 (1968).

ALLISON, A.C., DAVIES, A.J.S.: Nature 223, 330 (1971)

ALLISON, A.C.: p. 73 in "Immunopotentiation". 1 Vol., Amsterdam: Elsevier Excerpta Medica 1973.

AMIEL, J.L., MERY, A.M., MATHE, G.: p. 197 in "Cell-bound Immunity with Special Reference to Antilymphocyte Serum and Immunotherapy of Cancer". Vol. 43, Liège: University of Liège 1967.

ANDERSON, J., MOLLER, G., SJOBERG, O.: Europ. J. Immunol. 2, 349 (1972a).

ANDERSON, J., SJOBERG, O., MOLLER, G.: Transplant. Rev. 11, 131 (1972b).

BACH, J.F., DARDENNE, M., GOLDSTEIN, A.L., GUHA, A., WHITE, A.: Proc. Nat. Acad. Sci., USA, 68, 2734 (1971).

BAGSHAWE, K.D., GOLDING, P.R.: Immunity and Tolerance in Oncogenesis. 1 Vol., Div. Cancer Res., Perugia: University of Perugia 1971.

BALDWIN, R.W.: Adv. Cancer Res. 18, 1 (1973).

BALDWIN, R.W., BOWEN, J.G., PRICE, M.R.: Brit. J. Cancer 28, 16 (1973).

BALDWIN, R.W., PIMM, M.V.: Int. J. Cancer 12, 420 (1973a).

BALDWIN, R.W., PIMM, M.V.: Nat. Cancer Inst. Monogr. 39, 11 (1973b).

BALNER, H., OLD, L.J., CLARKE, D.A.: Proc. Soc. Exp. Biol. Med. 109, 58 (1962).

BANSAL, S.C., SJOGREN, H.O.: Int. J. Cancer 9, 490 (1972).

BANSAL, S.C., SJOGREN, H.O.: Int. J. Cancer 11, 163 (1973b).

BANSAL, S.C., SJOGREN, H.O.: Int. J. Cancer 12, 179 (1973c).

BANSAL, S.C., SJOGREN, H.O.: Fed. Proc. 32, 165 (1973a).

BANSAL, S.C., SJOGREN, H.O.: Israel J. Med. Sci., 10, 939 (1974).

BARTLETT, G.L.: Nat. Cancer Inst. Monogr. 35, 27 (1972).

BERNHEIM, J.L.: p. 473 in "Investigation and Stimulation of Immunity in Cancer Patients" (G. Mathé and R. Weiner, eds.). Vol. 47 R.R.C.R., Heidelberg: Springer-Verlag 1974.

BIOZZI, G., BENACERRAF, B., HALPERN, B.N.: Brit. J. Exp. Path., 34, 441 (1953).
BIOZZI, G., BENACERRAF, G., HALPERN, B.N.: Brit. J. Exp. Path., 36, 226 (1955).
BIOZZI, G., STIFFEL, D., MOUTON, Y., BOUTHILLIER, Y., DECREUSEFOND, C.: Immunology 14, 7 (1968).
BONMASSAR, E., BONMASSAR, A., VADLAMUDI, S., GOLDIN, A.: Cancer Res. 32, 1446 (1972).
BORNSTEIN, R.S., MASTRANGELO, M.J., SULIT, H., CHEE, D., YARBO, J.W., PREHN, L., PREHN, R.T.: Nat. Cancer Inst. Monogr. 39, 213 (1973).
BRAUN, W., ISHIZUKA, M., YAJIMA, Y., WEBB, D., WINCHURCH, R.: p. 139 in "Biological Effects of Polynucleotides" (R.F. Beers and W. Braun, eds.). 1 Vol., Heidelberg: Springer-Verlag 1971.
BRAUN, W., SHIOZAWA, C.: p. 338 in "Investigation and Stimulation of Immunity in Cancer Patients" (G. Mathé and R. Weiner, eds.). Vol. 47 R.R.C.R., Heidelberg: Springer-Verlag 1974.
BRETSCHER, P.: Transplant. Rev. 11, 217 (1972).
BRULEY-ROSSET, M., FLORENTIN, I., KHALIL, A.M., MATHÉ, G.: Int. Arch. Allergy and Applied Immunol., in press (1976).
CASTRO, J.E.: Europ. J. Cancer 10, 115 (1974).
CHASSOUX, D., SALOMON, J.C.: Int. J. Cancer, 16, 515 (1976).
CHEDID. L., LAMENSANS, A., PARANT, F., PARANT, M., ADAM, A., PETIT, J.F., LEDERER, E.: Cancer Res. 33, 2187 (1973).
CHEDID, L., PARANT, M., PARANT, F., GUSTAFSON, R.H., BERGER, F.M.: Proc. Nat. Acad. Sci., U.S.A., 69, 855 (1972).
CHIHARA, G., HAMURO, J., MAEDA, Y., ARAI, Y., FUKUOKA, F.: Cancer Res. 30, 2776 (1970).
CHIHARA, G., MAEDA, Y., HAMURO, J., SASAKA, T., FUKUOKA, F.: Sing. Nature 222, 687 (1969).
CONE, R.E., JOHNSON, A.G.: J. Exp. Med. 133, 665 (1971).
DE SOUSA, M.A.B., PARROT, D.M.V.: J. Exp. Med. 130, 671 (1969).
DONNER, M., VAILLIER, D.: p. 308 in "Investigation and Stimulation of Immunity in Cancer Patients" (G. Mathé and R. Weiner, eds.). Vol. 47 R.R.C.R., Heidelberg: Springer-Verlag 1974.
DRESSER, D.W.: Nature 191, 1169 (1961).
DRESSER, D.W.: Immunology 5, 378 (1962).
DRESSER, D.W.: Exp. Immunol. 3, 877 (1968a).
DRESSER, D.W.: Nature 217, 527 (1968b).
DRESSER, D.W.: The Immune Response Proc. Fourth Intern. Congress Pharmacol. 4, 192 (1970).
DRESSER, D.W., MITCHINSON, N.A.: Adv. Immunol. 8, 129 (1968).
DRESSER, D.W., PHILLIPS, J.M.: p. 3 in "Immunopotentiation". (Ciba Found. Symp. 18) 1 Vol., Amsterdam: Elsevier Excerpta Medica 1973.
DRESSER, D.W., TIEN-WEN TAO: Immunology 28, 443 (1975).

DRESSER, D.W., WORTIS, H.H., ANDERSON, H.R.: Clin. Exp. Immunol. 7, 817 (1970).

DUBOS, R.J., SCHAEDLER, R.W.: J. Exp. Med. 106, 703 (1957).

DUTTON, R.W., MISHELL, R.: J. Exp. Med. 126, 443 (1967).

FAUVE, R.M., HEVIN, M.B.: Ann. Inst. Pasteur, 120, 399 (1971).

FENYO, E.M., KLEIN, E., KLEIN, G., SWIECH, K.: J. Nat. Cancer Inst. 40, 69 (1968).

FINGER, E.H., BENEKE, G., EMMERLING, P.: Immunol. 154, 23 (1968).

FLORENTIN, I., HUCHET, R., BRULEY-ROSSET, M., HALLE-PANNENKO, O., MATHÉ, G.: Cancer Immunol. Immunoth., 1, 31 (1976).

FLOREY, H.W.: p. 40 in "General Pathology" (H.W. Florey, ed.). 4th ed., London: Immunol. Immunoth., 1, 31 (1976).

FLOERSHEIM, G.L.: Nature 216, 1235 (1967).

FLOREY, H.W.: p. 40 in "General Pathology" (H.W. Florey, ed.). 4th ed., London: Lloyd-Luke 1969.

FORD, W.L.: Cell Tissue kinetics, 2, 171 (1969a).

FORD, W.L.: Brit. J. Exp. Path., 50, 257 (1969b).

FREI, P.C., BENACERRAF, B., THORBECKE, G.J.: Proc. Nat. Acad. Sci. U.S.A., 53, 20 (1965).

FREUND, J.: Am. J. Clin, Path. 21, 645 (1951).

FREUND, J.: p. 46 in "The Nature and Significance of the Antibody Response (A.M. Pappenheimer, ed.). Vol. 1, New-York: Columbia Univ. Press (1953).

FROST, P., LANCE, E.M.: p. 29 in "Immunopotentiation". (Ciba Found. Symp. 18) 1 Vol., Amsterdam: Elsevier Excerpta Medica 1973.

GALL, D.: Immunology 11, 369 (1966).

GEFFARD, M., ORBACH-ARBOUYS, S.: Cancer Immunol. Immunoth., 1, 41 (1976).

GERSHON, R.K., KONDO, K., J. Immunol. 106, 1524 (1971a).

GERSHON, R.K., KONDO, K.: Immunology 21, 903 (1971b).

GERY, I., WAKSMAN, B.H.: J. Exp. Med. 136, 143 (1972).

GOLUB, E.S., WEIGLE, W.O.: J. Immunol 98, 1241 (1967).

GRESSER, I.: p. 327 in "Investigation and Stimulation of Immunity in Cancer Patients" (G. Mathé and R. Weiner, eds.). Vol. 47 R.R.C.R., Heidelberg: Springer-Verlag 1974.

GROOMS, G.A., MORTON, D.L.: Proc. Amer. Assoc. Cancer Res. (abstract No. 1160) 16, 261. (1975).

GUTTERMAN, J.U., MAVLIGIT, G., GOTTLIEB, J.A., BURGESS, M.A., McBRIDE, C.M., EINHORN, L., FREIREICH, E.J., HERSH, E.M.: New England J. Med. 291, 592 (1974).

HALL, J.G., MORRIS, B.: Quart. J. Exp. Physiol. 48, 235 (1963).

HALLE-PANNENKO, O., MATHÉ, G.: Transplant. Proc., in press (1976).

HALPERN, B.N., BIOZZI, G., STIFFEL, C.: p. 221 in "Role du Système Réticulo-Endothélial dans l'Immunité Antibactérienne et Anti-Tumorale" (B. Halpern, ed.). 1 Vol., Paris C.N.R.S. (1963).

HANNA, M.G. jr.: Seminars in Oncology 1, 319 (1974).

HANNA, M.G. jr., SNODGRASS, M.J., ZBAR, B., RAPP, H.J.: Nat. Cancer Inst. Monogr. 39, 71 (1973).

HANNA, M.G. jr., ZBAR, B., RAPP, H.J.: J. Nat. Cancer Inst. 48, 1441 (1972a).
HANNA, M.G. jr., ZBAR, B., RAPP, H.J.: J. Nat. Cancer Inst. 48, 1697 (1972b).
HELLSTROM, I., SJOGREN, H.O., WARNER, G., HELLSTROM, K.E.: Int. J. Cancer 7, 226 (1971).
HELLSTROM, I., WARNER, G.A., HELLSTROM, K.E., SJOGREN, H.O.: Int. J. Cancer 11, 280 (1973).
HIRSHAUT, Y., PINSKY, C., MARQUARDT, H., OETTGEN, H.F.: Proc. Am. Assoc. Cancer Res. (abstract No. 436) 14, 109 (1973).
HOWARD, J.G., CHRISTIE, G.H., SCOTT, M.T.: Cell. Immunol. 7, 290 (1973a).
HOWARD, J.G., SCOTT, M.T., CHRISTIE, G.H.: p. 101 in "Immunopotentiation". (Ciba Found. Symp. 18) Amsterdam: Elsevier Excerpta Medica 1973b.
HUNTER-CRAIG, I., NEWTON, K.A., WESTBURY, G., LACEY, B.W.: Brit. Med. J. 2, 512 (1970).
Immunopotentiation. A Ciba Foundation Symposium 18. 1 Vol. Amsterdam: Elsevier Excerpta Medica 1973.
ISHIZUKA, M., BRAUN, W., MATSUMOTO, T.: J. Immunol. 107, 1027 (1971).
JERNE, N.K., NORDIN, A.A., HENRY, C.C.: p. 109 in "Cell bound Antibodies" (B. Amos and H. Koprowski, eds.). Vol. 1, Philadelphia, Wistar Institute 1963.
JOHNSON, A.G., CONE, R.E., FRIEDMAN, H.M., HAN, I.H., JOHNSON, H.G., SCHMIDTKE, J.R., STOUT, R.D.: p. 157 in "Biological Effects of Polynucleotides" (R. Beers and W. Braun, eds.). Heidelberg: Springer-Verlag 1971.
JOHNSON, A.G., GAINES, S., LANDY, M.: J. Exp. Med. 103, 225 (1956).
JOLLES, P., MIGLIORE, D., BONHOMME, F.: Immunology 14, 159 (1968).
JOLLES, P., PARAF, A.: Chemical and biological basis of adjuvants. Vol. 13 Molecular biology biochemistry and biophysics. Heidelberg: Springer-Verlag 1973.
JOSEPH, R.R., BELPOMME, D., MATHÉ, G.: Brit. J. Cancer, 33, 567 (1976).
JUY, D., BONA, C., CHEDID. L.: C.R. Acad. Sci., Paris, 278, 2859 (1974).
KHALIL, A., BOURUT, C., HALLE-PANNENKO, O., MATHÉ, G., RAPPAPORT, H.: Biomedicine 22, 112 (1975).
KHALIL, A.M., MARTIN, M., BRULEY-ROSSET, M., MATHÉ, G.: Cancer Immunol. Immunoth., in press (1976a).
KHALIL, A.M., RAPPAPORT, H., MISSET, J.L.: Cancer Immunol. Immunoth., in press (1976b).
KIRCHNER, H., HOLDEN, H.T., HERBERMAN, R.B.: J. Immunol. 115, 1212 (1975).
KLEIN, E., HOLTERMANN, O.A.: Nat. Cancer Inst. Monogr. 35, 379 (1972).
KLEIN, E., HOLTERMANN, O.A., PAPERMASTER, B., MILGROM, H., ROSNER, D., KLEIN, L., WALKER, M.J., ZBAR, B.: Nat. Cancer Inst. Monogr. 39, 229 (1973).
KOGA, T., ISHIBASCHI, T., SUGIYAMA, K., TANAKA, A.: Int. ARch. Allergy 36, 233 (1969).
KUPERMAN, O., YASHPHE, D.J., SHARF, S., BEN-EFRAIM, S., WEISS, D.W.: Cell. Immunol. 3, 277 (1972).
LACOUR, F., SPIRA, A., LACOUR, J., PRADE, M.: Cancer Res. 32, 648 (1972).

LAMENSANS, A., MOLLIER, M.F., LAURENT, M.: Rev. Fr. Et. Clin. Biol. 13, 871 (1968).
LANDY, M., CHESSIN, L.W.: Antibioth. Chemoth. 15, 199 (1969).
LAWRENCE, W., SIMONSEN, M.: Transplantation 5, 1304 (1967).
LEVY, N.L.: Int. J. Cancer 11, 497 (1973).
LEVY, N.L., MAHALEY, M.S. jr., DAY, E.D.: Int. J. Cancer 10, 244 (1972).
LEVY, J.G., WHITNEY, R.B., SMITH, A.G., PANNO. L.: Brit. J. Cancer 30, 289 (1974).
LIKHITE, V.V.: Int. J. Cancer 14, 684 (1974).
LODD, N., CURTAIN, C.C., KIDSON, C.: Nature 214, 783 (1967).
LOUIS, J.A., CHILLIER, J.M., WEIGLE, W.O.: J. Exp. Med. 138, 1481 (1973).
MacKANESS, G.B.: p. 62 in "Infections Agents and Host Reactions" (Stuart Mudd ed.). Philadelphia: W.B. Saunders, Cy. Publ. 1970.
MacKANESS, G.B., LAGRANGE, P.H., ISHIBASHI, T.: J. Exp. Med. 139, 1540 (1974).
MacLENNAN, I.C.M., LOEWI, G.: Nature 219, 1069 (1968).
MacNEILL, T.A.: Immunology 18, 39 (1970).
MacNEILL, T.A., KILLEN, M.: Immunology 21, 751 (1971).
MAEDA, Y.Y., CHIHARA, G.: Nature 229, 634 (1971).
MAEDA, Y.Y., HAMURO, J., CHIHARA, G.: Int. J. Cancer 8, 41 (1971).
MARDINEY, M.R. jr., CHESS, L., BOCK, G.N., UNGARO, P.C., BUCHHOLZ, D.H.: p. 426 in "Investigation and Stimulation of Immunity in Cancer Patients" (G. Mathé and R. Weiner, eds.). Vol. 47 R.R.C.R., Heidelberg: Springer-Verlag 1974.
MARTYRE, M.C.: Biomedicine, in press (1976).
MARTYRE, M.C., WEINER, R., HALLE-PANNENKO, O.: p. 405 in "Investigation and Stimulation of Immunity in Cancer Patients" (G. Mathé and R. Weiner, eds.). Vol. 47 R.R.C.R., Heidelberg: Springer-Verlag 1974.
MARX, J.L.: Science 187, 1183 (1975).
MATHÉ, G.: in "European Society of Toxicology. Amsterdam: Excerpta Medica, 1976a, in press.
MATHÉ, G.: Rev. Fr. Et. Clin. Biol. 13, 881 (1968).
MATHÉ, G.: Adv. Cancer Research 14, 1 (1971).
MATHÉ, G.: p. 124 in "Complications of Cancer Chemotherapy" (G. Mathé and R.K. Oldham, eds.) Vol. 49 R.R.C.R., Heidelberg: Springer-Verlag 1974.
MATHÉ, G.: Cancer Immunol. Immunoth. 1, 3 (1976b).
MATHÉ, G., BINET, J.L., SEMAN, G., AMIEL, J.L., DAGUET, G.: p. 359 in "Tolerance Acquise et Tolerance Naturelle à l'égard de Substances Antigéniques Définies." Paris, C.N.R.S. 1963.
MATHÉ, G., DANTCHEV, D., POUILLART, P., FLORENTIN, I.: Presse Med. 1, 3135 (1972).
MATHÉ, G., DAUSSET, J., HERVET, E., AMIEL, J.L., COLOMBANI, J., BRULE, G.: J. Nat. Cancer Inst. 33, 193 (1964).
MATHÉ, G., HALLE-PANNENKO, O., FLORENTIN, I., BRULEY-ROSSET, M., KAMEL, M., HIU, I.J., BOURUT, C.: Europ. J. Cancer 11, 801 (1975).
MATHÉ, G., KAMEL, M., DEZFULIAN, M., HALLE-PANNENKO, O., BOURUT, C.: Cancer Res. 33, 1987 (1973).

MATHÉ, G., POUILLART, P., LAPEYRAQUE, F.: Brit. J. Cancer 23, 814 (1969).

MATHÉ, G., RAPPAPORT, H.: Histological typing of the neoplastic diseases of the haematopoietic and lymphoid tissues. 1 Vol. Geneva, World Health Organization (1976).

METCALF, D.: Brit. J. Cancer 15, 279, 769 (1961).

MIHICH, E.: Cancer Res. 29, 2345 (1969).

MITCHELL, M.S.: Biomedicine, in press (1976).

MITCHELL, M.A., KIRKPATRICK, D., MOKYR, M.B., GERY, I.: Nature, New Biol. 243, 216 (1973).

MODOLELL, M., LUCKENBACH, E.A., PARENT, M., MUNDER, P.G.: J. Immunol. 113, 395 (1974).

MOHYR, M.B., MITCHELL, M.S.: Cell. Immunol. 13, 264 (1975).

MOLLER, G., ANDERSON, J., SJOGREN, O.: Cell. Immunol. 4, 416 (1972).

MORTON, D.L., EILBER, F.R., JOSEPH. W.L., WOOD, W.C., TRAHAN, E., KETCHAM, A.S.: Ann. Surg. 172, 740 (1970a).

MORTON, D.L., EILBER, F.R., MALMGREN, R.A., WOOD, W.C.: Surgery 68, 158 (1970b).

NATHAN, C.F., KARNOVSKY, M.L., DAVID, J.R.: J. Exp. Med. 133, 1356 (1971).

NATHANSON, L.: Cancer Chemoth. Rep. 56, 659 (1972).

OLIVOTTO, M., BOMFORD, R.: Int. J. Cancer 13, 478 (1974).

PARR, I.: Brit. J. Cancer 26, 174 (1972).

PERLMANN, P., HOLM, G.: Immunol. 11, 117 (1969).

PERNIS, B., PARONETTO, F.: Proc. Soc. Exp. Biol., N.Y., 110, 390 (1962)..

PIMM, M.V., BALDWIN, R.W.: Nature 254, 77 (1975).

POUILLART, P., PALANGIE, T., SCHWARZENBERG, L., BRUGERIE, H., LHERITIER, J., MATHÉ, G.: Cancer Immunol. Immunoth., in press (1976).

POUILLART, P., PALANGIE, T., SCHWARZENBERG, L., BRUGERIE, H., LHERITIER, J., MATHÉ, G.: Biomedicine, 23, 469 (1975).

RAFF, M.C.: Transplant, Rev. 6, 52 (1971).

RAMON, G.: J. Med. Franç. 15, 381 (1926).

RAYNAUD, M., KOUZNETZOVA, B., BIZZINI, B., CHERMANN, J.C.: Ann. Inst. Pasteur 122, 695 (1972).

RIBI, E.E., MEYER, T.J., AZUMA, I., ZBAR, E.: Nat. Cancer Inst. Monogr. 39, 115 (1973).

RIOS, A., SIMMONS, R.L.: Transplantation 13, 343 (1972).

ROWLEY, D.A., FITCH, F.W., MOSIER, D.E., SOLLIDAY, S., COPPLESON, L.W., BRONW, B.W.: J. Exp. Med. 127, 985 (1968).

RUPP, J.C., MOORE, R.D., SCHOENBERG, M.D.: Arch. Path. 70, 43 (1960).

SARMA, P.S., SHIU, G., NEUBAUER, R.H., BARON, S., HUEBNER, R.J.: Proc. Nat. Acad. Sci. 62, 1046 (1969).

SAUTER, C., GIFFORD, G.E.: Nature 212,626 (1966).

SCHWARZENBERG, L., SIMMLER, M.C., PICO, J.L.: Cancer Immunol. Immunoth., 1, 69 (1976).

SCOTT, M.T.: Cell. Immunol. 5, 469 (1972a).

SCOTT, M.T.: Cell. Immunol. 5, 459 (1972b).

SCOTT, M.T.: Sem. Oncol. 1, 367 (1974).
SEDLACEK, H.H., MEESMANN, H., SEILER, F.R.: Proc. Amer. Assoc. Cancer Res. 16, 141 (abstract No. 64) (1975).
SIEGEL, B.V., MORTON, J.L.: Immunology 10, 559 (1966).
SJOGREN, H.O.: p. 51 in "The Role of Immunological Factors in Viral and Oncogenic Processes".(R.F. Beers, jr., R.C. Tilghman and E.G. Bassett, eds.). Baltimore: John Hopkins University Press 1974.
SMITH, G.V., MORSE, P.A. jr., DERAPS, G.D., RAJU, S., HARDY, J.D.: Surgery 74, 59 (1973).
SMITH, R.O., WOOD, W.B.: J. Exp. Med. 90, 567 (1949).
SPITZNAGEL, J., ALLISON, A.C.: J. Immunol. 104, 119 (1970b).
SPITZNAGEL, J., ALLISON, A.C.: J. Immunol. 104, 128 (1970a).
STEINKULLER, C.G., KRIGBAUM, L.G., WEISS, D.W.: Immunology 16, 225 (1969).
STEWART, T., TOLNAI, G.: Cancer 24, 117 (1969).
STJERNSWARD, J.: J. Nat. Cancer Inst. 36, 1189 (1966).
TAUB, R.N., KRANTZ, A.R., DRESSER, D.W.: Immunology 18, 171 (1970).
THOMAS, J.W., PLENDERLEITH, I.H., CLEMENTS, D., LANDI, S.: p. 315 in XIth International Cancer Congress, Florence October 1974. Vol. 1 Conferences, Symposia workshops.
TILL, J.E., McCULLOCH, E.A.: Rad. Res. 14, 213 (1961).
TRIPODI, D., PARKS, L.C., BRUGMANS, J.: New England J. Med. 289, 354 (1973).
UHR, J., MOLLER, G.: Adv. Immunol. 8, 81 (1968).
UNANUE, E.R., ASKONAS, B.A., ALLISON, A.C.: J. Immunol. 103, 71 (1969).
UNANUE, E.R., CEROTTINI, J.C.: Seminars in Hematology 7, 225 (1970).
VAN FURTH, R., COHN, Z.A., HIRSCH, J.G., HUMPHREY, J.H., SPECTOR, W.G., LANGEVOORT, H.L.: Bull. World Health Org. 46, 845 (1972).
WATSON, J., TRENKNER, E., COHN, M.: J. Exp. Med. 138, 699 (1973).
WEBB, D., BRAUN, W., PLESCIA, O.J.: Cancer Res. 32, 1814 (1972).
WESTPHAL, O., LUDERITZ, O., BISTER, F.: Z. Naturforsch. 7b, 148 (1952).
WHITE, R.G.: Brit. Med. Bull. 23, 39 (1967).
WILKINSON, P.C., O'NEILL, G.J., McINROY, R.J., CATER, J.C., ROBERTS, J.A.: p. 121 in "Immunopotentiation". (Ciba Found. Symp. 18). Amsterdam: Elsevier Excerpta Medica 1973.
WILLIAMS, A.C., KLEIN, E.: Cancer 25, 450 (1970).
WOLMARK, N., FISHER, B.: Cancer Res. 34, 2869 (1974).
WOODRUFF, M.F.A., DUNBAR, N.: p. 287 in "Immunopotentiation" (Ciba Found. Symp. 18). 1 Vol., Amsterdam, Elsevier Excerpta Medica 1973.
YOUNG, C.W., BITTAR, E.S.: Cancer Res. 34, 2675 (1974).
ZATZ, M.M., LANCE, E.M.: J. Exp. Med. 134, 224 (1971).
ZBAR, B.: p. 350 in "Investigation and Stimulation of Immunity in Cancer Patients" (G. Mathé and R. Weiner, eds.). Vol. 47 R.R.C.R., Heidelberg: Springer-Verlag 1974.
ZBAR, B., BERNSTEIN, I.D., RAPP, H.J.: J. Nat. Cancer Inst. 46, 831 (1971).
ZBAR, B., BERNSTEIN, I., TANAKA, T., RAPP, H.J.: Science 170, 1217 (1970).
ZBAR, B., RAPP, H.J., RIBI, E.A.: J. Nat. Cancer Inst. 48, 831 (1972).

Chapter 10
Conclusion: Immunotherapy and Immunorestoration Today in the Strategy of Cancer Treatment; Their Perspectives and that of Immunoprophylaxis

Apparently well-endowed with the data and concepts I have described, the medical oncologist is, in fact, still very poor in operational knowledge concerning the clinical use of active systemic immunotherapy, immunorestoration, local immunotherapy, and immunoprophylaxis.

He may be disturbed by some results which are, in the first approximation, contradictory: (a) agents which are, in certain circumstances, immunostimulating, may, in others, be immunosuppressive: in reality, there is no contradiction here, as immune reactions follow a kinetics (MATHÉ *et al.*, 1973, 1975a; DONNER and VAILLIER, 1974), and the results found may differ with the time of examination; moreover, there are factors which influence the type of immune responses, such as the dose, the route of administration, and the gap between the application of the adjuvant and that of the antigen (MATHÉ *et al.*, 1973, 1975a); (b) agents which, in some circumstances, help to reject tumours, may, in others, protect them by immunological enhancement, the mechanism of which is variable and probably complex (see Chapter 2); (c) systemic immunotherapy has been presented, from most experimental and clinical data, as the treatment of paucicellular (residual) tumours (MATHÉ, 1968), while some authors apply it to any presentation of any neoplasia (GUTTERMAN *et al.*, 1974a); (d) the simultaneous combination of immunotherapy and of oncostatic treatment, such as chemotherapy, is used by some (ISRAEL, 1974), whereas we use it according to strict rules, because chemotherapy applied just after BCG application can induce an intense immunodepression (MATHÉ *et al.*, 1974a, 1974b).

In practice, the medical oncologist is not torn between these contradictions. He has the knowledge given him in most experimental data and can follow a few simple rules, which enable him to avoid worsening the prognosis of the patient, and may help him to improve it. The first rule is that he should become less timorous: his intervention during the evolution of the tumour must be more frequent and must commence earlier than was the case before immunotherapy was introduced. He must teach and repeat that the indication for active systemic immunotherapy is the residual imperceptible disease left by surgery, radiotherapy, or chemotherapy, of which disease he can kill "the last cell" (MATHÉ, 1967). He should also make it clear that the ideal indication for active immunotherapy is not the voluminous terminal solid tumour; such tumours should be referred back to the surgeon or to the radiotherapist for reduction so that immunorestoration may be attempted, which, if successful, may render the patient sensitive to chemotherapy and/or immunotherapy (MATHÉ, 1966, 1975; MATHÉ *et al.*, 1976a).

For immunoprophylaxis, he should see the population concerned still earlier, before a tumour is detected, and he should try to organize with paediatricians, controlled trials on populations with a high cancer risk.

It must be remembered that the medical oncologist is not a clairvoyant: to modulate the immune condition of the patient, he must first evaluate it. He, therefore, needs a laboratory and, today, each hospital must have a cancer immunology laboratoy (see SIMMLER *et al.*, 1976a, 1976b; BRULEY-ROSSET *et al.*, 1976a) (see Appendix).

10.1. Evaluation of the Immune Status of the Patient

The first move for any oncologist, when examining a new patient, is to study his immune responses, particularly with the use of in vivo skin test methods and, if possible, some in vitro techniques. These will be considered in the Appendix.

The patient will then be classified: (a) totally immunodepressed, (b) presenting a dissociated insuffiency; or (c) presenting a normal immunoresponse.

10.2. Attempts at Immunorestoration in Case of Immunodepression

10.2.1. General Immunity

If the patient's responses are negative or very poor or dissociated, he has to become the object of a trial of immunorestoration, in order to make him a good responder (SIMMLER *et al.*, 1976b, 1976c), even if the envisaged treatment is chemotherapy: with POUILLART *et al.* (1976a), we have seen that, in patients with bronchial cancer, this treatment is much more efficient when skin tests are positive than when they are negative.

This necessity for normal immune responses, spontaneous or after restoration, is still more essential for systemic immunotherapy, as has been well demonstrated, experimentally by LEVY *et al.* (1974), and clinically (in melanoma patients with perceptibel tumours, submitted to chemotherapy interspersed with immunotherapy) by GUTTERMAN *et al.* (1974a), and by THOMAS *et al.* (1974).

It goes without saying that, if an observed immunodepression follows a course of chemotherapy, it is usually enough to wait a few weeks for spontaneous restoration (MATHÉ *et al.*, 1970; HERSCH *et al.*, 1971).

It should be recalled that immune-inhibited patients may become responders, in other words, they may be (or may seem to be) immunorestored by chemotherapy (pers. obs., see Chapter 6).

After local radiotherapy the immune responses must be carefully evaluated because this form of treatment induces a very intense immunodepression as we have shown experimentally with MARTIN *et al.* (1975a), and this is of long duration (see Fig. 6.3) as STJERNSWARD (1974) observed in patients. In such a case, an attempt at immunorestoration is indicated.

10.2.2. General Immunity and Sensitivity to Adjuvant(s) or Antigen(s) to be Used for Local Immunotherapy

We have seen that for local immunotherapy to be active, the patient has not only to present general immune responses, but he must respond specifically to the adjuvant(s) or antigen(s) that will be injected into the tumour; this has to be checked by specific skin tests. If the patient does not react when examined, he should be sensitized to the antigen used, whether it has an adjuvant action or not.

This sensitization is sufficient if the patient is not totally immunodepressed; if he is, he should first be submitted to a general immunorestoration (Fig. 10.1).

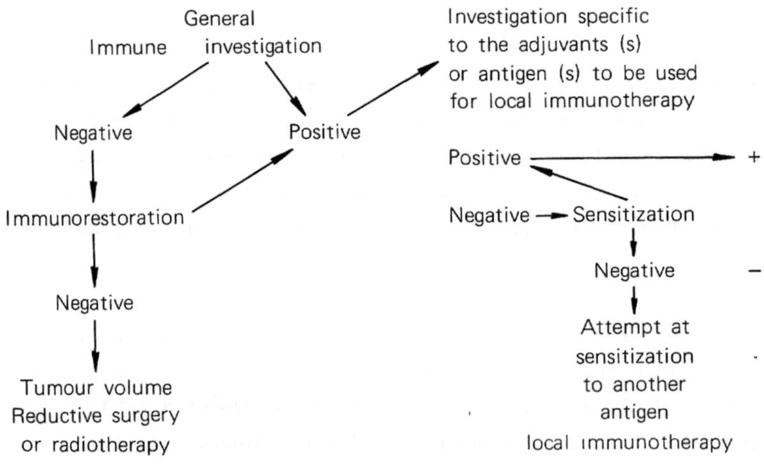

Fig. 10.1. Two examples of immunopreparation for local immunotherapy

10.3. The Patient is an Immunoresponder (without or after Immunorestoration)

10.3.1. Residual Imperceptible Disease

In this condition, the neoplasia is no longer perceptible because a primary or a series of primary treatments have been attempted to eradicate it, but the patient is supposed, from the staging of the disease, to carry a few cells that constitute what we call "the residual imperceptible disease" (see MATHÉ and CATTAN, 1976).

In such cases, systemic complementary chemotherapy, the results of which have been moderate or disappointing for 20 years (see MATHÉ and KENIS, 1975) but which now seem more promising (JAFFE *et al.*, 1974; BERLIN, 1974), is first indicated if the tumour is known to be chemotherapy-sensitive; in any case, systemic active immunotherapy is indicated, primarily in the case of a non-chemotherapy sensitive neoplasia, or complementary to systemic chemotherapy for 6 months to 2 years.

It is the indication of the residual imperceptible disease that has given the most convincing results in active systemic immunotherapy: in acute lymphoid leukaemia (MATHÉ *et al.*, 1968a, 1969a, 1976b), in acute myeloid leukaemia (POWLES *et al.*, 1973; VOGLER and CHAN, 1974; GUTTERMAN *et al.*, 1974b; MATHÉ *et al.*, 1975b), in chronic myeloid leukaemia (SOKAL, 1975), in leukaemic lymphosarcoma (MATHÉ *et al.*, 1975c, 1975d), in melanoma (BLUMING *et al.*, 1972; GUTTERMAN *et al.*, 1973, 1975a; MORTON, 1976) and in bronchus cancer (POUILLART *et al.*, 1976 not publ.).

Though there is an agreement on the indications for active systemic immunotherapy in this condition, which a few years ago was the only one considered as reasonable for this treatment, based on our experimental data (MATHÉ, 1968), a few points deserve discussion.

When to apply systemic active immunotherapy to residual disease left after chemotherapy?

We recognized the necessity for the patient to have a working immune machinery when immunotherapy is applied. We also saw that, in experimental animals carrying grafted tumours, there is an optimal time gap between chemotherapy and immunotherapy. This interval (CURRIE and BAGSHAWE, 1970) is probably necessary for the immune responses, which are suppressed by chemotherapy, to be spontaneously restored.

This post-chemotherapy immune restoration has been studied in man and the results, from the small amount of work that has been carried out, are probably valuable in determining the time for immunotherapy to be applied after chemotherapy. In our clinical experience, this time gap is about 15 days, after a cycle of 6 days combining 2 to 4 drugs (MATHÉ and KENIS, 1975). Some authors have described a rebound phenomenon following immunosuppressive chemotherapy cycles (CHEEMA and HERSH, 1971; SERROU *et al.*, 1974). Immune recovery, in such rebound cases, has also been said to be observed a few weeks after continuous long-term (3 years) chemotherapy (BORELLA and WEBSTER, 1971; GREEN and BORELLA, 1973).

Chemotherapy itself may have an immuno-stimulating effect, as shown by Lagrange and Mackaness (LAGRANGE *et al.*, 1974; MACKANESS *et al.*, 1974). These authors observed that cyclophosphamide, in certain conditions, is able to augment cell-mediated immunity, by releasing T-cells from the inhibiting influence of the humoral response. However, this occurs when the drug is given before the antigen; when it is given after, as is the case with cancer patients, T-cells become more susceptible to it.

Is it reasonable to intersperse chemotherapy and active systemic immunotherapy, or to apply them simultaneously?

Based on our first experimental results on L1210 leukaemia (MATHÉ, 1968; MATHÉ *et al.*, 1969b), from special experiments on sequential combinations of chemotherapy and active systemic immunotherapy in which BCG preceding cyclophosphamide diminished its antitumour action and induced a very strong immunosuppression (MATHÉ *et al.*, 1974a, 1974b), and from a clinical trial in which vincristine was administered during immunotherapy and diminished its effect (MATHÉ *et al.*, 1971), we have, over a long period of time, clinically applied immunotherapy exclusively following chemotherapy, when immune functions, suppressed by or during chemotherapy, are restored.

However, there have been some authors (POWLES, 1974; GUTTERMAN *et al.*, 1974b, 1975b) who have interspersed chemotherapy and systemic immunotherapy, and have obtained satisfactory results.

We have seen that several adjuvants act on stem cells, accelerating lymphocyte restoration (Fig. 9.1) (MATHÉ, 1974), and pushing the CFU in the cycle (POUILLART *et al.*, 1975, 1976b) (Table 9.I.). This may explain why we have been able, by administering BCG during the intervals between cycles of an intermittent chemotherapy, to shorten these intervals by accelerating haematological restoration (Fig. 3.48) (MATHÉ, 1974). Thus, such interspersing or combining of adjuvants and chemotherapy may enable us to give higher doses of cytostatics.

Moreover, authors who intersperse or mix chemotherapy and immunotherapy, if they reduce the efficiency of the lymphoid machinery, may take sufficient advantage of macrophage activation by the adjuvant (see Chapter 9), which may, in turn, add to the action of chemotherapy.

One hastens to add that, although most drugs used in chemotherapy are strongly immunosuppressive (MATHÉ and KENIS, 1975; LEVENTHAL *et al.*, 1974), there are some which are only minimally so. This is the case with imidazole carboxamide (DTC) (BRUCKNER *et al.*, 1974), which does not interfere with induction of delayed hypersensitivity to DNCB, does not suppress the primary immune responses to Vi antigen, and does not interfere with the secondary response to tetanus toxoid. The combination of this drug with immunotherapy is not theoretically unreasonable, and the remarkable results obtained by GUTTERMAN *et al.* (1974a) on melanoma patients treated with BCG combined with DTC may be explained by the properties of this drug. Hence, the interest in searching for non-immunosuppressive chemotherapy compounds: we have recently found, among a series of nitrosoureas, one derivative, RFCNU, (chloro-2-ethyl)-1 ribofuranosyl-isopropylidene-2'-3' paranitro-benzoate-5')-3 nitrosourea, which is strongly active against L 1210 leukaemia, and which does not affect immune functions (MATHÉ *et al.*, 1976c).

There is one more theoretical argument in favour of the possible interspersion or simultaneous combination of chemotherapy and immunotherapy: it is a fact that some drugs, or combination of drugs, may affect humoral-mediated immunity more than the cellular type. This has been shown to be true in the case of the combination of methotrexate followed, after 1 hour, by 5-fluorouracil, which selectively suppresses humoral antibody production without affecting cell-mediated immunity; the order and timing of administration is critical for this selective effect (DI LORENZO *et al.*, 1974). Cyclophosphamide is also known to affect antibody production, at certain doses, more than cell-mediated immunity (TURK and POULTER, 1972; TURK *et al.*, 1972), but great care must be taken in practice, because it is with this drug that we have obtained a very intense immunodepression in mice when it was administerd after BCG (MATHÉ *et al.*, 1974a).

Moreover, there is, as illustrated by Figures 6.1 and 3.12, an important dose-effect relationship (MATHÉ *et al.*, 1968b; THOMPSON *et al.*, 1972), with an optimal dose that exerts a strong antileukaemic effect and no immune depressive action (MATHÉ *et al.*, 1968b).

Finally, those who administer BCG and chemotherapy compounds simultaneously should not ignore other (non-immune) interactions between both types of agents. FARQUHAR (1974) demonstrated that the activities of several drugs metabolizing hepatic microsomal enzymes can be inhibited by BCG. This is particularly important in the use of DTC and cyclophosphamide, the effect of which depends partly on microsomal degradation.

In conclusion, at the present time, the use of the sequence (a) chemotherapy to reduce the neoplastic cell population as much as possible, (b) the time gap for chemotherapy-suppressed immune machinery restoration, (c) active systemic immunotherapy, is perfectly justified by most experimental data, and has proved to be safe and active in several clinical trials. The interspersion and simultaneous combination of chemotherapy and active systemic immunotherapy, if open to criticism in most circumstances, may be theoretically justified in special and very sophisticated conditions: these conditions must be precisely defined in man by "immunocompetent" specialists before they are used, because immunotherapy should not be applied without due care, as chemotherapy has been for too long (see MATHÉ and KENIS, 1975).

Is it reasonable to apply active systemic immunotherapy after surgery or radiotherapy as it is after chemotherapy?

This question may concern patients seen in the first perceptible phase of a tumour which has been submitted to radical surgery or radiotherapy.

The experimental work of BELEHRADEK *et al.* (1972) has not only shown that it is reasonable to apply immunotherapy after surgery, but that it is advisable to operate on the tumours before immunotherapy; they have shown that cell-mediated immunity can increase a few days after excision (BELEHRADEK *et al.*, 1972).

If surgery has left a few milligrams or grams of neoplastic tissue, clips can be left in to guide the application of a complementary radiotherapy.

The question arises here of the opportunity of applying systemic immunotherapy after radiotherapy. We have shown, experimentally (MARTIN *et al.*, 1975a), and STJERNSWARD (1974) has shown clinically, that radiotherapy, even when applied locally, leaves intense depression of the immune machinery. I quoted above an experiment by YRON *et al.* (1973) who failed to obtain a significant increase in the effect of radiotherapy on solid tumours by adding MER secondarily and suspected that this failure was due to the immunodepressive effect of irradiation.

However, in our laboratory, with MARTIN *et al.* (1975b), we have observed, in a strain of mice presenting long-term immunosuppression after radiotherapy, a remarkably favourable effect on Lewis tumours by immunotherapy applied after irradiation (Fig. 3.10). This result is very important for clinical cancer treatment. Active immunotherapy may be indicated after radiotherapy. The precaution to be taken is to verify that the immune responses have recovered (spontaneously or after immune restoration) before starting immunotherapy.

10.3.2. Perceptible Tumours Not Curable by Surgery Radiotherapy or Chemotherapy

Unfortunately, such tumours, which should no longer be seen, and would not if the patients had been submitted to the right strategy from the onset of disease, are still the lot of medical oncologists.

10.3.2.1. Indications for Local Immunotherapy

Local immunotherapy may be efficient in the case of perceptible accessible tumours which are non-operable, not radiosensitive, and not amenable to chemotherapy. A few tumours which have these characteristics are multiple melanoma, mycosis fungoides, and skin multiple metastasis of breast carcinoma or of other cancers.

One stumbling block of local immunotherapy use is that most of these tumours have been irradiated or reappear in irradiated sites; in such cases, local immunotherapy is not, or is very rarely, active (personal experience).

10. 3.2.2. Indications for Systemic Immunotherapy

A question which is still frequently asked is when to use active systemic immunotherapy in the case of perceptible, possibly voluminous tumours.

Here one is mainly dealing with neoplasias, often in relapse, for which radical surgery and/or radiotherapy are insufficient or conventionally "contra-indicated".

Once more, applying systemic active immunotherapy seems useless in the light of most experimental data, which showed that it is only able to kill a few cells (MATHÉ, 1968; PARR, 1972), and that, blocking factors are found in the serum of animals presenting perceptible (voluminous) tumours (HELLSTRÖM *et al.*, 1971, 1973; BALDWIN, 1973).

However, if most clinical data of the beneficial effect of active immunotherapy on perceptible tumours are far from being convincing, there are some which are impressive (GUTTERMAN *et al.*, 1974a; THOMAS *et al.*, 1974).

There are also, in the literature, about 130 cases of so-called spontaneous regressions of perceptible tumours (EVERSON and COLE, 1966), the reason for which may be a "spontaneous immunotherapy" due to some unknown adjuvants, for example micro-organisms. However, this could have been a local infection, in other words, a local immunotherapy.

A theoretical argument which upholds this possible indication for systemic immunotherapy in the case of perceptible tumours is that the effects of adjuvants do not only affect lymphocytes and macrophages via specific mechanisms. They also affect macrophages via a non-specific lytic action (BRULEY-ROSSET *et al.*, 1976b). The latter is not known to be suppressed by blocking factor(s), and we know that voluminous tumours that are intensively infiltrated by macrophages, such as Burkitt's lymphosarcoma (Fig. 2.3), have a good prognosis. It would be of interest to search for a correlation between histiocytic infiltration before or after immunotherapy, and the results of this treatment.

Here we have to recall other means which may be indicated for making active immunotherapy efficient. Since the blocking phenomenon, whatever its explanation, is bound to the large volume of the neoplasia, tumour volume reducing surgery (MATHÉ, 1966; MATHÉ and KENIS, 1975; MATHÉ *et al.*, 1976a) may be attempted (Fig. 10.2), which, even if it leaves one gram of neoplastic cells, may restore the tumour-host relationship, possibly by suppressing this immune blockage (see Chapter 2). When the neoplasia is chemotherapy-sensitive, oncostatics may be administered between tumour volume-reducing surgery and active immunotherapy (Figs. 10.3 and 10.4).

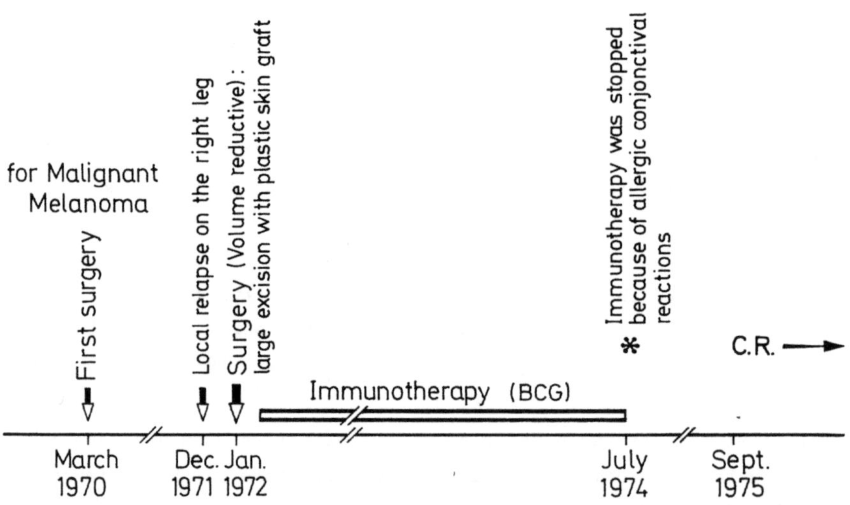

Fig. 10.2. Immunotherapy after tumour volume-reducing surgery in the case of a relapse melanoma (MATHÉ et al., 1976a)

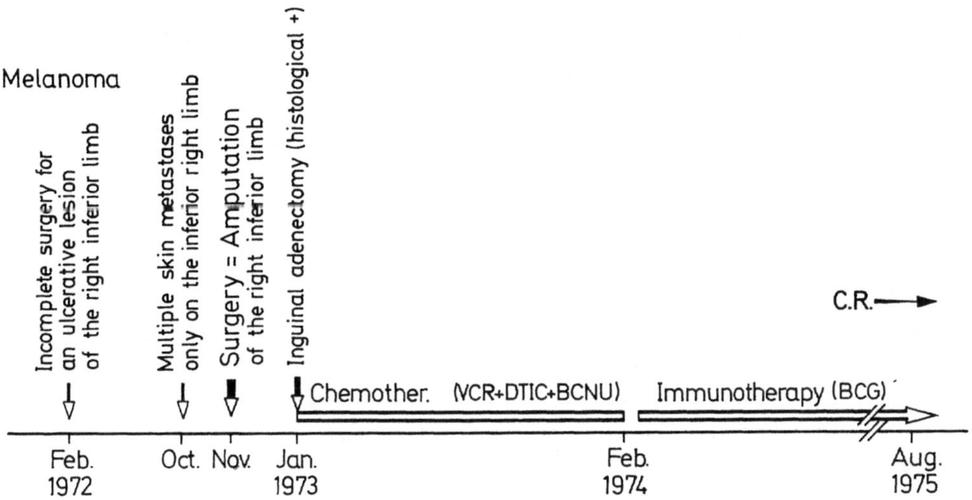

Fig. 10.3. Immunotherapy preceded by chemotherapy after tumour volume-reducing surgery in multiple skin metastasis of melanoma (MATHÉ et al., 1976a)

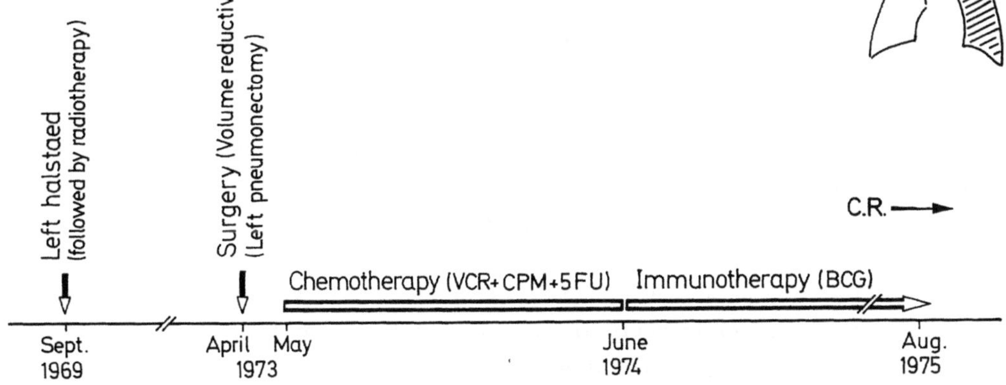

Fig. 10.4. Immunotherapy preceded by chemotherapy following tumour volume-reducing surgery in lung metastasis of a mammary carcinoma (MATHÉ et al., 1976a)

10.3.2.3. Combinations of Several Types of Immunotherapy

GRANT *et al.* (1974) have combined, in melanoma patients, the application of BCG on scarifications to the intratumoural injection of this mycobacterium and/or the administration of autologous irradiated tumour cells and/or the infusion of anti-tumour antibody; they report remarkable, but very short, regressions of the perceptible neoplasia(s).

SEIGLER *et al.* (1972, 1973) have tried what they call "multistage immunotherapy". First, they made the patients tuberculin-positive, if they were not so already. They then injected BCG into the tumours. They obtained autologous lymphocytes, which they then incubated with PHA, BCG, and irradiated melanoma cells. These lymphocytes were injected into the tumours into which they finally injected a mixture of neuraminidase-treated irradiated, autologous tumour cells and BCG. The authors claim that they have obtained about 20% complete remissions, without effect on visceral metastasis.

This result is not different from that obtained by MORTON (1971, 1972) and his collaborators (1970a, 1970b, 1971), and the procedure seems to be too sophisticated to bring consistent progress to immunotherapy; only the first stage is of great interest.

10.3.3. Surgery, Radiotherapy, Chemotherapy or Combination Curable Tumours

In such cases, it is ethical to begin treatment with either surgery, radiotherapy, chemotherapy or their combination which is known to effect cures (MATHÉ and CATTAN 1976).

However, curability by conventional treatment is far from 100%, even in patients for whom radical surgery radiotherapy, chemotherapy, or a combination is indicated.

Moreover, it has been shown recently that improvement in the rate of survival for several localizations of cancer has been much smaller during the 1960's than during the 1940's and the 1950's (CUTLER *et al.*, 1975).

We saw in a preceding paragraph that immunotherapy is able to improve curability when applied to the possible residual disease that the other forms of treatment may have left, which has been demonstrated for a few neoplasias and has to be studied in all tumours.

One is, therefore, curious to know if it would not be reasonable to start the treatment of some tumours by systemic immunopotentiation or by local immunotherapy.

As far as systemic immuno-intervention is concerned, it has been shown experimentally that, when applied before local radiotherapy (MOROSON and SCHECHTER, 1976), or chemotherapy (GIULIANI *et al.*, 1974), it potentiates their effects. Our experiments on its application before surgery are urgently needed.

As far as local immunotherapy is concerned, the courageous trials of KLEIN and coworkers (WILLIAMS and KLEIN, 1970), who used it exclusively to treat skin cancers, which are 90% curable by conventional means, and who obtained the same results with it, should prompt us to think about the use of this form of immunomanipulation as the first and, perhaps, the only treatment of other small cancers.

RAZA (pers. comm.) submitted their melanoma patients with poor prognosis to local BCG prior to surgery; they were free of disease at the time of publication. Since we know that this neoplasia is immunotherapy-sensitive (GUTTERMAN *et al.*, 1973; HOLMES *et al.*, 1975), and since its prognosis is not excellent after conventional treatment, at least in certain presentations (MATHÉ and CATTAN, 1976), it would not be unreasonable to choose this tumour in a controlled trial combining, first, local immunotherapy, and then conventional treatment.

It is also tempting to have a trial using local immunotherapy as the first or only treatment of small tumours, the conventional treatment of which leaves terrible functional or aesthetic sequelae, e.g. breast cancers. Local BCG was used by SPARKS *et al.* (1973) before surgical excision of a rat metastazing grafted (mammary) carcinoma: survival was considerably increased compared to controls (which were only operated on). This experimental result should make us venture to put forward this variety of cancer for a trial. It would, perhaps, be ethically and deontologically indicated to start with benign breast tumours, such as adenomas; WILLIAMS and KLEIN (1970) treated pre-malignant lesions of the skin, such as keratosis and leukoplasia with success.

10.4. Immunoprophylaxis of Cancer (Fig. 10.5.)

We have seen that vaccination by BCG against tuberculosis has raised hopes that infection by this mycobacterium is able, even in very small doses, and very early in life, to assure prophylaxis of childhood leukaemia (DAVIGNON *et al.*, 1970, 1971a, 1971b; ROSENTHAL *et al.*, 1972, 1974; HEMS and STUART, 1971).

The question then arises as to whether other modalities of BCG application (larger doses, repetition, etc.) could increase the prophylactic potential of BCG and other immunity adjuvants. This question will only be answered by randomized prospective trials. Not only leukaemia but all types of children's and adults' cancers should be considered.

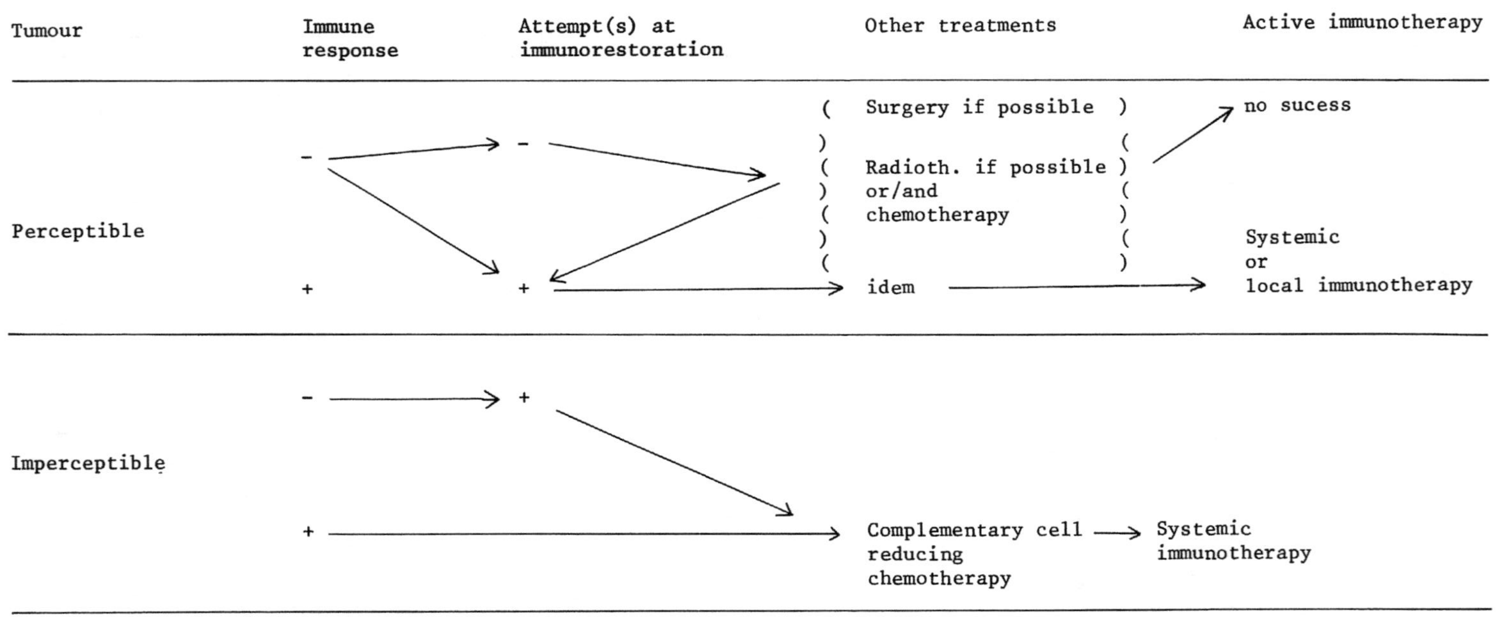

Fig. 10.5. Summary of indications of immunorestoration and active immunotherapy

References

BALDWIN, R.: Adv. Cancer Res., 18, 1 (1973).

BELEHRADEK, J., BARSKI, G., THONIER, M.: Int. J. Cancer 9, 461 (1972).

BERLIN, N.I. (ed.): A report to the profession from Breast Cancer Task Force, 30th September 1974, Department of Health Education and Welfare, Bethesda, USA.

BLUMING, A.Z., VOGEL, C.L., ZIEGLER, J.L., MODY, N., KAMYA, G.: Ann. Intern. Med. 76, 405 (1972).

BORELLA, L., WEBSTER, R.G.: Cancer Res. 31, 420 (1971).

BRUCKNER, H.W., MOKYR, M.B., MITCHELL, M.S.: Cancer Res. 34, 181 (1974).

BRULEY-ROSSET, M., BOTTO, H.G., GAUTHIER, H., POUILLART, P., FLORENTIN, I., HUGUENIN, P.: Israel J. Med. Sci., in press (1976a).

BRULEY-ROSSET, M., FLORENTIN, I., KHALIL, A.M., MATHÉ, G.: Int. Arch. Allergy and Applied Immunol. in press (1976b).

CHEEMA, A.R., HERSH, E.M.: Cancer 28, 851 (1971).

CURRIE, G.A., BAGSHAWE, K.D.: Brit. J. Cancer 1, 541 (1970).

CUTLER, S.J., MYERS, M.H., GREEN, S.B.: New England J. Med. 293, 122 (1975).

DAVIGNON, L., LEMONDE, P., ROBILLARD, P., FRAPPIER, A.: Lancet 2, 638 (1970).

DAVIGNON. L., LEMONDE, P., ROBILLARD, P., FRAPPIER, A.: Lancet 1, 80 (1971a).

DAVIGNON, L., LEMONDE, P., ROBILLARD, P., FRAPPIER, A.: Lancet 1, 799 (1971b).

DI LORENZO, J.A., GRISWOLD, D.E., BAREHAM, C.R., CALABRESI, P.: Cancer Res. 34, 124 (1974).

DONNER, M., VAILLIER, D.: p. 308 in "Investigation and Stimulation of Immunity in Cancer Patients" (G. Mathé and R. Weiner, eds.). Vol. 47 R.R.C.R., Heidelberg: Springer-Verlag 1974.

EVERSON, T.C., COLE, W.H.: Spontaneous regression of cancer. 1 Vol. Philadelphia: Saunders, ed. (1966).

FARQUHAR, D.: Proc. Amer. Assoc. Cancer Res. 15, 128 (1974).

GRANT, R.M., COCHRAN, A.J., HOYLE, D., MACKIE, R., MURRAY, E.L., ROSS, C.: Lancet, 2, 1096 (1974).

GIULIANI, F., CASAZZA, A.M., DI MARCO, A.: Biomedicine 21, 439 (1974).

GREEN, A.A., BORELLA, L.: Blood 42, 99 (1973).

GUTTERMAN, J.U., HERSH, E.M., McCREDIE, K.B., RODRIGUEZ, V., BODEY, G.P., FREIREICH, E.J.: Proc. Amer. Assoc. Cancer Res. 16, 127 (abstract No. 505) (1975b).

GUTTERMAN, J.U., McBRIDE, C.M., FREIREICH, E.J., MAVLIGIT, G., FREI, E. III, HERSH, E.M.: Lancet 1, 1208 (1973).

GUTTERMAN, J.U., MAVLIGIT, G., GOTTLIEB, J.A., BURGESS, M.A., McBRIDE, C.M., EINHORN, L., FREIREICH, E.J., HERSH, E.M.: New England J. Med. 291, 592 (1974a).

GUTTERMAN, J., MAVLIGIT, G., KENNEDY, A., HERSH, E.: Proc. Amer. Assoc. Cancer Res. 16, 245 (abstract No. 1096) (1975a).

GUTTERMAN, J.U., RODRIGUEZ, V., MAVLIGIT, G., BURGESS, M.A., GEHAN, E., HERSH, E.M., McCREDIE, K.B., REED, R., SMITH, T., BODEY, G.P. sr., FREIREICH, E.J.: Lancet 2, 1405 (1974b).

HELLSTROM, I., SJOGREN, H.O., WARNER, G.A., HELLSTROM, K.E.: Int. J. Cancer 7, 226 (1971).

HELLSTROM, I., WARNER, G.A., HELLSTROM, K.E., SJOGREN, H.O.: Int. J. Cancer 11, 280 (1973).

HEMS, G., STUART, A.: 1, 183 (1971).

HERSH, E., WHITECAR, J.P., MacCREDIE, K.B., BODEY, G.P., FREIREICH, E.J.: New Engl. J. Med. 285, 1211 (1971).

HOLMES, E.C., EILBER, F.R., MORTON, D.L.: J. Amer. Med. Assoc. 232, 1052 (1975).

ISRAEL, L.: p. 486 in "Investigation and Stimulation of Immunity in Cancer Patients" (G. Mathé and R. Weiner, eds.). Vol. 47 R.R.C.R., Heidelberg: Springer-Verlag 1974.

JAFFE, N., FREI, E. III, TRAGGIS, D., BISHOP, Y.: New England J. Med. 291, 994 (1974).

LAGRANGE, P.H., MacKANESS, G.B., MILLER, T.E.: J. Exp. Med. 139, 1529 (1974).

LEVENTHAL, B.G., COHEN, P., TRIEM, S.C.: Israel J. Med. Sci. 10, 866 (1974).

LEVY, N.L., SEIGLER, H.F., SHINGLETON, W.W.: Cancer 34, 1548 (1974).

MacKANESS, G.B., LAGRANGE, P.H., ISHIBASHI, T.: J. Exp. Med. 139, 1540 (1974).

MARTIN, M., BOURUT, C., HALLE-PANNENKO, O., MATHÉ, G.: Biomedicine, 23, 337 (1975b).

MARTIN, M., HUCHET, R., FLORENTIN, I., HALLE-PANNENKO, O., JURCZYK-PROCYK, S., PRITCHARD, L., MATHÉ, G.: in Fourth Annual Conference of the International Society for Experimental Hematology, Yugoslavia, September 1975a.

MATHÉ, G.: La chimiothérapie des cancers (leucémies, hématosarcomes et tumeurs solides). 2nd ed., Paris: Expansion Scientifique 1966.

MATHÉ, G.: Presse Med. 75, 2591 (1967).

MATHÉ, G.: Rev. Fr. Et. Clin. Biol. 13, 881 (1968).

MATHÉ, G.: p. 124 in "Complications of Cancer Chemotherapy" (G. Mathé and R. Weiner, eds.). Vol. 49 R.R.C.R., Heidelberg: Springer-Verlag 1974.

MATHÉ, G.: in "Société Internationale de Chirurgie", Edinburgh: September 1975.

MATHÉ, G., AMIEL, J.L., SCHWARZENBERG, L., SCHNEIDER, M., CATTAN, A., SCHLUMBERGER, J.R., HAYAT, M., VASSAL, F. DE: Rev. Fr. Et. Clin. Biol. 13, 454 (1968a).

MATHÉ, G., AMIEL, J.L., SCHWARZENBERG, L., SCHNEIDER, M., CATTAN, A., SCHLUMBERGER, J.R., HAYAT, M., VASSAL, F. DE: Lancet 1, 697 (1969a).

MATHÉ, G., AMIEL, J.L., SCHWARZENBERG, L., SCHNEIDER, M., HAYAT, M., VASSAL, F. DE, JASMIN, C., ROSENFELD, C., POUILLART, P.: Europ. J. Clin. Biol. Res. 16, 216 (1971).

MATHÉ, G., BELPOMME, D., POUILLART, P., SCHWARZENBERG, L., HAYAT, M., VASSAL, F. DE, JASMIN, C., AMIEL, J.L., MISSET, J.L., MUSSET, M.: Biomedicine, 23, 465 (1975c).

MATHÉ, G., CATTAN, A.: Cancérologie générale et clinique à l'usage du praticien et de l'étudiant. 1 Vol. Paris: Expansion Scientifique 1976.

MATHÉ, G., GARNIER, H., KUSS, R., BINET, J.P., LE BRIGAND, H., GENIN, J., PETIT, J.Y., CONSTANS, J.P., MUSSET, M., SCHWARZENBERG, L., VASSAL, F. DE, DELGADO, M.: J. Chirurgie, in press (1976a).

MATHÉ, G., HALLE-PANNENKO, O., BOURUT, C.: Transplant. Proc. 6, 431 (1974a).

MATHÉ, G., HALLE-PANNENKO, O., BOURUT, C.: Europ. J. Cancer 10, 661 (1974b).

MATHÉ, G., HALLE-PANNENKO, O., FLORENTIN, I., BRULEY-ROSSET, M., KAMEL, M., HIU, I.J., BOURUT, C.: Europ. J. Cancer, 11, 801 (1975a).

MATHÉ, G., HAYAT, M., SERROU, B., MONTERO, J.L., MORUZZI, A., IMBACH, J.L.: in "European Society for Toxicology". Amsterdam: Excerpta Medica, in press (1976c).

MATHÉ, G., KAMEL, M., DEZFULIAN, M., HALLE-PANNENKO, O., BOURUT, C.: Cancer Res. 33, 1987 (1973).

MATHÉ, G., KENIS, Y.: Biomedicine 18, 181 (1973).

MATHÉ, G., KENIS, Y.: La chimiothérapie des cancers (leucémies, hématosarcomes et tumeurs solides). 3rd ed., Paris: Expansion Scientifique 1975.

MATHÉ, G., MUSSET, M., SCHWARZENBERG, L., HAYAT, M., VASSAL, F. DE, AMIEL, J.L., POUILLART, P., MISSET, J.L.: Biomedicine, 23, 291 (1975b).

MATHÉ, G., POUILLART, P., LAPEYRAQUE, F.: Brit. J. Cancer 23, 814 (1969b).

MATHÉ, G., POUILLART, P., SCHWARZENBERG, L., HAYAT, M., AMIEL, J.L., VASSAL, F. DE: Brit. J. Cancer 31, 96 (1975d).

MATHÉ, G., SCHNEIDER, M., SCHWARZENBERG, L.: Rev. Fr. Et. Clin. Biol. 13, 695 (1968b).

MATHÉ, G., SCHNEIDER, M., SCHWARZENBERG, L.: Europ. J. Cancer 6, 23 (1970).

MATHÉ, G., VASSAL, F. DE, DELGADO, M., POUILLART, P., BELPOMME, D., HAYAT, M., SCHWARZENBERG, L., AMIEL, J.L., MISSET, J.L., MUSSET, M., JASMIN, C.: Cancer Immunol. Immunoth. 1, 77 (1976b).

MOROSON, H., SCHECHTER, M.: Biomedicine, in press (1976).

MORTON, D.O.: J. Reticuloendothel. Soc. 10, 137 (1971).

MORTON, D.L.: Nat. Cancer Inst. Monogr. 35, 375 (1972).

MORTON, D.L., EILBER, F.R., HOLMES, E.C., SPARKS, F.C., RAMMING, K.P.: Cancer Immunol. Immunoth., 1, 93 (1976).

MORTON, D.L., EILBER, F.R., JOSEPH, W.L., WOOD, W.C., TRAHAN, E., KETCHAM, A.S.: Ann. Surg. 172, 740 (1970a).

MORTON, D.L., EILBER, F.R., MALMGREN, R.A., WOOD, W.C.: Surgery 68, 158 (1970b).

MORTON. D.L., HOLMES, E.C., EILBER, F., WOOD, W.: Ann. Inter. Med. 74, 587 (1971). (1971).

PARR, I.: Brit. J. Cancer 26, 174 (1972).

POUILLART, P., PALANGIE, T., SCHWARZENBERG, L., BRUGERIE, E., LHERITIER, J., MATHÉ, G.: Biomedicine, 23, 469 (1975).

POUILLART, P., PALANGIE, T., SCHWARZENBERG, L., BRUGERIE, H., LHERITIER, J., MATHÉ, G.: Cancer Immunol. Immunoth. in press (1976b).

POWLES, R.: Cancer 34, 1558 (1974).

POWLES, R.L., CROWTHER, D., BATEMAN, C.J.T., BEARD, M.E.J., MacELWAIN, T.J., RUSSEL, J., LISTER, T.A., WITHEHOUSE, J.M.A., WRIGLEY, P.F.M., PIKE, M., ALEXANDER, P., HAMILTON FAIRLEY, G.: Brit. J. Cancer 28, 365 (1973).

ROSENTHAL, S.R.: Bull. Inst. Pasteur, 70, 29 (1972).

ROSENTHAL, S.R., CRISPEN, R.G., THORNE, M.G., PIEKARSKI, N., RAISYS, N., RETTIG, P.: p. 421 in "Investigation and Stimulation of Immunity in Cancer Patients" (G. Mathé and R. Weiner, eds.). Vol. 47 R.R.C.R., Heidelberg: Springer-Verlag 1974.

SEIGLER, H.F., SHINGLETON, W.W., METZGAR, R.S., BUCKLEY, C.E. III, BERGOC, P.M.: Ann. Surg. 178, 352 (1973).

SEIGLER, H.F., SHINGLETON, W.W., METZGAR, R.S., BUCKLEY, C.E., BERGOC, P.M., MILLER, D.S., FETTER, B.F., PHAUP, M.B.: Surgery 72, 162 (1972).

SERROU, B., DUBOIS, J.B., SILVA, R.: Proc. Amer. Assoc. Cancer Res. 15, 125 (abstract No. 497) (1974).

SIMMLER, M.C., RAMEAU, G., CHOU, M.J., MATHÉ, G.: Israel J. Med. Sci., in press (1976a).

SIMMLER, M.C., SCHWARZENBERG, L., MATHÉ, G.: Nouv. Presse Med., in press (1975b).

SIMMLER, M.C., SCHWARZENBERG, L., MATHÉ, G.: Cancer Immunol. Immunoth, in press (1976c).

SOKAL, J.E.: Proc. Amer. Assoc. Cancer Res. 16, 64 (abstract No. 255) (1975).

SPARKS, F.C., O'CONNELL, T.X., LEE, Y.T.N.: Surg. Forum 24, 118 (1973).

STJERNSWARD, J.: Lancet 2, 1285 (1974).

THOMAS, J.W., PLENDERLEITH, I.H., CLEMENTS, D., LANDI, S.: p. 315 in "XIth International Cancer Congress," Vol. 1. Florence: Conferences, Symposia workshops, October 1974.

THOMPSON, R.B., ALBEROLA, V., MATHÉ, G.: Europ. J. Clin. Biol. Res. 17, 900 (1972).

TURK, J.L., PARKER, D., POULTER, L.W.: Immunology 23, 493 (1972).

TURK, J.L., POULTER, L.W.: Clin. Exp. Immunol. 10, 285 (1972).

VOGLER, W.R., CHAN, Y.K.: Lancet 2, 128 (1974).

WILLIAMS, A.C., KLEIN, E.: Cancer 25, 450 (1970).

YRON, I., WEISS, D.W., ROBINSON, E., COHEN, D., ADELBERG, M.G., MEKORI, T., HABER, M.: Nat. Cancer Inst. Monogr. 39, 33 (1973).

Chapter 11
Appendix: Towards Monitoring the Immunological Status and Its Modification Induced by Immunotherapy, Immunorestoration or Immunoprophylaxis

Most cancer patients, at least at certain times during the course of the disease, are immunologically abnormal. It is, therefore, interesting to repeatedly explore their immune functions. ABLIN *et al.* (pers. comm.) proposed submitting all patients to an "immunostaging" and he defined rather naive immunostages" based on too simple and too complicated parameters (Table 11.I). Thus if it is too early to classify the immune status of cancer patients into grades or types, as we saw in Chapter 4 when describing the immunodeficiency, it is reasonable to submit them to a battery of tests at any important phase of their disease, especially before choosing a course of treatment, and to give the immune condition as much attention as is given to histocytological typing and topographical staging (MATHÉ and CATTAN, 1976).

Furthermore, since immunomanipulation may produce both favourable and unfavourable effects in patients, depending on the many factors which I have analysed in this book, it appears to be necessary to use all available tests to detect and, if possible, measure most functions of the immune machinery before, during, and after systemic or local immunotherapy, immunorestoration, or attempts at immunoprophylaxis, as well as the use of other means of treatment, surgery, chemotherapy, and radiotherapy, which are able to modify the immune condition.

Unfortunately, the tests available for investigating immune reactions are not as simple in their techniques and in their significance as those used by the microscopist for typing, and by the clinician for topographical staging.

There are, furthermore, three approaches: (a) general responses versus responses to tumour-associated antigens; (b) in vivo versus in vitro tests; (c) cellular versus humoral immunity exploration.

Some tests work "satisfactorily" today and must be used routinely; others are still experimental and can only be practised in oncology laboratories.

Table 11.I. Immunostaging of patients with malignancy proposed by ABLIN *et al.*, (pers. comm.)

"Immunostage"	Status	Immunity Humoral	Cellular
1	Strong	IgM	Present
2	Moderate	IgG	Present
3	Poor	IgM	Absent
4	Anergic	"Blocking" Antibody	Absent

11.1 Exploration of General Responses

Table 11.II lists the tests used in our clinical immunology laboratory (I.C.I.G. BROCHURE, 1973-1974), in an attempt to obtain information on the general immune status of the host.

The *in vivo tests* consist of studying skin reactions to different antigens.

(a) Those reactions to *recall antigens*, such as tuberculin protein purified derivative (PPD), candida, mumps, streptokinase-streptodornase, which investigate memory, recognition and inflammatory responses (CLAUSEN and SØBORG, 1969; HERSH *et. al.*, 1971; SCHNEIDER *et al.*, 1974; BREARD *et al.*, 1974; OLDHAM *et al.*, 1976; SIMMLER *et al.*, 1976a, 1976b) are the ones most frequently used. Their sensibility is poor, but their great advantage is that they can be repeated as often as necessary.

It is recommended that at least three recall antigens be used for each investigation, because it has been shown that, in such conditions, 90-100% of the normal population respond to at least one (AL-SARRAF *et al.*, 1970; EILBER and MORTON, 1970).

(b) Reactions to *primary antigens* (not previously met by the organism) are used to investigate sensitization and inflammatory responses, and the agents used are dinitrochlorobenzene (DNCB) (EILBER and MORTON, 1970), picryl chlóride (BLUMING *et al.*, 1972), keyhole limpet haemocyanin (KLH) (CURTIS *et al.*, 1970; HERSH *et al.*, 1974); they are more sensitive than recall antigen tests, but their great inconvenience is that, with the exception of KLH, they cannot be repeated as indicators of a primary response.

Table 11.II. Institut de Cancérologie et d'Immunogénêtique, Hôpital Paul Brousse, Laboratory of clinical immunology

Patient:

No.

Date:

Immunological investigation

General responses	Results
1. In vivo tests: skin reactions	
To recall antigens	Induration (mm)
Purified Protein Derivative (PPD)	
Candida	
Mumps	
Streptokinase-streptodornase	
To primary antigens	
Dinitrochlorobenzene (DNCB)	
Picryl chloride	
Keyhole limpet haemocyanin (KLH)	
2. In vitro tests	
2.1. Routine counts in peripheral blood of	(.../mm^3)
Lymphocytes	
Monocytes	
Immunoblasts	
2.2. Immunological and cytochemical counts with cell surface markers of	(.../mm^3)
T-lymphocytes (spontaneous rosettes with sheep red blood cells)	
B-lymphocytes (membrane immunoglobulins)	
Monocytes (peroxydase positive)	
Null cells	
2.3. Functions of mononuclear cells:	
2.3.1. Cell-mediated immunity	
Lymphocyte transformation by	c.p.m.(1)
Mitogens	
Phytohemagglutinin (PHA)	
Pokeweed mitogen (PWM)	

(1) c.p.m. = counts per minute

Antigens	stimulation index $(SI)^{(1)}$ %
KLH	
PPD	
Histocompatibility antigens	
(mixed lymphocytes reaction)	
Lymphocytotoxicity	$\%^{(2)}$
Toward permanent cell lines	
(^{51}Cr release)	
Due to K cells (L 1210 leukaemia target cells)	
Suppressor cells (inhibition of PHA transformation)	%
Leukocyte-inhibiting factor (LIF) in serum	%
2.3.2. Humoral-mediated immunity	
Immunoglobulin serum concentrations	g/l
IgG	
IgA	
IgM	
Antibody titer	Titer
Against KLH	

---

(1)
$$SI = \frac{\text{mean of c.p.m. in stimulated triplicate}}{\text{mean of c.p.m. in unstimulated triplicate}}$$

(2)
$$\%\ \text{lysis} = \frac{\text{cpm } ^{51}Cr \text{ released in experiment group}}{\text{cpm } ^{51}Cr \text{ released from cells by 4 x freeze-thaw}} \quad \frac{\text{cpm } ^{51}Cr \text{ released in autologous group}}{\text{cpm } ^{51}Cr \text{ released in autologous group}} \times 100$$

DNCB is a good "primary" antigen for seaching for a decreased delayed skin hypersensitivity; about 100% of normal subjects can be sensitized to it (BROWN *et al.*, 1967; EILBER and MORTON, 1970). Picryl chloride is, according to BLUMING *et al.* (1972), more rarely positive in normal subjects, and it would, therefore, be of special interest to detect immunostimulation.

(c) *KLH* has the two advantages of inducing a primary response type, even if it is injected repeatedly, and of simultaneously serving as an antigen for the investigation of cell-mediated and humoral immunity (CURTIS *et al.*, 1970).

In vitro tests comprise some which are included in laboratory routine examination, such as lymphocyte and monocyte counting; counting of circulating immunoblasts has been shown to be of interest in Chapter 4 (SCHNEIDER *et al.*, 1974).

The counting of T- and B-lymphocytes and a more precise counting of monocytes among the circulating mononuclear cells is now possible.

(a) Cytochemistry enables monocytes to be counted with more accuracy, because they give positive peroxydase reaction (HAYHOE and FLEMANS, 1970).

(b) Immune markers on the cell surface distinguish T-lymphocytes which form spontaneous rosettes with sheep red blood cells (SRBC) (FROLAND, 1972; PAPAMICHAIL *et al.*, 1972), while B-lymphocytes synthetise membrane immunoglobulins, demonstrable by immunofluorescence (FROLAND and NATVIG, 1971; WILSON and NOSSAL, 1971), and form, according to STATHOPOULUS and ELLIOTT (1974), rosettes with mouse red cells. Table 11.III indicates other tests for surface immunological markers of mononuclear cells and underlines those which must be used in cancer laboratories (MATHÉ and BELPOMME, 1974; Report of WHO/IARC, 1974).

Two different techniques are available for use in the SRBC rosette test: the so-called active rosette test in which the erythrocytes are incubated for only a very short time with lymphocytes, and the "total" rosette assay in which the contact is longer (WYBRAN and FUDENBERG, 1973).

WARA *et al.* (1976) have recommended that the rosette test be done in the presence of thymosin. A noticeable increase of rosettes produced by this hormone indicates that it would be efficient in the treatment of immunodeficiency syndromes. It may also distinguish among so-called "null cells", those which are to become thymocytes.

Mononuclear cells that are peroxydase-negative and present neither T nor B markers are, in fact, called "null cells"(JOSEPH *et al.*, 1976). According to CEROTTINI (personal communication), they may include K cells. JOSEPH *et al.* (1976), in our laboratory, found that their number is increased in the blood during long-term immunotherapy.

(c) Lymphocyte transformation, evaluated with tritiated thymidine incorporation and induced with mitogens and/or antigens, should also be practised routinely. However, although those tests are available in many laboratories today, attention should be drawn to the distribution of values in the normal population (Fig. 11.1) before interpreting their results (SIMMLER *et al.*, 1976a, 1976b).

Among mitogens, the most commonly used are phyto-haemagglutinin (PHA) concanavalin A (con A) and pokeweed mitogen (PWM), the first acting mainly on T-lymphocytes, the second identically on T-and B-cells (VALENTINE and LAWRENCE, 1971; DUBOIS *et al.*, 1973).

Table 11.III. Cell membrane markers of T- and B-lymphocytes and mononuclear macrophages used in clinical immunology laboratory

Markers	Tests	T	B	Third population[a]	
Receptor(s) (?) for SRBC	E rosettes	+	0	0	FROLAND, 1972 PAPAMICHAIL *et al.*, 1972
Receptor(s) (?) for MRBC	E rosettes	0	+ (on some)	0	STATHOPOULOS and ELLIOTT, 1974
Fc receptor(s) for ag-Ig complexes	EA rosettes [b]	0[c]	+ or 0[d] (on some)	+ (on some)	HAINZ, 1968 BASTEN *et al.*, 1972 SHEVACH *et al.*, 1973 HALLBERG *et al.*, 1973 FROLAND and NATVIG, 1973 BELPOMME *et al.*, 1974
Fc receptor(s) for Agg-IgG	Immunofluorescence with heat Agg-IgG	0	+	+(?)	DICKLER and KUNKEL, 1971
Activated C_3 (d or b) receptors	EAC rosettes	0	+ (on some)	+	HAINZ, 1968 BIANCO *et al.*, 1970 BASTEN *et al.*, 1972 SHEVACH *et al.*, 1973 ROSS *et al.*, 1973
Membrane Ig	Membrane Immunofluorescence	0	+	0	FROLAND and NATVIG, 1971 WILSON and NOSSAL, 1971
Specific T ag(s)	Cytotoxicity or immunofluorescence	+	0	0	TOBEN and COOPER, 1972 AISENBERG *et al.*, 1973 BROWN *et al.*, 1974
Specific B ag(s)	"	0	+	0	

Helix hemagglutinin Pomatia	Immunofluor-escence	+	0	0	HAMMARSTRÖM *et al.*, 1973
Receptors for Epstein-Barr Virus (EBV)	Rosettes	0	+	0	JONDAL and KLEIN, 1973

[a] Containing phagocytic and non-phagocytic monocytes as well as K cells.

[b] Using SRBC, OXRBC, or human RBC.

[c] + on educated T-lymphocytes.

[d] + using (EA sh.) sheep red blood cells sensitized by antibodies, or (EA ox.) ox red blood cells sensitized by antibodies.

- using (EA hu.) human red blood cells sensitized by antibodies.

SRBC = Sheep red blood cells

MRBC = Mice red blood cells

OXRBC= Ox red blood cells

ag-Ig = antigen-antibodies complexes.

Agg-IgG = heat aggregated IgG.

Ig = immunoglobulin.

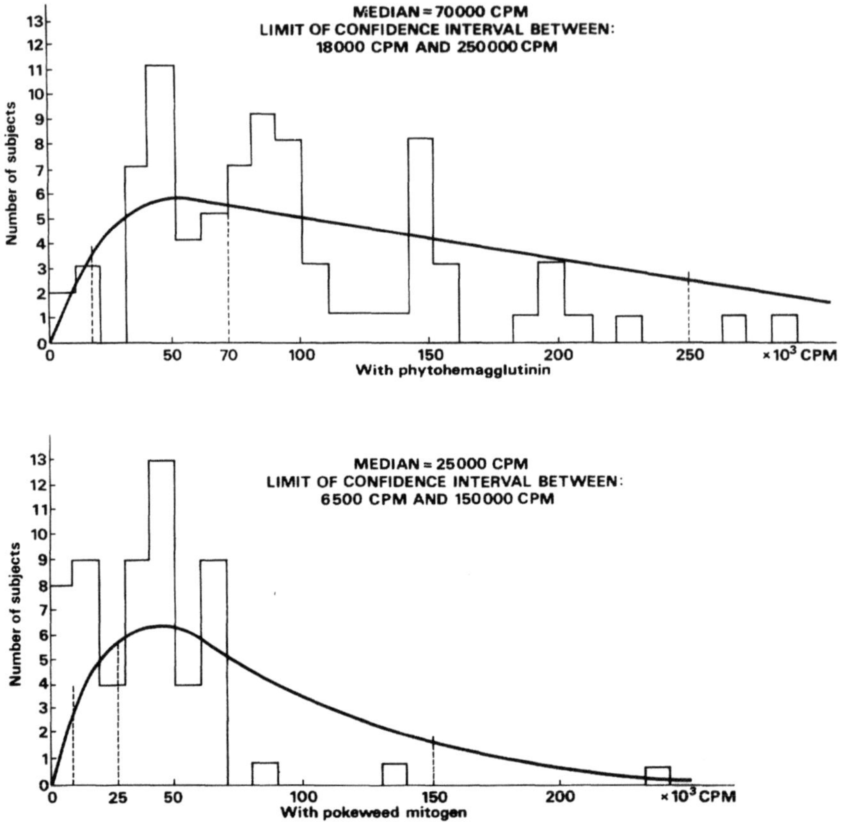

Fig. 11.1. Distribution of values in 80 normal subjects in logarithm normal for PHA and PWM transformations (SIMMLER et al., 1976a, 1976b)

Among antigens, KLH (CURTIS *et al.*, 1970) and PPD (DUBOIS *et al.*, 1973) are of interest because they allow, as seen above, comparison of skin reactions and lymphocyte transformation.

Another lymphocyte transformation test is based on the effect of a transplantation antigen which is nearer to tumour-associated antigen than KLH or PPD. This test is called the mixed lymphocyte reaction in which living lymphocytes of the subject to be tested are cultured with allogeneic mitomycin or irradiation sterilized lymphocytes (THURMAN *et al.*, 1973; BACH *et al.*, 1969).

(d) Recently, more tests have become available to evaluate lymphocyte functions: lymphocytotoxicity towards an established lymphoblastoid cell line (e.g., F 265), the lysis being evaluated by the release of ^{51}Cr carried by the target cells

(ROSENBERG *et al.*, 1974), and probably due to a mixture of B- and T-lymphocytes; cytotoxicity due to K cells, with a recent method using L1210 murine leukaemia cells as the target and a rabbit antitumour serum (CEROTTINI and BRUNNER, 1974) or another, using antibody-coated Chang cells (McLENNAN and LOEWI, 1968).

The search for suppressor cells (GERSHON *et al.*, 1972; WALDMAN *et al.*, 1974; GEFFARD and ORBACH, 1976) is still technically in the experimental stage, although its theoretical basis is simple: it might consist of inhibiting PHA transformation of normal lymphocytes by the population in which suppressor cells are sought.

Lymphocyte responses may be studied in the serum by the search for macrophage-inhibiting factor (MIF) (COHEN *et al.*, 1974) and/or for leucocyte-inhibiting factor (LIF) (ROSSO DI SAN SECONDO *et al.*, 1974), which are two lymphokines released during cell-mediated immunity reactions against various unknown antigens. The tests consist of evaluating migration inhibition (by the studied serum) of macrophages or leucocytes out of capillary tubes (Fig. 11.2). MIF (COHEN *et al.*, 1974) and LIF (BRULEY-ROSSET *et al.*, 1976) have been found in the serum of patients with various cancers.

(e) Finally, the humoral branch of immunity can be evaluated by the levels of the different immunogobulins in the serum, the immunodiffusion method (LAURELL's or MANCINI's, (see HARTMANN, 1971), which gives IgG, IgA, and IgM concentrations, being the only satisfactory one, and by the measurement of serum antibodies, those against

Fig. 11. 2. Migration patterns of human blood leucocytes in normal human sera (A + B) and sera of cancer patients (C + D)

KLH (CURTIS *et al.*, 1970) being the most interesting in as much as their level can be compared with the skin reactions and lymphocyte transformation against this antigen.

The correlation between different tests has been studied. SIMMLER *et al.* (1976a) showed that there is a good correlation between the mitogen response to PHA in vitro, and the in vivo reactions to recall antigens in patients with solid tumours before treatment (Table 6.XII). This correlation is decreased by chemotherapy, so that in vivo responses are more often reduced than in vitro transformation to mitogens (chemotherapy seems to affect a peripheral event more than a central proliferation of immune cells). SIMMLER *et al.* (1976a) also showed, in another group of patients, that there is no correlation between in vivo and in vitro responses to PPD, although a good correlation is found between in vitro transformation by mitogens PHA and PWM, and by PPD.

The correlation between skin tests and serum LIF activity is variable, for example, with the type of cancer: it is good for epidermoid bronchial and breast carcinoma patients; it is nil for subjects suffering from glioblastoma as observed in our laboratory by BRULEY-ROSSET *et al.*, (1976).

When the responses of skin tests and of in vitro tests, which must be made simultaneously, are dissociated, the results of the in vitro assays should be considered as the more accurate; it has been shown that skin tests may become positive in anergic patients after treatment with proteolytic enzymes such as protease I (THORNES, 1974). These observations should make us wary when interpreting the results of these tests as strict indicators of cell-mediated immunity.

11.2. Attempts at Evaluating Responses to Tumour-Associated Antigens

Both in vivo and in vitro approaches have been attempted in evaluating host responses against tumour-associated antigens.

In vivo tests have been proposed in which skin reactions are searched for palpation and measured at the site of i.d. injections of crude or solubilized extracts of autologous neoplasias (HUGHES and LITTEN, 1964; STEWART, 1969; FASS *et al.*, 1970; BLUMING *et al.*, 1971; OREN and HERBERMAN, 1971; HERBERMAN *et al.*, 1973; HERBERMAN, 1974). Reactions considered as positive have sometimes been observed in patients with leukaemias or Burkitt's lymphosarcoma, more frequently in remission than in the perceptible phase in the disease (BLUMING *et al.*, 1971; HERBERMAN *et al.*, 1973), in ocular melanoma (HERBERMAN *et al.*, 1974), and in carcinomas (same authors).

If positivity is not constant with tumour extracts, negativity is no more constant with normal tissue extracts. Bacterial contamination is one of the most critical factors (STEWART, 1969). Viral contamination must also be suspected. Non-specific reactions may depend on the protein concentration of the extract or may be directed against organ antigens as in the case of melanoma (BLUMING *et al.*, 1974).

The use of allogeneic tumour cells or cells from established cultured lines is still more open to criticism: a study which we conducted in acute leukaemia patients using such material gave results which could not be correlated to any parameters of the disease (OLDHAM *et al.*, 1976). Reactions against histocompatibility antigens, or cell line contaminating virus such as Epstein-Barr Virus (EBV), should be expected which could mask reactions against any tumour-associated antigen.

In vitro methods have been developed by which the immune reaction(s) against tumour-associated antigens can be studied (PERLMANN and HOLM, 1969; OETTGEN *et al.*, 1972; BLOOM and GLADE, 1971; OETTGEN and HELLSTROM, 1973; BANSAL, quoted by BANSAL and SJOGREN, 1974; MATHÉ and WEINER, 1974; BERKELHAMMER, 1974).

Several methods of detecting cell-mediated immunity have been developed.

(a) Lymphocyte transformation by autologous tumour cells (LTATC) has been claimed in acute leukaemia (FRIDMAN and KOURILSKY, 1969; LEVENTHAL *et al.*, 1972), as has an increase of the transformation level by immunization with tumour cells (POWLES *et al.*, 1971; GUTTERMAN *et al.*, 1973). It has also been described in various sarcomas (STJERNSWÄRD *et al.*, 1973; VANKY and STJERNSWÄRD, 1971; STJERNSWÄRD *et al.*, 1972), and melanoma (NAGEL *et al.*, 1971).

Both lymphocytes (CHESS *et al.*, 1972) and tumour cells (HALTERMAN *et al.*, 1973) can be frozen and stored for use on several assays.

This test is simple. However, it should be applied only to autologous (or isogeneic) tumour cells, as any histo-incompatible material induces lymphocyte stimulation. Moreover, though it is operational in many laboratories, care should be taken when accepting an increase of thymidine incorporation as significant (this is used to measure lymphocyte DNA synthesis) because of frequently observed non-specific fluctuations (SCHWEITZER *et al.*, 1973).

Extracted antigen(s) (after hypertonic or hypotonic lysis) may replace tumour cells (GUTTERMAN *et al.*, 1972a).

The following two tests have been described in Chapter 2.

(b) Colony inhibition assays, whatever the technique (HELLSTRÖM, 1967; HEPPNER and KOPP, 1971), utilize the ability of tumour cells to form colonies; the tested lymphocytes are added to the target cells and the mixture is incubated for several days, after which the number of colonies in the presence of sensitized lymphocytes is compared to the number found in the presence of control non-sensitized lymphocytes and to the number found when the immune serum, in which blocking factors are sought, is added one hour before addition of lymphocytes and removed just prior to this addition (Fig. 2.20).

Colony inhibition can be increased by prior stimulation of effector lymphocytes (HAYRY and DEFENDI, 1970; GOLUB *et al.*, 1973).

This test seems to be difficult to reproduce in many laboratories and its results are difficult to interpret, because of the great number of different parameters and cell populations involved.

(c) In cytotoxicity assays, cell lysis rather than cytostatism is evaluated.

The cellular technique (TAKASUGI and KLEIN, 1970) utilizes smaller quantities of material and a shorter incubation period than the colony-inhibition technique. The number of tumour cells remaining in the presence of effective lymphocytes is compared with the number in their absence. TAKASUKI *et al.* (1973) have developed a method of quantifying the microcytotoxicity assay by electron-image analysis.

A simpler and cheaper alternative consists of labelling tumour cells with an isotope and of quantifying its release as a measure of target cells damaged by lymphocytes. Many isotopes have been used: ^{3}H-thymidine, ^{125}I-iododeoxyruridine, ^{3}H-uridine, ^{3}H-proline, technetium 99m (BERKELHAMMER, 1974). ^{51}Cr assay (BRÜNNER *et al.*, 1968; HERBERMAN *et al.*, 1973) has the advantage that the maximum isotope release occurs in a few hours, as compared with several days for other isotopes. This test can be used on tumour cells which have not been grown in tissue culture. The release of this isotope from the proteins of the cells is generally considered as a good sign of cell death. Its disadvantages are that the level of specific reactivity is often low and the level of spontaneous release is often high.

The results given by these tests are the object of two main criticisms, other than technical ones. The first concerns the identity of the effector cell: it seems that only the short-term ^{51}Cr release assay would detect T-dependent effector cells (PLATA *et al.*, 1974), whereas the long-term incubation tests reveal not only the action of these lymphocytes but also that of thymus-independent cells (HERSEY *et al.*, 1973; O'TOOLE *et al.*, 1973). The second concerns specificity: although positive results were found in many types of tumours by different

N *et al.*, 1973) either by the chromium release assay or by the micro- assay, no correlation with the status of the disease was ever found, results are very often obtained with normal populations of lymphocytes.

n inhibition assays. Macrophage-inhibiting factor (MIF) activity was after exposure of lymphocytes from tumour-bearing animals to either : cells (KRONMAN *et al.*, 1969; ANDERSON *et al.*, 1973) or soluble tumour LTZER *et al.*, 1972). ANDERSEN *et al.* (1970) applied the system to man, that they found inhibition in the case of malignant mammary tumours and ase of benign breast neoplasia.

inhibitory factor (LIF) assay has also been proposed by BENDIXEN and SØBORG ROCKLIN *et al.* (1974): it is based on the inhibition of migration of human ut of capillary tubes in the presence of specific tumour associated The battery of tests that we use at present is given in Table 11.IV.

Institut de cancérologie et d'immunogénétique, Hôpital Paul-Brousse,
f clinical immunology

Patient:

No.

Date:

investigations

tumour associated antigen(s)

---

ests: skin reactions	Results: induration
) extracts of autologous neoplasia)	mm
) extracts of (allogeneic) cells)	

of) ( ) ( ) (

Table 11.IV. (continued)

2. In vitro tests	
2.1.1. Lymphocyte transformation by	c.p.m.
Autologous tumour cells	
fresh (with fresh AB serum	
fresh) with patient's serum	
frozen (with fresh AB serum	
frozen) with patient's serum	
Antigens extracted from tumour cells	
after hypertonic lysis (with fresh AB serum	
after hypertonic lysis)with patient's serum	
after hypotonic lysis (with fresh AB serum	
after hypotonic lysis)with patient's serum	
2.1.2. Leukocyte inhibitory factor (LIF) activated by:	%(1)
Tumour cells	
Antigens extracted from tumour cells	
after hypertonic lysis (with foetal calf serum	
after hypertonic lysis)with patient's serum	
after hypotonic lysis (with foetal calf serum	
after hypotonic lysis)with patient's serum	
2.1.3. Lymphocytotoxicity	%
Toward a permanent tumoral cell line	
(^{51}Cr release)	
with anti-cell line serum	(1)
with patient's serum	(2)
2.2. Humoral mediated immunity	
2.2.1. Immunofluorescence to detect antibodies against autologous tumour cells	% of fluorescent cells
Indirect technique	
2.2.2. Other assays	10^{-12} or ng
Direct technique Radioimmunoassy	

(1) Percentage of lysis by patient's lymphocytes with anti-cell line serum versus foetal calf serum.

(2) Percentage of lysis by patient's serum versus feotal calf serum.

MIF and LIF, although they are produced by a specific immunological reaction, are not specific for the target cells used. This explains the possible use of their search as a test for general immune capacity, and as a test for the "specific" one: the tests are called direct if factors are studied directly in the serum; the tests are called indirect if they are searched after in vitro reactivation of lymphocytes by tumoural antigens

The results of these tests, when used for "specific" tumour immunity, must be interpreted with caution in as much as non-specific toxicity of many neoplasia extracts has been observed (ROSENBERG, 1973).

Although it is too early to draw conclusions about the validity of migration inhibition tests, the following should be underlined: the lymphocytes tested have to be sensitized to give positive results; thus, assays can be done with histo-incompatible cells of the same antigenic type as that of the tumour; the patients are supposed to be immunized against tumour associated antigens and not against histocompatibility antigens; the time of incubation of the test is actually short enough to avoid an in vitro sensitization against these histocompatibility antigens.

Humoral-mediated tumour immunity has not been attempted as often as cell-mediated immunity because immunoglobulins are considered to assume a less important role.

Immunofluorescence, widely used for detection of TAA (LEVY *et al.*, 1972), is less easily utilized in the search for antibodies (Figs. 11.3 and 11.4). The direct technique involves labelling each test antiserum with fluorescein and reacting the antiserum with the tumour cells. The indirect technique, which is more practical because it enables unlabelled antiserum to react with tumour cells and then uses fluorescein-labelled anti-immunoglobulin serum to react with the test serum, is less specific.

Cytotoxic assays (Fig. 11.5) were developed by HELLSTRÖM and HELLSTRÖM (1971) and specially used by LEWIS (1967) and by his collaborators (1969) in melanoma patients: using fresh tumour cells and an exogenous source of complement, this author claimed that he was able to detect autologous antibodies. However, since sera do not lyse cells

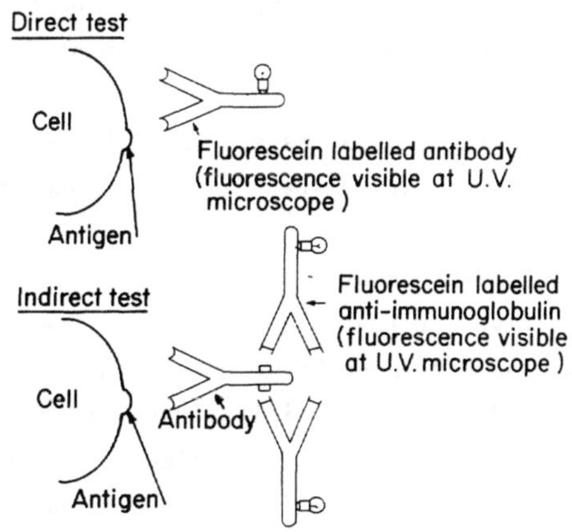

Fig. 11.3. Scheme of the immunofluorescence test

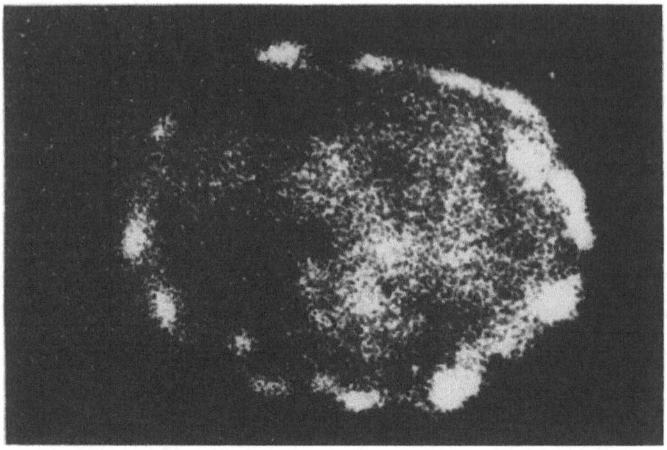

Fig. 11.4. Membrane immunofluorescence on a living cell revealing surface tumour-associated antigens

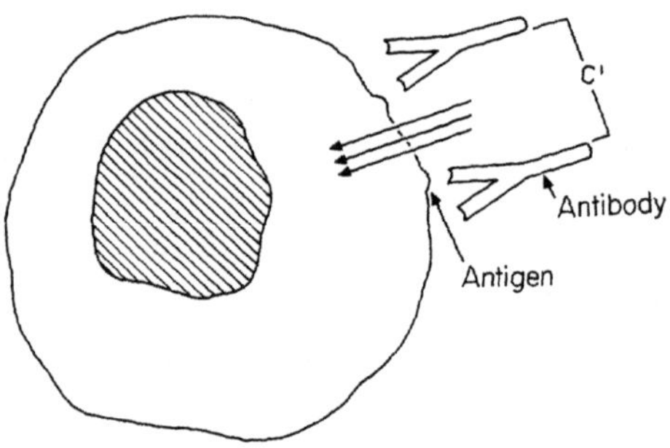

Fig. 11.5 Cytotoxicity test: the lesion is induced by the complex C'-antibody fixed on a cell surface

well, and since some contain hetero-antibodies, it is difficult to obtain the appropriate source of complement for each tumour. LEVY *et al.* (1972) described a very useful method of obtaining complement free of these antibodies.

Complement fixation assays (Fig. 11.6) (OLD *et al.*, 1968) have been rather disappointing as far as specificity is concerned (DORÉ *et al.*, 1967).

Radio-immuno-assay may one day supplant all other techniques, because it is the most sensitive serological test, and a great deal of experience has been acquired in the study of carcino-embryonic antigen(s) (CEA) (THOMSON *et al.*, 1969; ALPERT, 1972). It is a modified co-precipitation-inhibition technique, involving the combination of antibodies with labelled antigens and the competitive binding of antigens in the test serum. A triple isotope double-antibody radio-immuno-assay has been developed

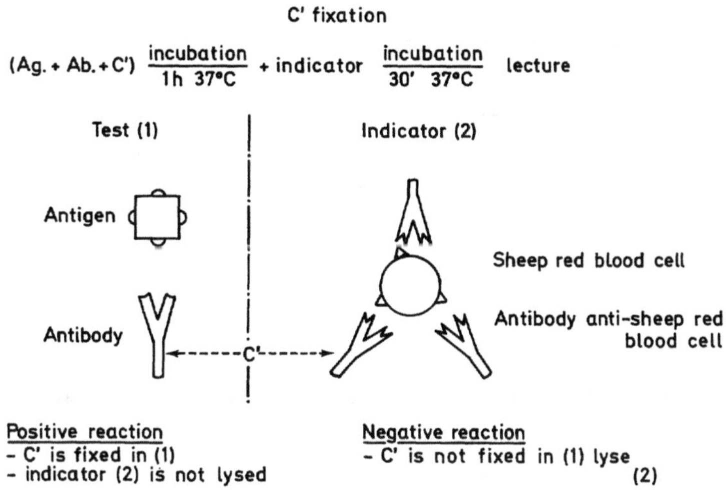

Fig. 11.6. Complement fixation test

(COLIGNAN *et al.*, 1972) for carcino-embryonic antigen(s), which requires small amounts of serum.

No results have yet been obtained for tumour antigens other than CEA. However, reactions against these antigens may have to be considered in evaluating the host immune behaviour with regard to cancer cells.

In conclusion, the study of serum antibody remains largely undeveloped, mainly because present concepts have persuaded us, as seen in Chapter 2, that the significance of complement-dependant immunoglobulin-antitumour effect is small, if not negligible, although IKONOPISOV *et al.* (1970) and CURRIE *et al.* (1971) showed its slight increase in melanoma patients treated with tumour extracts.

This is not a reason to neglect the search for the presence of antibodies, because we have seen their role in K-cell lysis, and because antigen-antibody complexes, and as perhaps certain antibodies are involved in the blocking of cell-mediated immunity (see Chapter 2).

The assays used to detect serum blocking factors are those utilized in evaluating cell-mediated immunity: the specific suppression of their responses is sought.

Several authors have claimed that they have proved the suitability of inhibiting lymphocyte tranformation in the serum of cancer patients (MAVLIGIT and ALEXANIAN,

1974), in addition to colony inhibition (Fig. 2.20) (HELLSTRÖM and HELLSTROM, 1971), cytotoxicity (BALDWIN *et al.*, 1973), and migration inhibition by lymphocytes (VANKY *et al.*, 1971; GUTTERMAN *et al.*, 1972b).

Such results must be regarded with caution. The significance of the blocking phenomenon is far from clear, and the correlation between in vitro demonstration of serum blocking factor(s) and in vivo events is still unknown (BERKERLHAMMER, 1974; LEVY and LECLERC, 1975).

References

AISENBERG, A.C., BLOCH, K.J., LONG, J.C., COLVIN, R.B.: Blood 41, 417 (1973).

AL-SARRAF, M., WONG, P., SARDESAI, S., VAITKEVICIUS, V.K.: Cancer 26, 262 (1970).

ALPERT, E.: Nat. Cancer Inst. Monogr. 35, 415 (1972).

ANDERSEN, V., BJERRUM, O., BENDIXEN, G., SCHIØDT, T., DISSING, I.: Int. J. Cancer 5, 357 (1970).

ANDERSON, J.M., KELLY, F., WOOD, S.E., RODGER, K.D., FRESNEY, R.I.: Brit. J. Cancer 28 (suppl. 1), 83 (1973).

BACH, M.L., BACH, F.S., JOO, P.: Science 166, 1520 (1969).

BALDWIN, R.W., EMBLETON, M.J., JOWES, J.J.P., LANGMAN, M.J.S.: Int. J. Cancer 12, 73 (1973).

BANSAL, S.C., SJOGREN, H.O.: Israel J. Med. Sci. 10, 939 (1974).

BASTEN, A., SPRENT, J., MILLER, J.F.: Nature, New Biol. 235, 178 (1972).

BELPOMME, D., DANTCHEV, D., DU RUSQUEC, E., GRANDJON, D., POUILLART, P., SCHWARZENBERG, L., AMIEL, J.L., MATHÉ, G.: Biomedicine 20, 109 (1974).

BENDIXEN, G., SØBORG, M.: J. Immunol. 104, 1551 (1970).

BERKELHAMMER, J.: Seminars in Oncology 1, 397 (1974).

BIANCO, C., PATRICK, R., NUSSENZWEIG, V.: J. Exp. Med. 132, 702 (1970).

BLOOM, E.R., GLADE, P.R., (eds.).: In vitro methods of cell-mediated immunity. 1 Vol., New-York: Academic Press 1971.

BLUMING, A., VOGEL, L., ZIEGLER, J.L.: p. 415 in "Investigation and Stimulation of Immunity in Cancer Patients" (G. Mathé and R. Weiner, eds.). Vol. 47 R.R.C.R., Heidelberg: Springer-Verlag 1974.

BLUMING, A.Z., VOGEL, C.L., ZIEGLER, J.L., MODY, N., KAMYA, G.: Ann. Intern. Med. 76, 405 (1972).

BLUMING, A.Z., ZIEGLER, J.L., FASS, L., HERBERMAN, R.B.: Clin. Exp. Immunol. 9, 713 (1971).

BREARD, J., WEINER, R.S., OLDHAM, R.K., MATHÉ, G.: p. 56 in "Complications of Cancer Chemotherapy". (G. Mathé and R.K. Oldham, eds). Vol. 49 R.R.C.R., Heidelberg: Springer-Verlag 1974.

BROWN, G., GREAVES, M.F., LISTER, T.A., RAPSON, N., PAPAMICHAEL, M.: Lancet 2, 753 (1974).

BROWN, R.S., HAYNES, H.A. FOLEY, T.H., GODWIN, H.S., BERARD, C.W., CARBONE, P.P.: Ann. Intern. Med. 67, 291 (1967).
BRULEY-ROSSET, M., BOTTO, H.G., GAUTHIER, H., POUILLART, P., FLORENTIN, I., HUGUENIN, P.: Israel J. Med. Sci. in press (1976).
BRUNNER, K.T., MAUEL, J., CEROTTINI, J.C., CHAPUIS, B.: Immunology 14, 181 (1968).
BUBENIK, J., PERLMANN, P., HELMSTEIN, K., MOBERGER, G.: Int. J. Cancer 5, 310 (1970).
CEROTTINI, J.C., BRUNNER, K.T.: Adv. Immunol. 18, 67 (1974).
CHESS, L., BOCK, G.N., MARDINEY, M.R.: Transplantation 14, 728 (1972).
CHU, E.H.Y., STJERNSWARD, J., CLIFFORD, P., KLEIN, G.: J. Nat. Cancer Inst. 39, 595 (1967).
CLAUSEN, J.E., SØBORG, M.: Acta Med. Scand. 186, 227 (1969).
COHEN, S., BEN FISHER, S., YOSHIDA, T., BETTIGOLE, R.E.: New Engl. J. Med. 290, 882 (1974).
COLIGNAN, J.E., EGAN, M.L., TODD, C.W.: Nat. Cancer Inst. Monogr. 35, 427 (1972).
CURRIE, G.A., LEJEUNE, F., FAIRLEY, G.H.: Brit. Med. J. 2, 305 (1971).
CURTIS, J.E., HERSH, E.M., HARRIS, J.E., MacBRIDE, C., FREIREICH, E.J.: Clin. Exp. Immunol. 6, 473 (1970).
DICKLER, H.B., KUNKEL, H.G.: J. Exp. Med. 136, 191 (1971).
DORE, J.F., MOTTA, R., MARHOLEV, L., HRSAK, I., COLAS DE LA NOUE, H., SEMAN, G., VASSAL, F. DE, MATHÉ, G.: Lancet 2, 1396 (1967).
DUBOIS, M.J.G.J., HUISMANS, D.R., SCHELLEKENS, P.Th.A., EIJSVOOGEL, V.P.: Tissue Antigens 3, 402 (1973).
EILBER, F.R., MORTON, D.L.: Cancer 25, 362 (1970).
FRIDMAN, W.H., KOURILSKY, F.M.: Nature 224, 277 (1969).
FROLAND, S.S.: Scand. J. Immunol. 1, 269 (1972).
FROLAND, S., NATVIG, J.B.: Nature, New Biol. 234, 251 (1971).
FROLAND, S.E., NATVIG, J.B.: Transplant Rev. 16, 114 (1973).
GEFFARD, M., ORBACH-ARBOUYS, S.: Cancer Immunol. Immunoth. 1, 41 (1976).
GERSHON, R.K., COHEN, P., HENCIN, R., LIEBHABER, S.: J. Immunol. 108, 586 (1972).
GOLUB, S.H., HEWETSON, J.F., SVEDMYR, E.A.J., KLEIN, G., SING, S.: Nat. Cancer Inst. Monogr. 37, 25 (1973).
GUTTERMAN, J.U., HERSH, E.M., McCREDIE, K.B., BODEY, G.P. sr., RODRIGUEZ, V., FREIREICH, E.J.: Cancer Res. 32, 2524 (1972b).
GUTTERMAN, J.U., MAVLIGIT, G., MacCREDIE, K.B., BODEY, G.P. sr., FREIREICH, E.J., HERSH, E.M.: Science 177, 1114, (1972a).
GUTTERMAN, J.U., MAVLIGIT, G., MacCREDIE, K.B., FREIREICH, E.J., HERSH, E.M.: Int. J. Cancer 11, 521 (1973).
HAINZ, H.: Science 162, 1281 (1968).
HALLBERG, T., HAEGERT, D., CLEIN, G.P., COOMBS, R.R.A.: Int. Arch. Allergy 44, 500 (1973).
HALTERMAN, R.H., JOHNSON, G.E., LEVENTHAL, B.G.: Nat. Cancer Inst. Monogr. 37, 149 (1973).

HAMMARSTROM, S., HELLSTROM, V., PERLMANN, P., DILLNER, M.L.: J. Exp. Med. 138, 1270 (1973).
HARTMANN, L.: Techniques modernes de laboratoire et explorations fonctionelles. Vol. 1, Paris: Expansion Scientifique 1971.
HAYHOE, F.G.J., FLEMANS, R.J.: Atlas de cytologie hématologique. 1 Vol., Paris: Editions Medicales Flammarion 1970.
HAYRY, P., DEFENDI, V.: Science 168, 133 (1970).
HELLSTROM, I.: Int. J. Cancer 2, 65 (1967).
HELLSTROM, I., HELLSTROM, K.E.: p. 409 in "In vitro Methods in Cell-Mediated Immunity" (B.R. Bloom and P.R. Glade, eds.). 1 Vol. New-York: Academic Press 1971.
HELLSTROM, I., HELLSTROM, K.E., BILL, A.H., PIERCE, C.E., YANG, J.P.S.: Int. J. Cancer 6, 172 (1970).
HELLSTROM, I., HELLSTROM, K.E., SJOGREN, O., WARNER, G.A.: Int. J. Cancer 7, 1, (1971).
HEPPNER, G.H., KOPP, J.S.: Int. J. Cancer 7, 26 (1971).
HERBERMAN, R.B.: p. 140 in "Investigation and Stimulation of Immunity in Cancer Patients" (G. Mathé and R. Weiner, eds.). Vol. 47 R.R.C.R., Heidelberg: Springer-Verlag 1974.
HERBERMAN, R.B., ROSENBERG, E.B., HALTERMAN, R.H., McCOY, J.L., LEVENTHAL, B.G.: Nat. Cancer Inst. Monogr. 35, 259 (1973).
HERSEY, P., MacLENNAN, I,C.M., CAMPBELL, A.C., HARRIS, R., FREEMAN, C.B.: Clin. Exp. Immunol. 14, 159 (1973).
HERSH, E.M., FREIREICH, E.J., McCREDIE, K.D., GUTTERMAN, J.U., BODEY, G.P. sr., WHITECAR, J.P., MAVLIGIT, G., CHEEMA, A.R.: p. 25 in "Investigation and Stimulation of Immunity in Cancer Patients" (G. Mathé and R. Weiner, eds.). Vol. 47 R.R.C.R., Heidelberg: Springer-Verlag 1974.
HERSH, E., WHITECAR, J.P., MacCREDIE, K.B., BODEY, G.P., FREIREICH, E.J.: New Engl. J.Med. 285, 1211 (1971).
HUBER, H., POLLEY, M.J., LINSCOTT, W.D., FUDENBERG, H.H., MULLER-EBERHARD, H.J.: Science 162, 1281 (1968).
HUGHES, L.E., LYTTON, B.: Brit. Med. J. 1, 209 (1964).
IKONOPISOV, R.L., LEWIS, M.G., HUNTER-CRAIG, I.D., BODENHAM, D.C., PHILLIPS, T.M., COOLING, C.I., PROCTOR, J., HAMILTON FAIRLEY, G., ALEXANDER, P.: Brit. Med. J. 2, 752 (1970).
INSTITUT DE CANCEROLOGIE ET D'IMMUNOGENETIQUE, Brochure 1973-1974, Villejuif, France.
JONDAL, M., KLEIN, G.: J. Exp. Med. 138, 1365 (1973).
JOSEPH, R.R., BELPOMME, D., MATHÉ, G.: Brit. J. Cancer in press (1976).
KRONMAN, B.S., WEPSIC, H.T., CHURCHILL; W.H. jr., ZBAR, B., BORSOS, T., RAPP, H.J.: Science 165, 296 (1969).
LEVENTHAL, B.G., HALTERMAN, R.H., ROSENBERG, E.B., HERBERMAN, R.B.: Cancer Res. 32, 1820 (1972).

LEVY, N.L., AMOS, D.B., SOLOVIEFF, G.V., DOS REIS, A.P.: Nat. Cancer Inst. Monogr. 35, 5 (1972).

LEVY, J.P., LECLERC, J.C.: Biomedicine 23, 249 (1975).

LEWIS, M.G.: Lancet 2, 921 (1967).

LEWIS, M.G., IKONOPISOV, R.L., NAIRN, R.C., PHILLIPS, T.M., FAIRLEY, G.H., BODENHAM, D.C., ALEXANDER, P.: Brit. Med. J. 3, 547 (1969).

MacLENNAN, I.C.M., LOEWI, G.: Nature 219, 1069 (1968).

MATHÉ, G., BELPOMME, D.: Biomedicine 20, 81 (1974).

MATHÉ, G., CATTAN, A.: Cancérologie générale et clinique à l'usage du praticien et de l'étudiant. 1 Vol., Paris: Expansion Scientifique 1976.

MATHÉ, G., WEINER, R. (eds.).: Investigation and Stimulation of Immunity in Cancer Patients. Vol. 47 R.R.C.R., Heidelberg: Springer-Verlag 1974.

MAVLIGIT, G., ALEXANIAN, R.: Proc. Amer. Assoc. Cancer Res. 15, 62 (abstract No. 248) (1974).

MELTZER, M.S., OPPENHEIM, J.J., LITTMAN, B.H., LEONARD, E.J., RAPP, H.J.: J. Nat. Cancer Inst. 49, 727 (1972).

NAGEL, G.A., PIESSENS, W.F., STILMANT, M.M., LEJEUNE, F.: Europ. J. Cancer 7, 41 (1971).

OETTGEN, H.F., BEAN, M.A., KLEIN, G.: Cancer Res. 32, 2845 (1972).

OETTGEN, H.P. HELLSTROM, K.E.: p. 951 in "Cancer Medicine" (E. Frei and J. Holland, eds.). 1 Vol., Philadelphia: Lea and Febiger 1973.

OLD, L.J., BOYSE, E.A., GEERING, G., OETTGEN, H.F.: Cancer Res. 28, 1288 (1968).

OLDHAM, R.K., WEINER, R.S., MATHÉ, G., BREARD, J., SIMMLER, M.C., CARDE, P., HERBERMAN, R.: Int. J. Cancer 17, 326 (1976).

OREN, M.E., HERBERMAN, R.B.: Clin. Exp. Immunol. 9, 45 (1971).

O'TOOLE, C., PERLMANN, P., WIGZELL, H., UNSGAARD, B., ZETTERLUND, C.G.: Lancet 1, 1085 (1973).

PAPAMICHAIL, M., HOLBOROW, E.J., KEITH, H.I., CURREY, H.K.F.: Lancet 2, 64 (1972).

PERLMANN, P., HOLM, G.: Adv. Immunol. 11, 117 (1969).

PLATA, F., GOMARD, E., LECLERC, J.C., LEVY, J.P.: J. Immunol. 112, 1477 (1974).

POWLES, R.L., BALCHIN, L.A., FAIRLEY, G.H., ALEXANDER, P.: Brit. Med. J. 1, 486 (1971).

Report of a W.H.O./Iarc-sponsored Workshop on human B- and T-cells, London, July 15-17th 1974. Scand. J. Immunol. 3, 521 (1974).

ROCKLIN, R.E.: J. Immunol. 112, 1461 (1974).

ROSENBERG, S.A.: Nat. Cancer Inst. Monogr. 37, 139 (1973).

ROSENBERG, E.B., McCOY, J.L., GREEN, S.S., DONNELLY, F.C., SIWARSKI, D.F., LEVINE, P.H., HERBERMAN, R.B.: J. Nat. Cancer Inst. 52, 345 (1974).

ROSS, G.D., RABELLINO, E.M., POLLEY, M.J., GREY, H.M.: J. Clin. Invest. 52, 377 (1973).

ROSSO DI SAN SECONDO, V., SCALAMOGNA, M., SIRCHIA, G.: Tissue Antigens 4, 383 (1974).

SCHNEIDER, M., MATHÉ, G., SCHWARZENBERG, L., POUILLART, P., WEINER, R., AMIEL, J.L., HAYAT, M., JASMIN, C., VASSAL, F. DE: p. 42 in "Investigation and Stimulation of Immunity in Cancer Patients" (G. Mathé and R. Weiner, eds.). Vol. 47 R.R.C.R., Heidelberg: Springer-Verlag 1974.

SCHWEITZER, M., MELIEF, C.J.M., EIJSVOOGEL, V.P.: Int. J. Cancer 11, 11 (1973).

SHEVACH, E., JAFFE, E.S., GREEN, I.: Transplant. Rev. 16, 3 (1973).

SIMMLER, M.C., RAMEAU, G., CHOU, M.J., MATHÉ, G.: Israel J. Med. Sci. in press(1976a).

SIMMLER, M.C., SCHWARZENBERG, L., MATHÉ, G.: Cancer Immunol. Immunoth. in press (1976b).

STATHOPOULOS, G., ELLIOTT, E.V.: Lancet 1, 600 (1974).

STEWART, T.H.M.: Cancer 23, 1368 (1969).

STJERNSWARD, J., JONDAL, M., VANKY, F., WIGZELL, H., SEALLY, R.: Lancet 1, 1352 (1972).

STJERNSWARD, J., VANKY, F., KLEIN, E.: Brit. J. Cancer 28 (suppl. 1) 72 (1973).

TAKASUGI, M., KLEIN, E.: Transplantation 9, 219 (1970).

TAKASUGI, M., MICKEY, M.R., TERASAKI, P.I.: Nat. Cancer Inst. Monogr. 37, 77 (1973),

THOMSON, D.M.P., KRUPEY, J., FREEDMAN, S.O., GOLD, P.: Proc. Nat. Acad. Sci. USA 64, 161 (1969).

THORNES, R.D.: Lancet 2, 382 (1974).

THURMAN, G.B., STRONG, D.N., AHMED, A., GREEN, S.S., SELL, K.W., HARTZMAN, R., BACH, F.H.: Clin. Exp. Immunol. 15, 289 (1973).

TOBEN, H.R., COOPER, M.D.: Clin. Res. 20, 797 (1972).

VALENTINE, F.T., LAWRENCE, H.S.: Adv. Int. Med. 17, 51 (1971).

VANKY, F., STJERNSWARD, J.: Israel J. Med. Sci. 7, 211 (1971).

VANKY, F., STJERNSWARD, J., KLEIN, G., NILSONNE, U.: J. Nat. Cancer Inst. 47, 95 (1971).

WALDMANN, T.A., BRODER, S., BLAESE, R.M., DURM, M., BLACKMAN, M., STROBER, W.: Lancet 2, 609 (1974).

WARA, D.M., AMMAN, A.J.: Ann. New-York Acad. Sci. in press (1976).

WILSON, J.D., NOSSAL, G.J.V.: Lancet 2, 788 (1971).

WYBRAN, J., FUDENBERG, H.: J. Clin. Invest., 52, 1026 (1973).

Subject Index

Recent Results in Cancer Research

Sponsored by the Swiss League against Cancer. Editor in Chief: P. Rentchnick, Genève

44 Special Topics in Carcinogenesis. Edited by E. GRUNDMANN (Symposium).
45 KOLDOVSKÝ, P.: Carcinoembryonic Antigens.
46 Diagnosis and Therapy of Malignant Lymphoma. Edited by K. MUSSHOFF (Symposium).
47 Investigation and Stimulation of Immunity in Cancer Patients. Edited by G. MATHÉ and R. WEINER (Symposium).
48 Platinum Coordination Complexes in Cancer Chemotherapy. Edited by T. A. CONNORS and J. J. ROBERTS (Symposium).
49 Complications of Cancer Chemotherapy. Edited by G. MATHÉ and R. K. OLDHAM (Symposium).
50 Cancer Registry. Edited by E. GRUNDMANN and E. PEDERSEN (Symposium).
51 Gliomas. Current Concepts in Biology, Diagnosis and Therapy. Edited by J. HEKMATPANAH (Symposium).
52 The Ambivalence of Cytostatic Therapy. Edited by E. GRUNDMANN and R. GROSS (Symposium).
54 Malignant Bone Tumors. Edited by E. GRUNDMANN.
Special Supplement: Biology of Amphibian Tumors. Edited by M. MIZELL.

Further publications by Georges Mathé

Physiologie normale et pathologique due métabolisme de l'eau (with J. MABURGER). 1 Vol. Paris: Editions Médicales Flammarion 1952.

Aspects histologiques et cytologiques des leucémies et hématosarcomes (with G. SÉMAN). 1 Vol. Paris: Maloine 1963.

L'aplasie myélo-lymphoide de l'irradiation totale (with J. L. AMIEL). 1 Vol. Paris: Gauthier-Villars 1965.

Bone marrow transplantation and white cell transfusions (with J. L. AMIEL and L. SCHWARZENBERG). 1 Vol. Springfield: Ch. C. Thomas 1971.

La chimiothérapie des cancers (leucémies, hématosarcomes, tumeurs solide). 1st ed. 1966, 2nd ed. 1969, 3rd ed. 1975 (with Y. KENIS). Paris: Expansion Scientifique.

Cancer Chemotherapy (leukaemias, haematosarcomas and solid tumours) (with Y. KENIS). Heidelberg-New York: Springer-Verlag 1976.

Histocytological typing of the neoplastic diseases of the hemato-poietic and lymphoid tissues (with H. RAPPAPORT). 1 Vol. Geneva: World Health Organisation, in press 1975.

Introduction à l'immunothérapie active des cancers: immunopévention et immunorestauration. Paris: Expansion Scientifique Française 1976.

La greffe. Aspects biologiques et cliniques (with J. L. AMIEL). 1. Vol. Paris: Masson 1952.

Scientific Basis of Cancer Chemotherapy. Recent Results in Cancer Research. Vol. 21. Heidelberg-New York: Springer-Verlag 1969.

Aseptic Environments and Cancer Treatment. Recent Results in Cancer Research. Vol. 29. Heidelberg-New York: Springer-Verlag 1970.

Advances in the Treatment of Acute (Blastic) Leukemias. Recent Results in Cancer Research. Vol. 30. Heidelberg-New York: Springer-Verlag 1970.

Adriamycin (with S. K. CARTER, A. DI MARCO, M. GHIONE and I. H. KRAKOFF). 1 Vol. Heidelberg-New York: Springer-Verlag 1972.

Natural history and modern treatment of Hodgkin's disease (with M. TUBIANA). Series Hematologica, 2 Vol. Copenhagen: Munksgaard 1973.

Nomenclature, Methodology and Results of Clinical Trials in Acute Leukaemias (with P. POUILLART and L. SCHWARZENBERG). Recent Results in Cancer Research. Vol. 43. Heidelberg-New York: Springer-Verlag 1973.

Investigation and Stimulation of Immunity in Cancer Patients (with R. WEINER). Recent Results in Cancer Research. Vol. 47. Heidelberg-New York: Springer-Verlag 1974.

Complications of Cancer Chemotherapy (with R. K. OLDHAM). Recent Results in Cancer Research. Vol. 49. Heidelberg-New York: Springer-Verlag 1974.

Lymphocytes and Macrophages in Cancer Patients (with I. FLORENTIN and M. C. SIMMLER). Recent Results in Cancer Research. Heidelberg-New York: Springer-Verlag, in press 1976.